C0-AWO-024

Energy Crisis
Volume 3
1975-77

Energy Crisis
Volume 3
1975-77

Edited by Lester A. Sobel

Contributing editors: Mary Elizabeth Clifford, Joseph Fickes, Hal Kosut, Christopher Hunt, Chris Larson

Indexer: Grace M. Ferrara

Facts On File
119 West 57th Street, New York, N.Y. 10019

Energy Crisis
Volume 3
1975-77

Published by Facts on File, Inc.,
119 West 57th Street, New York, N.Y. 10019.

Library of Congress Catalog Card No. 74-75154
ISBN 0-87196-280-2

9 8 7 6 5 4 3 2 1
PRINTED IN
THE UNITED STATES OF AMERICA

Contents

Introduction

THE POSSIBILITY OF AN "ENERGY CRISIS" has been discussed by scientific, economic and political thinkers for more than half a century. In 1938, President Franklin D. Roosevelt ordered a study of U.S. energy resources, and he warned the following year that these resources "are not inexhaustible." "As early as 1954," the 50th American Assembly reported, "it became an important national objective to develop nuclear power as a source of electricity." The U.S. "need was not seen as imminent," an Assembly report said, but "for other parts of the world—without our rich fossil fuel supplies—the day of dependence on [nuclear] fussion was expected to come much sooner."

The Arab oil embargo of 1973—accompanied by shortages of gasoline and heating oil and followed by the quadrupling of the price of foreign oil—began to impress on Americans the seriousness of the situation. Yet the shortages ended—at least temporarily—and the U.S. and the other developed nations accommodated themselves to the necessity of paying very much higher prices for the oil they used and, all too often (as it is frequently admitted), wasted.

But "the energy crisis," Barry Commoner wrote in the Feb. 2, 1976 issue of *The New Yorker,* "illuminates the world's most dangerous political issues as it wrenches into open view the brutality of national competition for resources, the festering issues of economic and social injustice and the tragic absurdity of war." He indicated that "the crisis forces us to make long-avoided choices" that require a "reexamining [of] the precepts of the economic system that now govern how energy is produced and used."

Rep. Mike McCormack (D, Wash.) had observed in a speech in

Pittsburgh Nov. 17, 1975 that "one of the most dangerous aspects of the energy crisis is that a large portion of our fellow citizens do not understand it. Indeed, a surprising portion of Americans deny that an energy crisis exists, and many who do believe that it has been contrived by evil powers which could easily and quickly undo their nefarious deeds, that is, solve the energy crisis by magic. . . ." Mansfield said:

. . .Out of the confused debate that has developed with respect to the energy crisis, perhaps only one concept is universally accepted. That is that we must not waste our energy; that we must reduce our national annual energy growth rate. What isn't yet apparent, however, even to many sincere and concerned policy-makers, is that the total energy consumption of our nation must continue to increase for as far as we can see into the future, even if we establish extraordinarily successful spartan conservation programs.

The report of the Ford Foundation Energy Policy Project assumes, as a reasonable scenario for the future, limiting the growth of total energy consumption for this nation to 2% per year. This would be a dramatic reduction in our contemporary growth rate of about 3.6%, but it may be a long-range attainable goal which might not seriously harm the economy.

It is important to understand, however, that a 2% growth rate means doubling our energy consumption in about 35 years. Within this overall growth, there will inevitably be a dramatic shift from the use of petroleum and natural gas to the production of electricity from coal and nuclear fission. Here the annual growth rate must be about 5% per year to maintain an overall 2% rate. This means doubling electric energy generation by about 1990.

Unfortunately, most of the debate on the energy crisis, in spite of the perils we are facing, has centered around such subjects as import tariffs, quotas, gasoline taxes and prices, allocation programs, regulations and incentives. While all this is important and has been the subject of intense Congressional activity, obsession with it tends to distort our perspective. I do not take this matter lightly, but acting as if these problems constitute the energy crisis is somewhat like wrestling for deck chairs on the *Titanic*.

The stark realities are that our oil production is down about 10% in the last three years; our natural gas production is down almost as much; our coal production has increased less than 10% in this period; the percent of petroleum we import has now increased to about 41%; . . . and a handful of anti-nuclear activists are advocating a moratorium on nuclear energy production—our only hope, along with coal, for energy self-sufficiency during this century.

The challenge we face today is to reject these attempts to confuse us, and to overcome the paralysis that seems to grip us, and to develop now a systems approach to an integrated national energy policy, not only to eliminate waste and conserve energy wherever practical, but also to produce the energy that we will need in the future. . . .

"The energy crisis," according to Sen. Adlai E. Stevenson 3d (D, Ill.), ". . . is still a phrase in search of meaning." He told the U.S. Senate April 9, 1976 that "as I see it, the energy crisis consists of at least three interlocking crises: a crisis of energy prices; a crisis of supply; and, finally, a crisis of national security.'' Stevenson continued:

. . . By no coincidence the quadrupling of oil prices in 1973 was followed by the worst inflation in this century and the worst recession since the Great Depression. Demand fell—but prices continued to rise. Employment fell—but wages rose at higher rates. The inflation of 1973 and 1974 was caused by energy prices more than by anything else; and that inflation, in a turnabout of all the old rules of economic behavior, caused the recession that followed.

The rising price of energy—touched off by the rising price of oil—rippled out through the economy to drive up the cost of every product and service. The plight of cities, industries and farms, crises in themselves, can be traced in large part to the spiraling cost of energy. Energy-induced inflation may have cost consumers $150 billion in purchasing power during 1974 and 1975, not all of it recaptured by the American economy. That reduced purchasing power guaranteed a recession. . . .

. . . [E]nergy conservation is one of the most promising and neglected means of solving the supply problem. We are wasting upwards of 25% of the energy used to heat and cool our buildings and run our factories. A dollar invested in conservation saves twice as much energy as a dollar invested in production can produce. But major investments in conservation are not being made. . . .

Foreign oil can be interdicted at its source—and also in transit. Thirty-six percent of the world's oil, 60% of Europe's, must pass through the Straits of Hormuz at the mouth of the Persian Gulf. That passage could be closed, I am told, by the mysterious sinking of one or two tankers. The transportation of oil and gas can be interdicted at its source, at the mouth of the Persian Gulf, the Red Sea, around the periphery of Africa, in the Eastern Mediterranean, in the North Sea, indeed at any point in the world where oil is either produced or transported. . . . The Soviet Union is acquiring the power to interdict transportation of the world's oil supplies. . . .

Confronted by the knotty complexities of the energy crisis, the political left responds with proposals for oil company divestiture. But the advocates of divestiture can offer no assurance that smaller, more numerous producing and refining units would sell energy at a price other than the price established by the cartel. Smaller units of production may be in a weaker position from which to bargain with foreign producers for supplies at reasonable prices.

The political right, on the other hand, offers us 19th century *laissez faire*, embodied in clichés uncritically. But in this world, unregulated domestic energy prices rise and fall with the prices of the OPEC [Organization of Petroleum Exporting Countries] cartel. In this world, producers curtail supplies of energy at home and abroad to increase prices. And in this world, energy is not a luxury but as essential to our way of life as air and water are to our bodies. . . .

In the week of March 8, 1976 the U.S., for the first time, imported more oil (8,196,000 barrels a day) than it produced (8,013,500 barrels a day).

Shortly thereafter, Sen. Dale Bumpers (D, Ark.) held hearings of the Senate Interior & Insular Affairs Committee on U.S. energy policy. In his opening remarks July 22, 1976, Bumpers said that these ''facts should be highlighted'':

First: Foreign imports of crude oil are running at about 40% of our domestic needs. Prior to the 1973 embargo, 33% of our needs were supplied by foreign sources;

Second: In 1970 we paid about $3 billion for foreign oil; in 1975, after a 500% increase in the price of oil, the cost soared to $27 billion, and next year the cost may reach $35 billion;

Third: In spite of the critical shortages caused by the OPEC embargo, we are actually importing a greater percentage of our oil from these less-than-reliable sources than we did at the time of the embargo.

Summarized, our appetite for energy has not been diminished by either higher prices or greater dependence on foreign oil. The plain and simple truth of the matter is that too many in this country exhibit too little concern for the fact that energy has become our Achilles' heel. . . .

The "geopolitical" implications of the increased U.S. dependence on oil imports was analyzed by Melvin A. Conant, former assistant administrator of the Federal Energy Administration for international energy affairs, in testimony before the Senate committee Feb. 1, 1977. Conant noted that U.S. and European energy consumption had doubled and Japanese energy consumption had tripled during 1960-76. "It was in this period," he said, "that the U.S. ceased to be able to meet its energy requirements from indigenous resources and ceased to be able to function as the emergency oil supplier to our Japanese and NATO [North Atlantic Treaty Organization] allies. During this period while the Middle East and Africa remained the single most important source of oil for Western Europe and Japan, their importance in meeting U.S. requirements nearly tripled. Throughout 1960-1976, only the USSR was and remained energy self-sufficient and hence had no supply vulnerability." Conant continued:

The events surrounding the Arab-Israel War of October 1973 dramatized the potential divisions that competition for access to Middle Eastern oil can exert on our relations with our major industrial allies. The unwillingness of some European allies to grant us staging facilities for our resupply effort to Israel arose from the fact that they feared that making such facilities available could endanger the flow of oil from the oil-producing states with potentially devastating effects on their national economies. . . . [O]ur allies have lived with energy dependency far longer than we have. Their perspective, therefore, is different from our own. . . . Although the industrial countries weathered the OPEC embargo of 1973–74 without irreperable damage to the Western alliance, the crisis demonstrated that the different degrees of our oil dependence upon the Middle East and the greater importance to the key producers of the European and Japanese markets will continue to be the root of our energy and related differences with our industrial allies.

Since the end of the embargo, the level of competition for "access" to Middle Eastern oil has intensified as each of the major industrial countries has come to the realization that with current projections of energy needs over the next decade, the importance of "access" to Middle Eastern crude oil reserves will remain paramount. Indeed the importance of Middle Eastern crude can be demonstrated by reference to the following statistics. While OPEC countries account for approximately 60% of world oil production . . . and over 70% of world oil reserves, the oil-producing nations of the Persian Gulf account for 40% of world oil production and over 50% of world oil reserves. . . .

. . . [T]he possible consequences to our alliance structures arising from increased allied competition compel us to comprehend the variety of issues and the opportunities for forceful initiatives which are embedded in this question of "access" to vital energy supplies. Each of us is vulnerable to the constraints of the geopolitics of energy. . . . [F]or many years ahead the *location* and the sovereign *control* over oil entering world trade will continue to be in the hands of the non-industrial world. Among the free nations, virtually only the U.S. still has options which it can exercise to limit its supply vulnerability to the political and economic actions of key producers.

In this regard, I would like to emphasize that it is virtually impossible for even the U.S. to become energy independent before the turn of the century. This is not to imply that our strategic vulnerability could not be drastically reduced by the rapid development of our coal, nuclear and other alternative energy sources, rather it is merely to emphasize that oil will remain the dominant fuel for the industrialized world for the rest of the century.

. . . Although we share the previous and new administrations' concerns over the serious safety and environmental issues raised by the rapid development of our domestic coal and nuclear power resources, we must stress in the strongest possible terms the fact that the failure to develop energy alternatives in time and on a sufficient scale will enhance the importance of imported oil for all the industrial economies. Oil and oil alone, will make up for shortfalls in our efforts.

Indeed, the failure to make the hard decisions in the domestic arena will complicate our relations with our industrial allies. The U.S. may well contain great unrealized energy reserves whose exploitation could conceivably meet our needs and in time even some of our allies! Unless we determine soon whether this can be the case, the Europeans and Japanese will have little choice but to continue to expand the nuclear sectors of their economies and to attempt to the degree possible to diversify the source of their fossil fuel imports. But they shall remain crucially dependent on the Middle East.

In this regard, Soviet/European and Soviet/Japanese energy deals and negotiations give cause for concern. The USSR has a marked energy advantage over the U.S. through its far lesser dependence upon imported oil and its potential role as a major exporter of natural gas to Europe and Japan. The Soviet Union can probably maintain energy autarky within this century. The strategic significance of such an accomplishment could be monumental.

In the meantime, the commanding position of the USSR as a leading exporter of gas to Europe and probably to Japan, and also as the "transit" area for large volumes of Middle East gas to Europe, could give the USSR unprecedented political leverage. . . .

The major facts about the world energy situation during the period from 1969 through mid-1975 are recorded in the two FACTS ON FILE books *Energy Crisis, Volume 1, 1969–73*, and *Volume 2, 1974–75*. The events of the following two years are detailed in this book, which brings the narrative to about the midpoint of 1977. The material in this book is based largely on the record compiled by

FACTS ON FILE in its weekly coverage of world events. Much of the subject matter is controversial, but a conscientious effort was made to record all developments without editorial bias and to produce a balanced and reliable reference tool.

LESTER A. SOBEL

New York, N.Y.
September, 1977

U.S. Policy

Situation Deteriorates

U.S. policy during the period 1975–77 was based on the knowledge, supported by a growing mass of statistics, that American use of fuel was growing—after a two-year decline—even as U.S. energy resources were diminishing.

Energy use statistics. Petroleum remained the nation's most important single energy source; it provided 46% of the energy used in the U.S. in 1975, a 0.2% increase over 1974, according to statistics released by the Interior Department's Bureau of Mines April 4, 1976.

A decline of 2.5% was recorded in total U.S. energy use in 1975. It was the second consecutive annual decline and first two-year decline since 1945–46. The figure for the 1975 use of heat, light and power in all forms was 70,580 trillion BTUs (British thermal units), according to revised data (made public in 1977). This compared with 1973's record 74,555 trillion BTUs.

The bureau attributed the 1975 decline to high fuel prices, the slow economy, energy conservation efforts and a relatively mild winter.

A drop in industrial fuel consumption, of 6%, was the largest component in the decline in total energy use in 1975. Household and commercial use of energy fell 2%. Fuel use by electric utilities increased about 0.5%, as it did also in transportation.

Nuclear power contributed 2.3% of the nation's total energy in 1975, up from 1.7% in 1974. Use of bituminous coal and lignite increased 1.7%. There were declines in use of anthracite coal (down 7.8%), natural gas (7.2%), petroleum-products (1.9%) and water power (4%).

The production of bituminous coal and lignite rose 6.1% in 1975 to a record 640 million tons, of which 404 million tons were consumed by electric utilities. Domestic crude-oil production fell 4.5% to 3.1 billion barrels. Production of natural gas for domestic markets was down 6.9% to 20.1 trillion cubic feet.

But in 1976, total U.S. energy use rose by 4.8% to 73,999 BTUs—still below the 1973 record—the Bureau of Mines reported March 31, 1977.

According to bureau figures, household and commercial energy use rose 7.3% during 1976, reflecting the impact of abnormally cold weather in the last quarter. Industrial energy use increased only 3.3%. Energy required to generate electricity by utilities was up 5.6% from 1975. Consumption for transportation rose 4.2%, with gasoline use up 4.5%.

Petroleum remained the nation's largest energy source, supplying 47.2% of energy needs. Natural-gas usage declined 1% to account for 27.3% of the total.

Slight increases were reported for usages of coal and lignite, to an 18.4% share, and nuclear power, to a 2.8% share. Hydroelectric-power usage was down an unspecified amount due to drought in the West.

Imports of crude oil and petroleum products increased 21.4% during the year to a total of 2.6 billion barrels at an estimated cost of $32 billion. Imports accounted for 40.6% of U.S. oil use.

Oil reserves shrink. According to American Petroleum Institute (API) statistics released March 30, 1976, proved U.S. reserves of crude oil dropped by about 1.6 billion barrels during 1975, despite a 17.5% increase in the number of wells drilled.

The proved crude reserves were estimated at 32.7 billion barrels as of Dec. 31, 1975. The 1974 figure was 34.3 billion barrels. An increase of 1.3 billion barrels from both new oil fields and extension or development of known reservoirs was noted in the revised estimate of 1975 proved reserves. This was negated by the withdrawal during 1975 of 2.9 million barrels for domestic production.

API President Frank N. Ikard noted that the proved reserves of crude oil had declined by more than 6.3 billion barrels, or 16.2%, since the end of 1970, when the estimate was 39 billion barrels.

Zarb reviews energy situation. Speaking on the second anniversary of the Arab oil embargo, Federal Energy Administrator Frank G. Zarb had said Oct. 20, 1975 that the country produced less oil and natural gas than in 1973 and was more dependent on imports.

In what a Federal Energy Administration news released called a mixed review of energy developments, Zarb said U.S. dependence on OPEC-produced oil was now more than 60%, up more than 11% from the figure it had been in 1973. (The American Petroleum Institute Sept. 17 said crude oil output in the week ended Sept. 12 had fallen to 8.24 million barrels a day, the lowest level since Aug. 1966.) Natural gas production was running 8% below 1973 levels and shortages were expected for the winter. Coal production was up slightly, and a total output of 640 million tons was projected for 1975 compared with 590 million tons in 1973.

Among the positive aspects cited by Zarb was a decline in petroleum consumption of 1.9 million barrels below the rate projected in 1974. (According to the American Petroleum Institute Nov. 10, consumption of petroleum products for the first 10 months of 1975 was 1.7% below the figure for the comparable period of 1974.) The average number of drilling rigs in operation in 1974 was 24% above the figure for the previous year, and 19.6 more wells were drilled. (The International Association of Drilling Contractors said Sept. 3 that the industry had 1,684 rotary drilling rigs in operation, the highest number since 1963.) Zarb also noted that fuel economy in automobiles had gone up, with 1976 models testing at 12.8% better than those of the previous year and 26.6% better than for 1974.

U.S. output decline blamed on price curbs. In Senate debate Dec. 17, 1975, Sen. Dewey Bartlett (R, Okla.) again blamed price restrictions for causing a decline in U.S. oil output by reducing the incentive for drilling for oil domestically. Bartlett's statement, as it appeared in the Congressional Record, included these assertions:

"In 1955 when the U.S. consumption of petroleum was 8.9 million barrels per day, there were on the average 2,700 rigs actively drilling in the United States. Now with consumption at nearly 17 million barrels per day, there are only 1,800 active rigs. In order to proportionately provide the same effort as there was in 1955, we should now have nearly 5,200 rigs drilling. Thus, our domestic drilling effort must nearly triple, and we are far behind.

"The Independent Petroleum Association of America has presented an interesting analysis of this same question which shows that during the period from 1960 to 1973, the domestic petroleum industry invested on the average $2.5 billion per year to drill wells; in 1974, expenditures were $4.5 billion—nearly an 80 percent increase. Yet, this level of investment is still inadequate. As the IPAA states:

"To reverse the downtrend in domestic oil production and raise 1985 output to desired levels, we must almost double the 1974 expenditures for drilling in the year 1976–1980 and more than double the . . . expenditures in the succeeding five years, 1981–1985. . . ."

Sen. Dale Bumpers (D, Ark.) came to similar conclusions in remarks in the Senate the same day. He said:

"In January 1973, the price of oil in this country was $3.40. In July of this year, 1975, the price was $10.60, or over 3 times the price in 1973.

"This occurred, Mr. President, at a time when the price of oil in this country was supposed to be controlled.

"What has happened during that period, while the price of oil was more than tripling?

"Here are the facts:

"In 1972, there were 27,291 wells drilled in this country—a low water mark, admittedly.

"In 1973, it dropped again, to 26,592 wells, or a decrease of 2.5 percent.

"But in 1974, following the first year of the Arab oil embargo, there were 31,698 wells drilled, or an increase of 19.2 percent.

"In 1975, the estimate is that there will be 34,214 wells drilled, still another increase of 7.9 percent over the wells drilled in 1974.

"That shows conclusively that people are indeed looking for oil, and the reason they are looking for it is because they can do so profitably and they will continue to do so. . . ."

Weekly oil imports set new record. U.S. imports of petroleum and petroleum products climbed during the week ended March 4, 1977 to record levels for the third week in a row, according to American Petroleum Institute figures released March 9. Total imports of 10.3 million barrels a day eclipsed the mark of 10.1 million barrels a day set the week before.

While imports of crude oil dipped in the most recent week to 6.6 million barrels a day from 6.95 million barrels, product imports rose to a record 3.69 million barrels a day from 3.13 million barrels a day during the week that ended Feb. 25.

A record 1.17 million barrels a day of middle distillates, including home heating oil, was imported during the week that ended March 4. The sharp increase from the previous week's figure of 790,000 barrels a day was attributed to orders placed in early February when the weather was extremely cold.

Commerce Department officials reported March 28 that imported petroleum had accounted for 42% of domestic oil use during 1976; however, during the first two months of 1977, foreign oil supplies accounted for an average 45% of consumption and in one week during February, half of the oil consumed in the nation came from foreign sources. Prior to the quadrupling of oil prices in late 1973, foreign oil supplies had accounted for 25%-30% of total U.S. use.

Officials also noted that oil prices, as well as volume, had increased during February. A barrel of crude oil imported to the U.S. in February cost $13.02, up from $12.53 in January. The U.S. oil bill in February was $3.32 billion, $178.9 million more than in January.

Refinery output hits 11-year low—Revised figures showed that U.S. refineries had produced an average of 7.95 million barrels of oil a day in February, the first drop below the eight-million mark since 1966, the American Petroleum Institute said March 9.

McKetta's review & analysis. Dr. John J. McKetta, chemical engineering professor at the University of Texas and former (1970–71) chairman of the Advisory Comittee on Energy to the Secretary of the Interior, reviewed and analyzed the U.S. energy situation in answer to questions submitted by Dresser Industries, Inc. McKetta's views, as revealed in his answers and inserted in the Congressional Record Feb. 17, 1977 by Rep. Robert Krueger (D, Tex.):

In 1974, the U.S. produced 9 million barrels of oil per day. The figure declined to 8.5 million in 1975, and declined again in 1976, even while demand was rising at higher than expected rates. We aren't gaining in production—we are continuing to lose ground. We aren't moving toward energy self-sufficiency; we are moving farther away from it. Just a few years ago, we were importing about 25 percent of the oil used in this country. In 1976, we imported approximately 45 percent, and the figure is still growing. Our increasing dependence on imported oil brings greater risks of another embargo, and more intimidation in the conduct of our foreign policy. This endangers and jeopardizes our nation.

Since 1956, our total energy demand growth rate has outstripped our total energy production growth rate, causing a disturbingly high growth rate in energy imports.

Q. What about energy sources other than oil and gas? Can we count on them to close the gap between our production and demand?

U. S. Oil Imports

[Thousands of barrels per day]

4 weeks ending	May 7, 1976	May 9, 1975	May 10, 1974	May 11, 1973	Percentage change 1976/75	1976/74	1976/73
Crude oil	4,785	3,476	3,562	3,240	+37.7	+34.3	+47.7
Motor gasoline	103	103	241	66		−57.3	+56.1
Distillate	87	160	261	229	−45.6	−66.7	−62.0
Residual	877	942	1,690	1,564	−6.9	−48.1	−43.9
Jet fuel—Naphtha	23	20	10	23	+15.0	+130.0	
Jet fuel—Kerosene	71	147	136	190	−51.7	−47.8	−62.6
Other products	259	279	511	467	−7.2	−49.3	−44.5
Total products	1,420	1,651	2,849	2,539	−14.0	−50.2	−44.1
Total imports	6,205	5,127	6,411	5,779	+21.0	−3.2	+7.4
Year to date (19 reports)							

4 weeks ending	May 7, 1976	May 9, 1975	May 10, 1974	May 11, 1973	Percentage change 1976/75	1976/74	1976/73
Crude oil	4,993	3,682	2,722	2,885	−35.6	+83.4	+73.1
Motor gasoline	127	133	178	65	−4.5	−28.7	+95.4
Distillate	178	247	316	340	−27.9	−43.7	−47.7
Residual	1,268	1,369	1,689	2,095	−7.4	−24.9	−39.5
Jet fuel—Naphtha	25	30	11	19	−16.7	+127.3	+31.6
Jet fuel—Kerosene	91	135	141	162	−32.6	−35.5	−43.8
Other products	310	315	454	397	−1.6	−31.7	−21.9
Total products	1,999	2,229	2,789	3,078	−10.3	−28.3	−35.1
Total imports	6,992	5,911	5,511	5,963	+18.3	+26.9	+17.3

Source: API Weekly Statistical Bulletin.

U. S. Refinery Operations

[Thousands of barrels per day]

4 week ending	May 7, 1976	May 9, 1975	May 10, 1974	May 11, 1973	Percentage change 1976/75	1976/74	1976/73
Operable capacity	15,139	15,027	14,230	13,618	+0.7	+6.4	+11.2
Input to crude oil processing units	12,997	12,235	12,341	NA	+6.2	+5.3	NA
Percent capacity utilized [1]	85.9	81.4	86.7	NA			
Crude oil runs	12,661	11,765	11,835	12,106	+7.6	+7.0	+4.6
Percent capacity utilized [2]	83.6	78.3	83.2	88.9			
Production:							
Motor gasoline	6,598	6,009	6,264	6,516	+9.8	+5.3	+1.3
Distillate	2,589	2,439	2,594	2,489	+6.2	−.2	+4.0
Residual	1,270	1,212	873	891	+4.8	+45.5	+42.5
Kerosene	143	158	116	197	−9.5	−23.3	−27.4
Jet fuel—Naphtha	206	199	189	183	+3.5	+9.0	+12.6
Jet fuel—Kerosene	691	680	645	684	+1.6	+7.1	+1.0
Year to date (19 reports) week ending							

4 week ending	May 7, 1976	May 9, 1975	May 10, 1974	May 11, 1973	Percentage change 1976/75	1976/74	1976/73
Operable capacity	15,120	14,989	14,149	13,535	+.9	+6.9	+11.7
Input to crude oil processing units	13,161	12,506	11,980	NA	+5.2	+9.9	NA
Percent capacity utilized [1]	87.0	83.4	84.7	NA			
Crude oil runs	12,795	12,011	11,470	12,171	+6.5	+11.6	+5.1
Percent capacity utilized [2]	84.6	80.1	81.1	89.9			
Production:							
Motor gasoline	6,549	6,232	6,036	6,234	+5.1	+8.5	+5.1
Distillate	2,737	2,625	2,561	2,743	+4.3	+6.9	−.2
Residual	1,336	1,303	953	971	+2.5	+40.2	+37.6
Kerosene	173	194	164	259	−10.8	+5.5	−33.2
Jet fuel—Naphtha	184	176	215	192	+4.5	−14.4	−4.2
Jet fuel—Kerosene	737	676	621	712	+9.0	+18.7	+3.5

[1] New API definition.
[2] Old API definition.

Source: API Weekly Statistical Bulletin

A. Our projection regarding other U.S. energy sources has declined substantially between 1970 and 1975. In 1970, we estimated that all of our energy sources could provide 95 quadrillion British Thermal Units by 1985. Our 1975 estimates show that we were overly optimistic and now we expect only 62 Q's* from all energy sources by 1985.

While our predictions of oil and gas production by 1985 have been revised downward by 3 Q's, predictions of supplies of other energy sources, such as coal, nuclear, geothermal, solar and synthetics, have been revised downward by almost 30 Q's. The outlook now is that coal will be the only energy source other than oil and gas that can make a significant contribution to our energy supplies over the next ten years.

Nuclear energy, for example, will provide only about 9 Q's by 1985, if all planned nuclear plants are actually built.

At the same time, 1975 predictions of 1985 energy demand have been revised upward from the 1970 estimates. . . .

Obviously, we must make every effort to develop all energy sources, but oil, gas and coal are the only ones that can make a significant contribution to our needs in the immediate future, or, at least, the next two decades. We have no choice but to concentrate our efforts on hydrocarbons, especially oil and gas, until other sources of energy are able to share the load—probably around the year 2000.

Q. What are the consequences of our rising energy imports?

A. First, we must conclude that our increasing dependence upon imported oil could result in a profound and dangerous weakness in the U.S. position abroad. Today, for example, we face the dilemma of supporting Israel, with whom we have religious and friendly ties, or supporting the Arab states, whose oil supply is vital to our nation.

Second, there is the specter of economic disaster. In 1976, the U.S. cost of imported energy liquids will be $37 billion. It was only $6 billion in 1973. If we project import trends based on our 1970 production and demand projections, we would need to import 8 billion barrels of oil during the year 2000. The cost of that amount of fuel could be over $200 billion. This is equivalent to 20 million new jobs at $10,000 per person per year. If we use 1975 projections for production and demand, the situation becomes even more disastrous.

We should also keep in mind that our rising bill for energy imports has the same effect as if our tax bill were increasing . . . the result is that industry has less funds to provide new jobs and consumers have less funds available to spend to support employment.

Q. Is our growing dependence on imported oil proof that the United States is running out of oil and gas?

A. No, the tragedy of our problem is that the crisis is developing even though expert scientific studies indicate there are sufficient domestic resources of crude oil and natural gas to last for the foreseeable future, and certainly until other energy sources, such as nuclear power, develop to the point of supplying the greater part of our needs. As of 1975, the U.S. had proved recoverable reserves of about 228 trillion cubic feet of gas and 34 billion barrels of oil. Estimates of undiscovered U.S. recoverable potentials go as high as 750 trillion cubic feet for gas and 104 billion barrels for oil.

Q. You mentioned earlier that in addition to oil and gas, coal must be used more extensively. Why is the use limited now?

A. In 1970 the Mine Safety Act and EPA regulations caused declines in the production and usage of coal. Twenty-two percent of the total mines were closed during 1970–71. Since that time, restrictions on the use of high sulfur coal have decreased usage. Production and usage have risen since 1971, but 1975 levels will have to triple by 1985 if we are to approach self-sufficiency.

Q. What must be done to reverse the energy situation in this country?

A. Simply stated, we must close the gap between supply and demand. More precisely, we must adopt an energy policy that establishes specific objectives and provides the incentives and controls required to achieve them.

Our strategy should be to increase domestic production and reserves of oil and gas over the short term, say to 1990, to gain time for development of our alternate energy resources.

Q. Specifically, what steps should be taken to accomplish this?

A. One of the first steps should be to return to a free market for energy. If we decontrol the price of oil and gas, we will have higher prices. Although basically unpopular higher prices would help correct our energy situation in two ways: first, they would slow demand; second, they would spur domestic exploratory drilling, which would eventually increase both our production and reserves. We know from the experiences of other countries that when gasoline costs more than a dollar a gallon, consumption drops substantially.

Higher prices for oil and gas would also encourage greater recovery of oil and gas from existing wells. Only about 35 percent of known oil deposits is recoverable using present technology. Higher prices would allow technological advances which, in turn, would contribute to improvement in recovery rates of existing oil reserves. Also, there are known natural gas deposits that are not now commercial; higher prices would change the economics, encourage drilling in those fields and help increase our gas production.

* 1 Q equals one quadrillion British Thermal Units. This is the energy in 1 trillion cu. ft. of gas or 46 million tons of coal or 180 million bbls. of oil or 243 million megawatt hours.

In addition, we need to triple the use of coal, because it represents our most abundant and easily accessible source of domestic energy.

Q. Do you have other thoughts about making the problem less severe?

A. In addition to accelerating our production, we need to conserve energy through more efficient use. One area subject to significant improvement is power generation. For example, the conversion of natural gas to electricity results in an energy loss of over 30 percent. Also, we should convert immediately to coal for electrical power generation, especially in states that do not produce large quantities of natural gas.

We should declare a moratorium on automotive catalytic converters and exhaust gas recirculation devices, except in the few cities with special environmental problems, and put lead back into gasoline so that we could return to the more efficient engines for automobiles.

We must become reasonable about environmental demands and establish trade offs between energy development and environmental constraints. We need to review many of the obstructive governmental regulations. We have to take a more rational approach to the prudent use of nuclear energy. Finally, research must be increased to develop every practical energy source as rapidly as possible.

Q. Is it too late to head off our energy shortage?

A. Yes, it is too late to completely eliminate the gap between supply and demand over the next 15 years. Some of the actions we just discussed, however, would decrease oil demand by about 2 million barrels per day and increase the oil supply by some 3 to 4 million barrels per day by 1985.

As I said at the outset, time is running out. No matter what we do, we are going to feel the adverse effect of the energy crisis on our living standards. . . .

Williams finds outlook grim. Sen. Harrison A. Williams Jr. (D, N.J.) told the Senate May 11, 1977 that "in regard to national energy security, most energy experts agree that we are in a far more precarious position today than we were at the time of the oil embargo in 1973. Today, after a major energy crisis and the proposals of Project Independence, we find that our imports of all petroleum products have increased to over 50% of the total supply in recent months." Williams continued:

"It is particularly disconcerting that OPEC-enforced increases in crude oil prices have had far-reaching and disruptive impact on our own economic stability. The 1973–74 embargo and the ensuing higher prices resulted in a loss of about 500,000 jobs and approximately $20 billion in our gross national product. The fourfold increase in OPEC oil prices had a tremendous inflationary impact upon our economy. There was an outflow of about $34 billion in 1976 to pay for imported oil, or $160 for each American, 11 times that in 1972. Three years after the supposed end of the crisis the American people continue to face the economic burdens it caused. Even more ominous is the recognition that due to our increased dependence on energy imports, another interruption of our energy supply would be far more disruptive to our economy than the disruption of 1973–74.

"It is equally alarming to see that during this critical period domestic oil production has reached its lowest level in 10 years. The average production rate in 1973 of crude oil and condensate was 9.2 million barrels per day. The current level is reportedly only 8.1 million barrels per day. This reduction will significantly offset the anticipated relief to be brought by the estimated 1.2 million barrels a day we hope will be provided by the Alaskan pipeline upon completion this year. Even the possible realization of an additional 2 million barrels a day from the development of the Outer Continental Shelf will not be sufficient to guarantee independence from foreign oil sources in the near future. It is rapidly becoming evident that as long as our Nation has a high dependency upon petroleum to meet its energy demands, it will also be forced to rely upon oil imports and be a major participant in the international energy trading system.

"Two important policies that the United States has attempted to implement as a result of the energy crisis are the development of alternate energy resources and a commitment to increased energy conservation. Both are clearly necessary, but the progress toward reducing dependence on foreign energy sources has been very disappointing.

"The potential for energy conservation in the United States is very clear. In 1960, the U.S. annual consumption of energy was 247 million BTU's per capita. By 1973, a peak consumption rate of 359 million BTU's per capita had been reached. This was followed by a period of short term conservation, necessitated by the energy being coupled with our domestic recession, and leading to a temporary decline in consumption. In 1975, per capita energy consumption was down to 335 million BTU's. Last year, however, saw a 1.8 percent growth in energy consumed—344 million BTU's per capita. It has been projected that by the end of 1977 we will hit the previous peak of 359 million BTU's. I might also add that our current consumption rate is more than double that of other large industrial nations.

"Our Nation consumes about 33 percent of the world's energy supply. Our large consumption and its relationship to our productive capacity clearly indicate some of the difficulties the United States will face in its pursuit of energy sufficiency. Much of this consumption of energy is unnecessary, and often wasteful. The new administration and Congress have the opportunity, perhaps here as nowhere else, to provide positive leadership to the American people in creating a national energy conservation ethic. By creating an energy conservation ethic, we can both reduce our current energy consumption and provide for an orderly assessment of the direction of our energy development program. This would begin a process of thoughtfully weighing our energy options and accurately evaluating how realistic and hazardous the current pursuit of energy sources actually is. As a first step, forceful conservation policies can insure that the enormous rate of additional consumption will decrease in a manner that will not seriously affect our economic stability. However, conservation alone will not be adequate to provide for our future energy needs.

"Concurrently, the development of alternate and additional energy resources must be continued and expanded, with constant surveillance of their environmental and economical impact. Coal, the most frequently cited alternative, lies in reserves that could reportedly meet energy needs for several hundreds years at the current rate of consumption. Although the development of our coal resources is critically important, the unfortunate technical facts regarding its use suggest that it also cannot completely solve our future energy problems. The processes of coal extraction and conversion are highly expensive, and their possible adverse effects on the environment could be tremendously costly to society in the long run.

"We must also realize that the coal reserves may not last as long as has been forecast. The President's Council on Environmental Quality has estimated that only 30 percent of the coal we extract actually reaches the final user. The rest is exhausted in processing, transportation, conversion to electricity, and electrical leakage. As final demand increases, the quantity of coal exhausted in meeting that demand will also rise. Available coal reserves will be more rapidly depleted, and the problem of finding other sources will still be unsolved. A growth rate of energy consumption of only 7 percent annually would mean doubling energy use every 10 years. This would quickly shorten the time that coal could meet our national energy needs.

"The possibility of increasing our Nation's nuclear energy dependency has also been presented as a solution to the problem of energy self-sufficiency. However, it appears that even if all necessary safeguards are implemented, and the critical environmental and safety issues are finally resolved, nuclear energy will not be a panacea. The Symposium on the Exploration of Uranium Ore Deposits held in Vienna in April 1976 found that expected nuclear power production until the end of this century will require an estimated 4 million short tons of uranium. To meet this demand, discoveries of new uranium deposits will have to rise from the current average of 80,000 short tons per year to about 300,000 short tons per year by 1990, and to almost double that in the following decade. At best nuclear power could supply half of our electricity by 2000, which would represent about one-fourth of our total national energy input.

"Our own domestic reserves of uranium are limited in supply. While the element itself is relatively common in some parts of the Earth's crust, future availability of processed uranium for commercial use will depend heavily on its price, currently over $40 a pound and increasing. If the United States irrevocably commits itself to nuclear energy, we may be required because of the rising worldwide demand for energy to accept some dependency on foreign supply at increased prices.

"Other potential sources of energy are now being investigated; among them are nuclear fusion, solar and wind energy, biological energy conversion, and development and use of geothermal resources. The costs of such research will be high, but it is clear that they are necessary. In addition, the capital investment for these technologies will remain very expensive. However, without an adequate energy supply that the alternatives could provide, the United States will find itself confronted with a serious threat to its economic base, and the imposition of severe hardships on the lives of its citizens.

"Another critical nuclear energy problem has arisen because of the significant increase in the export of nuclear materials and technology throughout the world. At the present time Iran, South Korea, Brazil, Argentina, and several other developing nations are in the process of investing several billions of dollars in nuclear development. The number of nations which export is growing and competition to sell nuclear technology is increasing. In addition, as more facilities for the full nuclear cycle are being exported, commercial competition will reduce the ability of any individual exporting nation to impose safety and accountability standards on purchasers, and at the same time, the potential of importing nations to produce nuclear weapons will increase rapidly. . . ."

FPC issues electric power reports. U.S. electric utilities produced a record 2 036,-486,503 megawatt-hours of electricity in 1976, 6.2% more than in 1975, according to a Federal Power Commission (FPC) report released March 23, 1977. The installed generating capacity of the utilities reached a record level of 531,287 megawatts during the year, up 4.5% from 1975. (A megawatt-hour represented the total energy developed by a power of one million watts acting for one hour.)

To produce electric energy in 1976, utilities burned 448.1 million tons of coal; 555.4 million barrels of fuel oil, and 3,078 billion cubic feet of natural gas. Fuel consumption totals for 1975 had been: coal, 406 million tons; oil, 506.1 million barrels, and gas, 3,158 billion cubic feet.

Of the total net production by electric utilities in 1976, coal-fired plants were responsible for 46.34%, oil-fired plants 15.68%, gas-fired plants 14.45%, nuclear plants 9.38%, hydroelectric plants 13.92% and all other sources .19%.

The FPC March 25 issued similar overall data for the years 1971–76. They showed average annual increases of 4.8% in net electricity production during the five-year period. Installed generating capacity of electric utilities increased at an average annual rate of 7.6%.

According to figures for net electricity production by source of energy:

- The percentage share of nuclear power increased to 9.4% in 1976 from 2.4% in 1971.
- Production from oil, which was 13.6% of the total in 1971, increased to 16.8% in 1973, decreased to 15.1% in 1975 and increased again to 15.7% in 1976.
- Production from natural gas declined to 14.4% of the total in 1976 from 23.2% in 1971.

In other action, the FPC March 11 announced it had ended its round-the-clock monitoring of the natural gas and electric power emergency conditions that had beset the East and Midwest during January and February.

Ford Program

Energy targets reduced. Two goals of the Ford Administration's program to lessen U.S. dependence on foreign oil supplies were quietly abandoned by the White House, according to two news reports.

The Washington Post reported Sept. 21, 1975 that Gerald R. Ford's plan to produce synthetic fuel from coal and shale by 1985 that was the equivalent of 1 million barrels of oil imported daily had been drastically reduced by the White House Energy Resources Council. The new production target for "synfuel" was 350,-000 barrels, according to a report given congressmen by the council.

The staggering cost of the original proposal was given as one reason for scaling down the project. According to current estimates, the more modest synfuel program would cost about $7 billion, compared with more than $20 billion for the larger project.

Other reasons given for reducing the project's scope were environmental hazards associated with the production process, the social upheaval caused by locating large industries in remote Western areas where coal and shale were plentiful, and the shortage of water in these Western regions. (The fuel conversion process required large amounts of water.)

The report also noted that none of the three synthetic fuel projects started in the last 10 years was thriving:

■ The cost of removing oil from Canada's tar sands had tripled since 1970 and had caused several oil companies participating in the venture to withdraw or postpone their activities.

■ Construction of an oil shale plant in Colorado had been delayed because costs had risen from $300 million to $800 million in two years.

■ Applications submitted to the Federal Power Commission to extract a synthetic natural gas from coal had been deferred because of technological troubles and uncertainty about government policy on the decontrol of natural gas prices.

In another indication of scaled-down energy aims, the Journal of Commerce reported Aug. 19 that the Administration no longer found its 2-million-barrel-a-day reduction in oil imports a realistic goal for 1977; instead, the Administration sought to reduce imports (currently totaling 7 million barrels a day) by about 1.5 million barrels.

The more modest proposal was dependent upon Congressional approval of several controversial aspects of the Administration's energy plan, such as decontrol, the opening of production at naval oil reserves, and the conversion of oil-burning utilities to coal.

The continuing impasse over a national energy policy also had caused the White House to make an earlier revision in its most optimistic goal for lessening U.S. dependence on foreign oil. President Ford initially had called for a million-barrel-a-day cutback in oil imports by the end of 1975. That target was abandoned when Congress first balked at the President's legislative program.

Ford's energy message. President Ford Feb. 26, 1976 sent to Congress a special message on energy emphasizing the need to "regain our energy independence." "We must reduce our vulnerability to the economic disruption which a few foreign countries can cause by cutting off our energy supplies or by arbitrarily raising prices," he told Congress.

Among his major proposals was a request for presidential authority, subject to disapproval by Congress, to decide between two competing proposals for delivery of natural gas from Alaska's north slope. The authority for decision currently lay with the Federal Power Commission, subject to judicial review. Under the President's plan, the FPC would be requested to make recommendations on the choice by February 1977.

One of the proposals, from a consortium of U.S. and Canadian companies, was for a 6,280-mile pipeline from Alaska through Canada to points in the western and midwestern U.S. The other proposal, from El Paso Co., was for an 809-mile pipeline from Prudhoe Bay to the southern coast of Alaska. The line generally would parallel that of the Alaskan oil pipeline currently under construction. At the coast terminal, the gas would be liquefied for shipment by tankers to Southern California, where it would be regasified. (The Interior Department had approved the two competing plans Dec. 10, 1975. Both plans, estimated to cost about $7 billion, were described by Interior Secretary Thomas S. Kleppe as "technically and economically feasible.")

The President requested a provision to prohibit legal challenges to the gasline decision, once approved by Congress. The request was made in light of the requisite Canadian cooperation on the gasline, to bar possible delay from protracted litigation after the Canadian decision on a route, if it were made in favor of the Canadian route.

The message reiterated the Administration's proposal for deregulation of new natural gas. The President said the higher prices from removal of price controls would stimulate 25% more production by 1985.

The President announced in his message that he would take administrative action to limit imports of liquefied natural gas to one trillion cubic feet a year by 1985, or about 5% of total U.S. natural gas demand. Although little liquefied natural gas was currently imported, there were proposals to import it from Nigeria and Algeria, and the Administration was said to be wary of getting into a dependent situation to foreign suppliers.

Another major proposal in the message was for a $1 billion authorization for financial assistance to states and localities for coping with problems flowing from development of energy resources on federal lands, such as offshore oil reserves. The proposed grants, loans and

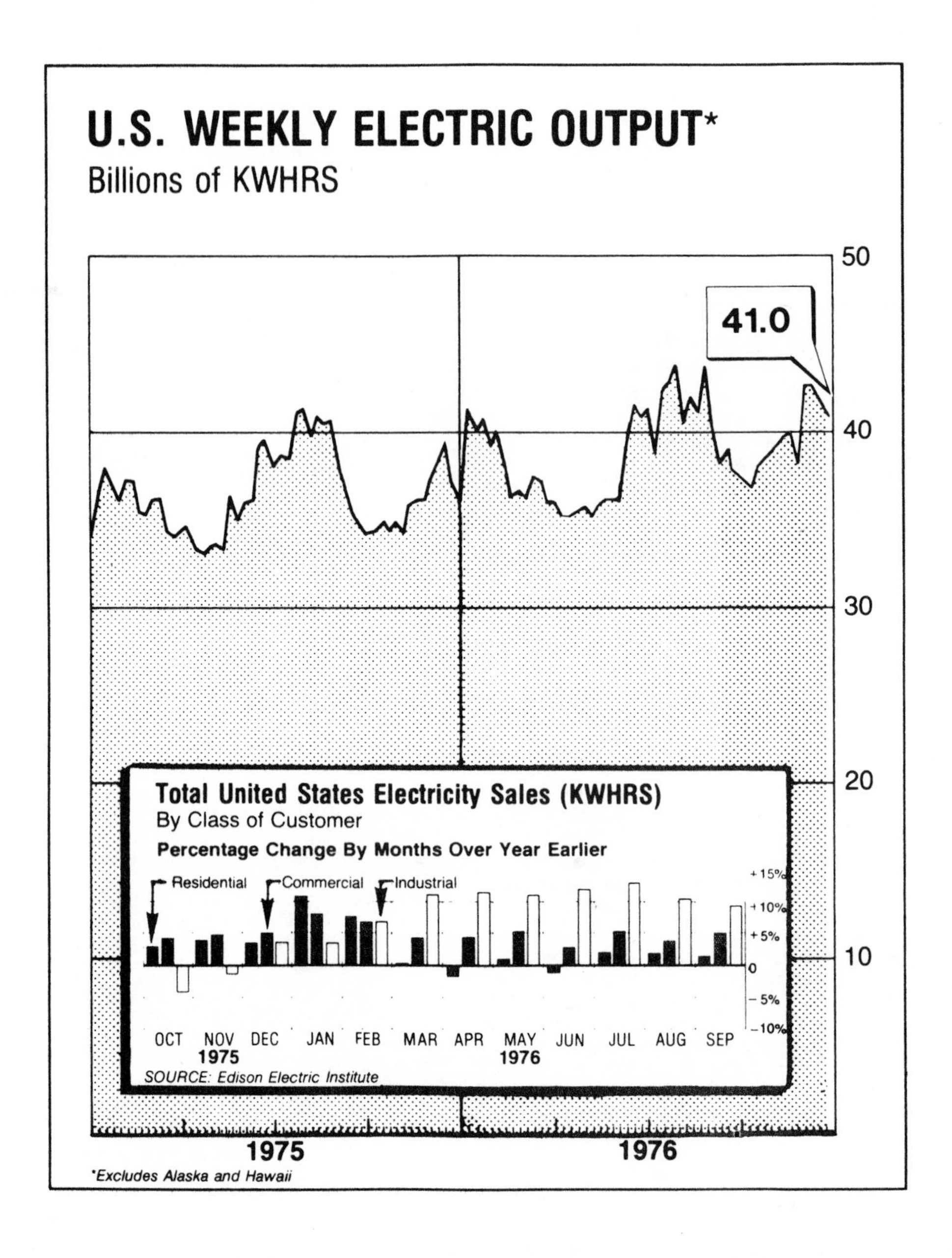
U.S. WEEKLY ELECTRIC OUTPUT*
Billions of KWHRS
41.0
50
40
30
20
10
Total United States Electricity Sales (KWHRS)
By Class of Customer
Percentage Change By Months Over Year Earlier
Residential
Commercial
Industrial
+15%
+10%
+5%
0
−5%
−10%
OCT NOV DEC JAN FEB MAR APR MAY JUN JUL AUG SEP
1975
1976
SOURCE: Edison Electric Institute
1975
1976
*Excludes Alaska and Hawaii

loan guarantees would be extended over a 15-year period for roads, schools, water facilities and other community development projects.

The President urged Congress to complete action on legislation to authorize production of oil from the naval petroleum reserves.

In the nuclear field, Ford reaffirmed the Administration's commitment for further development of the breeder-type of nuclear power reactor. The President also requested again legislation to open the uranium enrichment operation to private business.

For the International Atomic Energy Agency, the President sought a special contribution of up to $5 million worth of equipment, training and technical assistance to strengthen efforts against diversion of nuclear materials into weaponry.

Text of the President's message:

To the Congress of the United States:

A little over two years ago, the Arab embargo proved that our Nation had become excessively dependent upon others for our oil supplies. We now realize how critical energy is to the defense of our country, to the strength of our economy, and to the quality of our lives.

We must reduce our vulnerability to the economic disruption which a few foreign countries can cause by cutting off our energy supplies or by arbitrarily raising prices. We must regain our energy independence.

During the past year, we have made some progress toward achieving our energy independence goals, but the fact remains that we have a long way to go. However, we cannot take the steps required to solve our energy problems until the Congress provides the necessary additional authority that I have requested If we do not take these steps, our vulnerability will increase dramatically.

In my first State of the Union Address last year, I pointed out that our vulnerability would continue to grow unless a comprehensive energy policy and program were implemented. I outlined these goals for regaining our energy independence:

—First, to halt our growing dependence on imported oil during the next few critical years.

—Second, to attain energy independence by 1985 by achieving invulnerability to disruptions caused by oil import embargoes. Specifically, we must reduce oil imports to between 3 and 5 million barrels a day, with an accompanying ability to offset any future embargo with stored petroleum reserves and emergency standby measures.

—Third, to mobilize our technology and resources to supply a significant share of the free world's energy needs beyond 1985.

In pursuing these goals, we have sought to provide energy at the lowest cost consistent with our need for adequate and secure supplies We should rely upon the private sector and market forces since it is the most efficient means of achieving these goals. We must also achieve a balance between our environmental and energy objectives.

These goals were reasonable and sound a year ago and they remain so today.

Since January of 1975, this Administration has initiated the most comprehensive set of energy programs possible under current authority. This includes actions to conserve energy, to increase the production of domestic energy resources, and to develop technology necessary to produce energy from newer sources.

During this time, I have also placed before the Congress a major set of legislative proposals that would provide the additional authority that is needed to achieve our energy independence goals.

Thus far, the Congress has completed action on only one major piece of energy legislation—the Energy Policy and Conservation Act—which I signed into law on December 22, 1975. That law includes four of the original proposals I submitted to the Congress over a year ago. achieve our energy independence goals. posals still await final action by the Congress.

NATURAL GAS

The need for Congressional action is most critical in the area of natural gas. We must reverse the decline in natural gas production and deal effectively with the growing shortages that face us each winter.

Deregulating the price of new natural gas remains the most important action that can be taken by the Congress to improve our future gas supply situation. If the price of natural gas remains under current regulation, total domestic production will decline to less than 18 trillion cubic feet in 1985. However, if deregulation is enacted, production would be about 25 percent higher by 1985. Natural gas shortages mean higher costs for consumers who are forced to switch to more expensive alternative fuels and mean, inevitably, an increasing dependence on imported oil. Curtailment of natural gas to industrial users in the winters ahead means more unemployment and further economic hardships.

Therefore, I again urge the Congress to approve legislation that will remove Federal price regulation from new natural gas supplies and will provide the added short-term authorities needed to deal with any severe shortages forecast for next winter.

I also urge prompt action by the Congress on a bill I will be submitting shortly which is designed to expedite the selection of a route and the construction of a transportation system to bring the vast supplies of natural gas from the North Slope of Alaska to the "lower 48" markets. This legislation would make possible production of about 1 trillion cubic feet of additional natural gas each year by the early 1980s.

We expect imports of liquefied natural gas (LNG) to grow in the next several years to supplement our declining domestic supply of natural gas. We must balance these supply needs against the risk of becoming overly dependent on any particular source of supply.

Recognizing these concerns, I have directed the Energy Resources Council to establish procedures for reviewing proposed contracts within the Executive Branch, balancing the need for supplies with the need to avoid excessive dependence, and encouraging new imports where this is appropriate. By 1985, we should be able to import 1 trillion cubic feet of LNG to help meet our needs without becoming overly dependent upon foreign sources.

NUCLEAR POWER

Greater utilization must be made of nuclear energy in order to achieve energy independence and maintain a strong economy. It is likewise vital that we continue our world leadership as a reliable supplier of nuclear technology in order to assure that worldwide growth in nuclear power is achieved with responsible and effective controls.

At present 57 commercial nuclear power plants are on line, providing more than 9 percent of our electrical requirements, and a total of 179 additional plants are planned or committed. If the electrical power supplied by the 57 existing nuclear power plants were supplied by oil-fired plants, an additional one million barrels of oil would be consumed each day.

On January 19, 1975, I activated the independent Nuclear Regulatory Commission (NRC) which has the responsibility for assuring the safety, reliability, and environmental acceptability of commercial nuclear power. The safety record for nuclear power plants is outstanding. Nevertheless, we must continue our efforts to assure that it will remain so in the years ahead. The NRC has taken a number of steps to reduce unnecessary regulatory delays and is continually alert to the need to review its policies and procedures for carrying out its assigned responsibilities.

I have requested greatly increased funding in my 1977 budget to accelerate research and development efforts that will meet our short-term needs to:

—make the safety of commercial nuclear power plants even more certain;
—develop further domestic safeguards technologies to assure against the theft and misuse of nuclear materials as the use of nuclear-generated electric power grows;
—provide for safe and secure long-term storage of radioactive wastes;
—and encourage industry to improve the reliability and reduce the construction time of commercial nuclear power plants.

I have also requested additional funds to identify new uranium resources and have directed ERDA to work with private industry to determine what additional actions are needed to bring capacity on-line to reprocess and recycle nuclear fuels.

Internationally, the United States in consultation with other nations which supply nuclear technology has decided to follow stringent export principles to ensure that international sharing of the benefits of nuclear energy does not lead to the proliferation of nuclear weapons. I have also decided that the U.S. should make a special contribution of up to $5 million in the next 5 years to strengthen the safeguards program of the International Atomic Energy Agency.

It is essential that the Congress act if we are to take timely advantage of our nuclear energy potential. I urge enactment of the Nuclear Licensing Act to streamline the licensing procedures for the construction of new powerplants.

I again strongly urge the Congress to give high priority to my Nuclear Fuel Assurance Act to provide enriched uranium needed for commercial nuclear power plants here and abroad. This proposed legislation which I submitted in June 1975, would provide the basis for transition to a private competitive uranium enrichment industry and prevent the heavy drain on the Federal budget. If the Federal Government were required to finance the necessary additional uranium enrichment capacity, it would have to commit more than $8 billion over the next 2 to 3 years and $2 billion annually thereafter. The taxpayers would eventually be repaid for these expenditures but not until sometime in the 1990's. Federal expenditures are not necessary under the provisions of this Act since industry is prepared to assume this responsibility with limited government cooperation and some temporary assurances. Furthermore, a commtment to new Federal expenditures for uranium enrichment could interfere with efforts to increase funding for other critical energy programs.

COAL

Coal is the most abundant energy resource available in the United States, yet production is at the same level as in the 1920's and accounts for only about 17 percent of the Nation's energy consumption. Coal must be used increasingly as an alternative to scarce, expensive or insecure oil and natural gas supplies. We must act to remove unnecessary constraints on coal so that production can grow from the 1975 level of 640 million tons to over 1 billion tons by 1985 in order to help achieve energy independence.

We are moving ahead where legislative authority is available.

The Secretary of the Interior has recently adopted a new coal leasing policy for the leasing and development of more coal on Federal lands. To implement this policy, regulations will be issued governing coal mining operations on Federal lands, providing for timely development, and requiring effective surface mining controls which will minimize adverse environmental impacts and require that mined lands be reclaimed. As a reflection of the States' interests, the Department proposes to allow application on Federal lands of State coal mine reclamation standards which are more stringent than Federal standards, unless overriding National interests are involved.

I have directed the Federal Energy Administration and the Environmental Protection Agency to work toward the conversion of the maximum number of utilities and major industrial facilities from gas or oil to coal as permitted under recently extended authorities.

We are also stepping up research and development efforts to find better ways of extracting, producing, and using coal.

Again, however, the actions we can take are not enough to meet our goals. Action by the Congress is essential.

I urge the Congress to enact the Clean Air Act amendments I proposed which will provide the balance we need between air quality and energy goals. These amendments would permit greater use of coal without sacrificing the air quality standards necessary to protect public health.

OIL

We must reverse the decline in the Nation's oil production. I intend to implement the maximum production incentives that can be justified under the new Energy Policy and Conservation Act. In addition, the Department of the Interior will continue its aggressive Outer Continental Shelf development program while giving careful attention to environmental considerations.

But these actions are not enough. We need prompt action by the Congress on my proposals to allow production from the Naval Petroleum Reserves. This legislation is now awaiting action by a House-Senate Conference Committee.

Production from the Reserves could provide almost one million barrels of oil per day by 1985 and will provide both the funding and the oil for our strategic oil reserves.

I also urge the Congress to act quickly on amending the Clean Air Act auto emission standards that I proposed last June to achieve a balance between objectives for improving air quality, increasing gasoline mileage, and avoiding unnecessary increases in costs to consumers.

BUILDING ENERGY FACILITIES

In order to attain energy independence for the United States, the construction of numerous nuclear power plants, coal-fired power plants, oil refineries, synthetic fuel plants, and other facilities will be required over the next two decades.

Again, action by the Congress is needed.

I urge Congress to approve my October, 1975 proposal to create an Energy Independence Authority, a new government corporation to assist private sector financing of new energy facilities.

This legislation will help assure that capital is available for the massive investment that must be made over the next few years in energy facilities, but will not be forthcoming otherwise. The legislation also provides for expediting the regulatory process at the Federal level for critical energy projects.

I also urge Congressional action on legislation needed to authorize loan guarantees to aid in the construction of commercial facilities to produce synthetic fuels so that they may make a significant contribution by 1985.

Commercial facilities eligible for funding under this program include those for synthetic gas, coal liquefaction and oil shale, which are not now economically competitive. Management of this program would initially reside with the Energy Research and Development Administration but would be transferred to the proposed Energy Independence Authority.

My proposed energy facilities siting legislation and utility rate reform legislation, as well as the Electric Utilities Construction Incentives Act complete the legislation which would provide the incentives, assistance, and new procedures needed to assure that facilities are available to provide additional domestic energy supplies.

ENERGY DEVELOPMENT IMPACT ASSISTANCE

Some areas of the country will experience rapid growth and change because of the development of Federally-owned energy resources. We must provide special help to heavily impacted areas where this development will occur.

I urge the Congress to act quickly on my proposed new, comprehensive, Federal Energy Impact Assistance Act which was submitted to the Congress on February 4, 1976.

This legislation would establish a $1 billion program of financial assistance to areas affected by new Federal energy resource development over the next 15 years. It would provide loans, loan guarantees, and planning grants for energy-related public facilities. Funds would be repaid from future energy development. Repayment of loans could be forgiven if development did not occur as expected.

This legislation is the only approach which assures that communities that need assistance will get it where it is needed, when it is needed.

ENERGY CONSERVATION

The Nation has made major progress in reducing energy consumption in the last 2 years but greatly increased savings can yet be realized in all sectors.

I have directed that the Executive Branch continue a strong energy management program. This program has already reduced energy consumption by 24 percent in the past 2 years, saving the equivalent of over 250,000 barrels of oil per day.

We are moving to implement the conservation authorities of the new Energy Policy and Conservation Act, including those calling for State energy conservation programs, and labeling of appliances to provide consumers with energy efficiency information.

I have asked for a 63 percent increase in funding for energy conservation research and development in my 1977 budget.

If the Congress will provide needed legislation, we will make more progress. I urge the Congress to pass legislation to provide for thermal efficiency standards for new buildings, to enact my proposed $55 million weatherization assistance program for low-income and elderly persons, and to provide a 15-percent tax credit for energy conservation improvements in existing residential buildings. Together, these conservation proposals can save 450,000 barrels of oil per day by 1985.

INTERNATIONAL ENERGY ACTIVITIES

We have also made significant progress in establishing an international energy policy. The United States and other major oil-consuming nations have established a comprehensive long-term energy program through the International Energy Agency—IEA—committing ourselves to continuing cooperation to reduce dependence on imported oil. By reducing demand for imported oil, consuming nations can, over time, regain their influence over oil prices and end vulnerability to abrupt supply cutoffs and unilateral price increases.

The International Energy Agency has established a framework for cooperative efforts to accelerate the development of alternative energy sources. The Department of State, in cooperation with FEA, ERDA, and other Federal agencies, will continue to work closely with the IEA.

While domestic energy independence is an essential and attainable goal, we must recognize that this is an interdependent world. There is a link between economic growth and the availability of energy at reasonable prices. The United States will need some energy imports in the years ahead. Many of the other consuming nations will not be energy independent. Therefore, we must continue to search for solutions to the problems of both the world's energy producers and consumers.

The U.S. delegation to the new Energy Commission will pursue these solutions, including the U.S. proposal to create an International Energy Institute. This Institute will mobilize the technical and financial resources of the industrialized and oil-producing countries to assist developing countries in meeting their energy problems.

1985 AND BEYOND

As our easily recoverable domestic fuel reserves are depleted, the need for advancing the technologies of nuclear energy, synthetic fuels, solar energy, and geothermal energy will become paramount to sustaining our energy achievements beyond 1985. I have therefore proposed an increase in the Federal budget for energy research and development from $2.2 billion in 1976 to $2.9 billion in the proposed 1977 budget. This 30-percent increase represents a major expansion of activities directed at accelerating programs for achieving long-term energy independence.

These funds are slated for increased work on nuclear fusion and fission power development, particularly for demonstrating the commercial viability of breeder reactors; new technology development for coal mining and coal use; enhanced recovery of oil from current reserves; advanced power conversion systems; solar and geothermal energy development; and conservation research and development.

It is only through greater research and development efforts today that we will be in a position beyond 1985 to supply a significant share of the free world's energy needs and technology.

SUMMARY

I envision an energy future for the United States free of the threat of embargoes and arbitrary price increases by foreign governments. I see a world in which all nations strengthen their cooperative efforts to solve critical energy problems. I envision a major expansion in the production and use of coal, aggressive exploration for domestic oil and gas, a strong commitment to nuclear power, significant technological breakthroughs in harnessing the unlimited potential of solar energy and fusion power, and a strengthened conservation ethic in our use of energy.

I am convinced that the United States has the ability to achieve energy independence.

I urge the Congress to provide the needed legislative authority without further delay.

GERALD R. FORD.

THE WHITE HOUSE, *February 26, 1976.*

Nadar attacks Ford policy. Consumer advocate Ralph Nader sponsored a national conference in Washington May 20–21, 1976 on public policies to promote energy efficiency. In his introductory remarks, Nader attacked the Ford Administration for alleged failures in energy programs and policies:

> As Roger Sant, the Federal Energy Administration's departing energy conservation administrator, has said, by improving our energy efficiency "we could stop our energy growth right now," while our economy continued to expand. A man who talks like that can't expect to last long in the energy-growth-oriented Ford Administration and, sure enough, Sant is on his way out.
>
> There are basically two energy efficiency problems in this country today. One is the intolerable gap between our present efficiency of energy use and the energy efficiency which we know is economically and technically feasible. The second problem can be characterized as another kind of energy efficiency gap, the "White House energy efficiency gap."
>
> The White House energy efficiency gap is the gap between White House lip service to the goal of using energy more efficiently and actual Ford Administration policies regarding energy efficiency. . . .
>
> The only thing more scandalous than the massive waste of energy in our society, with its corrosive effect on our standard of living and our national security, is the footdragging of the White House when it comes to public policies that would promote energy efficiency. Three recent Ford Administration actions regarding energy policy illustrate how the White House has made energy efficiency a controversial issue rather than a consensus policy which accurately reflects our national interests.
>
> First, there is the Energy Research and Development Administration's budget. While the White House preaches energy efficiency, it practices atomic socialism. The ERDA budget is overwhelmingly devoted to the nuclear fuel cycle and the breeder reactor program. . . .
>
> The second illustration of the White House energy efficiency gap was provided by the Ford Administration's testimony on the proposed Energy Conservation Act of 1976, which would encourage energy efficiency improvements where they would have the most effect, in the vast population of existing homes, commercial buildings, and factories. Many

existing buildings are literally designed to waste energy. They are overlighted, overheated, overcooled, overventilated, and overexposed to the outdoor environment by poor design and inadequate insulation. If they were automobiles, we would call them "lemons," and there would be massive recalls for correction of defects.

The proposed Energy Conservation Act would do the next best thing. Among other things, it would provide loan guarantees and interest subsidies to make it easier for the owners of these structural lemons to borrow money to invest in energy efficiency improvements. It would also finance state energy conservation programs which would educate consumers on how to save energy and provide energy audits to guide wise investments in energy saving modifications by homeowners, businessmen, and manufacturers.

Surely such modest proposals to guide energy users to improved efficiency would be expected to gain the support of an Administration committed to a Project Independence program.

Yet when hearings were held on the Energy Conservation Act in February, the White House sent a Federal Energy Administration spokesman to tell the Congress that the Energy Conservation Act is "premature," and that any major public policy initiative in the area of energy efficiency "ought to be preceded by sound analysis of consumer and industrial behavior, which could give us some indication of the effectiveness of measures to promote energy conservation."

Just think about that advice for a moment. That's like telling a person who is hemorrhaging from an open wound to sit back and study the problem rather than stifling the bleeding . . .

In addition to the footdragging on energy efficiency initiatives represented by its budget priorities and its opposition to the Energy Conservation Act, a third White House policy which illustrates the White House energy efficiency gap is the Ford-Rockefeller proposal for a $100 billion Energy Independence Authority. Although the White House opposes the Energy Conservation Act's encouragement of investments in energy efficiency, it supports the EIA, which would allocate capital to synthetic fuels plants, uranium enrichment, atomic fuel reprocessing, and the federal purchase of atomic power plants for lease to electric utilities. This proposal should be called the "Energy Cartel Subsidy Act," since its basic purpose is to subsidize our domestic energy cartel.

The EIA is the preeminent example of the Ford Administration's general policy of favoring programs which divert scarce capital to energy producers as the primary answer to our energy problems. Although the departing Roger Sant has repeatedly testified and lectured that a barrel of oil saved is as good as a barrel produced, the White House isn't listening. . . .

In addition to reducing our need for imported oil, investments in energy efficiency can improve our economy by reducing inflation through improved economic efficiency and, in the opinion of many experts, increase employment because energy efficiency investments tend to create more jobs than energy supply investments.

The persistence of the White House support for policies that would divert capital into less productive energy supply investments in preference to policies that would promote energy efficiency investments is, in short, a policy of less bang for the buck. Such a policy is based on the White House's undue reliance on the energy supply industry for guidance on energy policy. It is time for the White House to hear from experts in the field of energy efficiency. . . .

Energy funds for fiscal '77. President Ford July 12, 1976 signed legislation appropriating $5.75 billion in fiscal 1977 for a variety of energy programs administered by the Energy Research and Development Administration (ERDA). The bill also provided $3.97 billion in fiscal 1977 for the Army Corps of Engineers, the Interior Department and a number of independent federal and regional agencies. The $3.97 billion was for public water and power projects.

The bill had cleared Congress June 29 when the Senate approved it by voice vote, the House 381–15. Congress beefed up the Administration's request for solar energy programs, appropriating a total of $290.4 million. (The Administration had requested $141.8 million for ERDA solar energy operating expenses; the bill allocated $258.5 million for this.) Congress approved without revision an Administration request for $630.26 million for operating expenses of nuclear-fission reactor programs. Included in that amount were funds for a demonstration breeder reactor in Tennessee. The total appropriation for nuclear-fission projects was $762.7 million. Nuclear-fusion projects received $426.5 million, with Congress increasing the Administration's request for ERDA operating expenses from $239.4 million to $275.0 million. Among the appropriations for operating expenses of other ERDA programs were (the Administration proposal, when different, is given in parenthesis):

- Geothermal energy, $53.2 million ($48.6 million).
- Environmental research and safety, $253 million ($239.5 million).
- High energy physics, $170 million ($167.5 million).

■ Basic energy sciences, $197.4 million ($182.8 million).

■ Nuclear materials security, $27.42 million ($25.74 million).

■ Nuclear weapons, $1.362 billion ($1.367 billion).

■ Space nuclear systems, $31 million.

■ Naval reactors, $191.5 million.

■ Nuclear explosives applications, $1.3 million.

■ Uranium enrichment, $925.175 million.

A $1.572-billion appropriation was included for plant and capital equipment expenditures by ERDA.

Incentives bill. President Ford Aug. 14, 1976 signed a bill raising the price of domestic oil and authorizing loan and other incentives to spur energy-conservation investments. The bill also extended the life of the Federal Energy Administration through Dec. 31, 1977, authorizing $227 million for that agency in fiscal 1977.

The FEA's statutory authorization had expired July 30. Ford that day signed an executive order permitting the agency to continue its work—technically, the order created a Federal Energy Office—pending enactment of the bill then being considered by Congress. The bill cleared Congress Aug. 10 when the House approved it, 293–88. The Senate had passed it Aug. 5 by voice vote.

The provisions dealing with oil:

■ Eliminated the price ceiling—$11.63 at the time of the bill's passage—on oil coming from "stripper" wells (those producing less than 10 barrels a day). Stripper wells accounted for about 12% of domestic oil production.

■ Permitted the administration to offer price incentives for oil drawn from wells by "tertiary" methods (technologically advanced methods used to extract oil that could not be pumped by conventional means).

■ Authorized a 10% annual increase in the price of domestic crude oil. Previous legislation allowed a 3% annual price increase, plus additional increases up to an overall maximum of 10% to offset inflation.

Energy conservation items included:

■ A requirement that the Department of Housing and Urban Development formulate energy-conservation construction standards. After approval by both houses of Congress, the standards were intended to be incorporated in local building standards.

■ Authorization of $2 billion in loan guarantees to help finance purchase of energy-conserving equipment by hospitals, universities, local governments, owners of large apartment buildings and small businesses. Individual loans would be limited to $5 million.

■ Authorization of $200 million over two years for HUD-administered pilot projects that would give loans, grants or loan guarantees to homeowners or renters who insulated their homes or installed storm windows. These subsidies or loans could cover up to 20% of the cost of a project but not more than $400.

■ Authorization of $200 million over three years to provide free insulation for the homes of low-income persons.

Bills not cleared. Congress, when it adjourned Oct. 2, 1976, had failed to clear many energy-related bills that had been on the agenda of either the Republican Administration or the Democratic leadership in Congress. Some bills were casualties of parliamentary delaying tactics in the last, time-pressed days of the session. Other bills that had Democratic backing were held back because they were opposed by President Ford and it was unlikely he would sign them. Conversely, some Administration bills never emerged from committee in the Democratic-controlled Congress.

Among energy-related measures not cleared by Congress in 1976:

■ A bill instituting daylight saving time seven months—rather than six—each year. The bill passed the Senate, but was defeated in the House Sept. 21.

■ A bill amending the 1970 Clean Air Act. The bill delayed imposition of auto-emission standards and contained provisions limiting permissible pollution in areas with clean air ("nondegradation"). Critics of the nondegredation section said it would impair the economy and prevent development of new energy supplies. The

Administration had opposed the provisions, as had business groups, electric utilities, oil companies, and real estate and construction spokesmen. The automobile industry—backed by the Administration—had said that even the eased emission-limit deadlines were too stringent. The bill, a compromise between House and Senate versions achieved in the last week of the session, was not voted on in the House. In the Senate it was killed by a filibuster led by Jake Garn (R, Utah).

■ A bill revising regulations on leasing to industry of outer-continental-shelf oil-and-gas rights. It had been opposed by the oil-and-gas industry and the Administration. A conference version was killed in the House Sept. 28.

■ A bill to regulate strip-mining and to set standards for reclamation of previously stripped lands. The bill, similar to legislation vetoed by Ford, was blocked in committee.

■ Bills to deregulate the price of natural gas. The House passed a bill providing for some deregulation, and the Senate considered a separate bill that retained regulation but provided for a substantial increase in price. No final action was taken.

■ A bill providing for the breakup—along functional lines (production, refining, marketing, and transportation)—of the major oil companies. The so-called "divestiture" bill was approved 8–7 by the Senate Judiciary Committee June 15, but advanced no further.

■ A bill permitting private industry to enrich uranium under specific contracts subject to congressional approval and with federal guarantees for investors. Uranium enrichment had been a monopoly of the federal government. The bill, backed by the Administration and the nuclear industry, passed the House. The Senate failed to vote on it.

■ A bill providing multibillion-dollar-loan guarantees for the development of synthetic fuels from coal, oil shale and other resources. The House rejected the bill Sept. 23. The Senate failed to act on similar legislation.

■ A bill regulating the export of nuclear materials to insure that they would not be reprocessed to make weapons. One bill setting strict standards was reported out of committee in the Senate, but not voted on. The Administration opposed that bill and favored another that only directed the Administration to negotiate safeguards to control the spread of nuclear-weapons capability. (Nuclear export controls were also incorporated in a bill dealing with the Arab boycott. That bill also died.)

Congress gets oil-stockpile plan. The Federal Energy Administration sent to Congress Dec. 15, 1976 a plan to stockpile oil for use in another embargo. The plan called for storing 500 million barrels of crude oil, the equivalent of a 90-day supply of imported oil, in underground salt domes in Texas and Louisiana that were near existing pipeline systems.

The $8-billion price tag for the program based on a plan to ask foreign countries and domestic oil companies to bid competitively for the supply contracts.

'Backsliding' scored. An energy task force sponsored by the Twentieth Century Fund Jan. 11, 1977 issued a report criticizing what it called "backsliding" instead of "progress" in U.S. energy programs. (The fund was a research foundation that conducted policy-oriented studies in the social sciences.)

Among the recommendations of the report: drafting of a U.S. energy policy; establishment of an emergency three-to-six-month stockpile of oil; implementation of cost-oriented guidelines for environmental protection efforts, and use of new incentive programs to encourage the conservation of energy, to expand the use of domestic coal reserves and to promote the development of new sources of energy.

Final Ford budget. Gerald R. Ford was defeated by Jimmy Carter Nov. 2, 1976 in his campaign for reelection as President. Ford submitted his final budget as President Jan. 17, 1977. The energy request of $14.2 billion was a 34% boost over the $10.6-billion figure for fiscal 1977. Most of the increase was directed toward production of uranium-enrichment facilities and an emergency oil-storage program authorized by Congress in 1975.

For the latter, $1.69 billion was requested, of which $1.3 billion was for purchase of the oil. The remainder would be for preparation of storage facilities, which were slated to be underground salt domes and mines mostly along the Gulf Coast of Texas and Louisiana.

The uranium-enrichment program would consume $1.2 billion in fiscal 1978 outlays for the purpose of upgrading three government-owned plants and expanding another.

Other projects included exploration of the national petroleum reserve in Alaska. (Jurisdiction over the reserve was scheduled to be transferred in June from the Navy to the Interior Department.) The budget item for it was $209.5 million.

A $55-million request was made for a program to encourage insulation of homes of low-income persons.

President Ford renewed a request first made in 1975 for an energy authority to provide loans and backing for energy projects facing financial problems. Financing authority of $100 billion was envisioned, with up to $10 billion of it for use in fiscal 1978. Since the authority's loans would be repaid, its impact on the fiscal 1978 budget was small, appearing as a $42-million operating loss.

A 28% increase was budgeted for energy research—$3.7 billion, up from the fiscal 1977 level of $2.9 billion. Nuclear research and development, at $2 billion compared to $1.5 billion in fiscal 1977, accounted for much of the spending. The $1.4 billion for non-nuclear research was a $200 million gain from fiscal 1977.

The Ford Administration sought a 24% increase in outlays to $736 million, from $595 million in fiscal 1977, for nuclear fission programs, mainly the controversial liquid-metal, fast-breeder reactor using plutonium, a major ingredient in the atomic bomb.

A 34% increase to $431 million was scheduled for nuclear fusion-power development, a relatively new technology being explored for commercial electrical power.

Research for more efficient use of coal and oil was budgeted at $500 million in fiscal 1978, up from $445 million. Research on energy conservation was to rise 12% to $140 million.

Ford proposed that research on solar heating and cooling be maintained at the current $61-million level and that research concerning the conversion of sun power directly into electricity be expanded 42% to the $173-million level.

1976 Political Positions

Energy policy was one of the major topics discussed during the 1976 Presidential election campaign.

Democratic platform. The Democratic national platform made the following statements on the energy issue:

Almost three years have passed since the oil embargo. Yet, by any measure, the nation's energy lifeline is in far greater peril today. America is running out of energy—natural gas, gasoline and oil.

The economy is already being stifled. The resulting threat of unemployment and diminished production is already present.

If America, as we know it, is to survive, we must move quickly to develop renewable sources of energy.

The Democratic Party will strive to replace the rapidly diminishing supply of petroleum and natural gas with solar, geothermal, wind, tide and other forms of energy, and we recommend that the federal government promptly expend whatever funds are required to develop new systems of energy.

We have grown increasingly dependent on imported oil. Domestic production, despite massive price increases, continues to decline. Energy stockpiles, while authorized, are yet to be created. We have no agreements with any producing nations for security of supply. Efforts to develop alternative energy sources have moved forward slowly. Production of our most available and plentiful alternative—coal—is not increasing. Energy conservation is still a slogan, instead of a program.

Republican energy policy has failed because it is based on illusions; the illusion of a free market in energy that does not exist, the illusion that ever-increasing energy prices will not harm the economy, and the illusion of an energy program based on unobtainable independence.

The time has come to deal with the realities of the energy crisis, not its illusions. The realities are that rising energy prices, falling domestic supply, increasing demand, and the threat to national security of growing imports, have not been contained by the private sector.

The Democratic energy platform begins with a recognition that the federal government has an important role to play in insuring the nation's energy future, and that it must be given the tools it needs to protect the economy and the nation's consumers from arbitrary and excessive energy price increases and help the nation embark on a massive domestic energy program focusing on conservation, coal conversion, exploration and development of new technologies to insure an adequate short-term and long-term supply of energy for the nation's needs. A nation advanced enough and wealthy enough to send a man to the moon must dedicate itself to developing alternate sources of energy.

Energy pricing. Enactment of the Energy Policy and Conservation Act of 1975 established oil ceiling prices at levels sufficient to maximize domestic production but still below OPEC equivalents. The act was a direct result of the Democratic Congress' commitment to the principle that beyond certain levels, increasing energy prices simply produce high-cost energy—without producing any additional energy supplies.

This oil-pricing lesson should also be applied to natural gas. Those now pressing to turn natural-gas price regulation over to OPEC, while arguing the rhetoric of so-called deregulation, must not prevail. The pricing of new natural gas is in need of reform. We should narrow the gap between oil and natural gas prices with new natural-gas ceiling prices that maximize production and investment while protecting the economy and the consumer. Any reforms in the pricing of new natural gas should not be at the cost of severe economic dislocations that would accelerate inflation and increase unemployment.

An examination must be made of advertising cost policies of utilities and the imposition of these costs on the consumer. Advertising costs used to influence public policy ought to be borne by stockholders of the utility companies and not by the consumers.

Domestic supply and demand. The most promising neglected domestic option for helping balance our energy budget is energy conservation. But major investments in conservation are still not being made.

The Democratic Party will support legislation to establish national building-performance standards on a regional basis designed to improve energy efficiency. We will provide new incentives for aiding individual homeowners, particularly average income families and the poor in undertaking conservation investments. We will support the reform of utility rate structures and regulatory rules to encourage conservation and ease the utility rate burden on residential users; farmers and other consumers who can least afford it; make more efficient use of electrical generating capacity; and we will aggressively pursue implementation of automobile efficiency standards and appliance labeling programs already established by Democratic initiative in the Energy Policy and Conservation Act.

Coal currently comprises 80% of the nation's energy resources, but produces only 16% of the nation's energy. The Democratic Party believes that the United States' coal production can and must be increased without endangering the health and safety of miners, diminishing the land and water resources necessary for increased food production, and sacrificing the personal and property rights of farmers, ranchers and Indian tribes.

We must encourage the production of the highest quality coal, closest to consuming markets, in order to insure that investments in energy production reinforce the economics of energy producing and consuming regions. Improved rail transportation systems will make coal available where it is actually needed, and will insure a rail transport network required for a healthy industrial and agricultural economy.

We support an active federal role in the research and development of clean burning and commercially competitive coal burning systems and technologies, and we encourage the conversion to coal of industrial users of natural gas and imported oil. Air quality standards that make possible the burning of coal without danger to the public health or degradation of the nation's clear air must be developed and implemented.

The Democratic Party wants to put an end to the economic depression, loss of life and environmental destruction that has long accompanied irresponsible coal development in Appalachia. Strip mining legislation designed to protect and restore the environment, while ending the uncertainty over the rules governing future coal mining, must be enacted.

The huge reserves of oil, gas and coal on federal territory, including the outercontinental shelf, belong to all the people. The Republicans have pursued leasing policies which give the public treasury the least benefit and the energy industry the most benefit from these public resources. Consistent with environmentally sound practices, new leasing procedures must be adopted to correct these policies, as well as insure the timely development of existing leases.

Major federal initiatives, including major governmental participation in early high-risk development projects, are required if we are to harness renewable resources like solar, wind, geothermal, the oceans, and other new technologies such as fusion, fuel cells and the conversion of solid waste and starches into energy. The Ford Administration has failed to provide those initiatives, and, in the process, has denied American workers important new opportunities for employment in the building and servicing of emerging new energy industries.

U.S. dependence on nuclear power should be kept to the minimum necessary to meet our needs. We should apply stronger safety standards as we regulate its use. And we must be honest with our people concerning its problems and dangers as well as its benefits.

An increasing share of the nuclear research dollar must be invested in finding better solutions to the problems of nuclear waste disposal, reactor safety and nuclear safeguards—both domestically and internationally.

Competition in the domestic petroleum industry. Legislation must be enacted to insure energy administrators and legislators access to information they need for making the kind of informed decisions that future energy policy will require. We believe full disclosure of data on reserves, supplies and costs of production should be mandated by law.

It is increasingly clear that there is no free, competitive market for crude oil in the United States. Instead, through their control of the nation's oil pipelines, refineries and marketing, the major oil producers have the capability of controlling the field and often the downstream price of almost all oil.

When competition inadequate to insure free markets and maximum benefit to American consumers exists, we support effective restrictions on the right of major companies to own all phases of the oil industry.

We also support the legal prohibition against corporate ownership of competing types of energy, such as oil and coal. We believe such "horizontal" concentration of economic power to be dangerous both to the national interest and to the functioning of the competitive system.

Improved energy planning. Establishment of a more orderly system for setting energy goals and developing programs for reaching those goals should be undertaken. The current proliferation of energy jurisdictions among many executive agencies underscores the need for a more coordinated system. Such a system should be undertaken, and provide for centralization of overall energy planning in a specific executive agency and an assessment of the capital needs for all priority programs to increase production and conservation of energy.

The international economy. Eight years of mismanagement of the American economy have contributed to global recession and inflation. The most important contribution a Democratic Administration will make to the returning health of the world economy will be to restore the health of our own economy, with all that means to international economic stability and progress. . . .

Energy. The United States must be a leader in promoting cooperation among the industrialized countries in developing alternative energy sources and reducing energy consumption, thus reducing our dependence on imports from the Middle East and restraining high energy prices. Under a Democratic Administration, the United States also will support international efforts to develop the vast energy potential of the developing countries.

We will also actively seek to limit the dangers inherent in the international development of atomic energy and in the proliferation of nuclear weapons. Steps to be given high priority will include: revitalization of the Nonproliferation Treaty, expansion of the International Atomic Energy Agency and other international safeguards and monitoring of national facilities, cooperation against potential terrorism involving nuclear weapons, agreement by suppliers not to transfer enrichment or reprocessing facilities, international assurance of supply of nuclear fuel only to countries cooperating with strict nonproliferation measures, subsidization of multinational nuclear facilities, and gradual conversion to international control of non-weapons fissionable material.

Republican platform. These were the positions on energy taken in the Republican national platform:

In 1973, Americans were shocked to discover that a plentiful supply of energy could no longer be assumed. Unfortunately, the Democrat majority in Congress still has not responded to this clear and urgent warning. The United States is now consuming more imported oil than it was three years ago and our dependence on foreign sources has continued to increase to the point where we now import more than 40% of our oil.

One fact should now be clear: We must reduce sharply our dependence on other nations for energy and strive to achieve energy independence at the earliest possible date. We cannot allow the economic destiny and international policy of the U.S. to be dictated by the sovereign powers that control major portions of the world's petroleum supplies.

Our approach toward energy self-sufficiency must involve both expansion of energy supply and improvement of energy efficiency. It must include elements that insure increased conservation at all levels of our society. It must also provide incentives for the exploration and development of domestic gas, oil, coal and uranium, and for expanded research and development in the use of solar, geothermal, co-generation, solid waste, wind, water, and other sources of energy.

We must use our non-renewable resources wisely while we develop alternative supplies for the future. Our standard of living is directly tied to a continued supply of energy resources. Without an adequate supply of energy, our entire economy will crumble.

Unwise government intervention in the marketplace has caused shortage of supply, unrealistic prices and increased dependence on foreign sources. We must immediately eliminate price controls on oil and newly-discovered natural gas in order to increase supply, and to provide the capital that is needed to finance further exploration and development of domestic hydrocarbon reserves.

Fair and realistic market prices will encourage sensible conservation efforts and establish priorities in the use of our resources, which over the long run will provide a secure supply at reasonable prices for all.

The nation's clear and present need is for vast amounts of new capital to finance exploration, discovery, refining, and delivery of currently usable forms of energy, including the use of coal as well as discovery and development of new sources. At this critical time, the Democrats have characteristically resorted to political demagoguery seeking short-term political gain at the expense of the long-term national interest. They object to the petroleum industry making any profit. The petroleum industry is an important segment of our economy and is entitled to reasonable profits to permit further exploration and development.

At the height of the energy crisis, the Republican Administration proposed a strong, balanced energy package directed at both expansion of supply and conservation of energy. The response from the Democrats in Congress was to inhibit expanded production through artificially set price and allocation controls, thereby preventing market forces from working to make energy expansion economically feasible.

Now, the Democrats proposed to dismember the American oil industry. We vigorously oppose such divestiture of oil companies—a move which would surely result in higher energy costs, inefficiency and undercapitalization of the industry.

Democrats have also proposed that the federal government compete with industry in energy development by creating a national oil company. We totally oppose this expensive, inefficient and wasteful intrusion into an area which is best handled by private enterprise.

The Democrats are playing politics with energy. If they are permitted to continue, we will pay a heavy price in lost energy and lost jobs during the decades ahead.

Immediate removal of counter-productive bureaucratic red tape will eliminate hindrances to the exploration and development of hydrocarbons and other energy resources. We will accelerate development of oil shale reserves, Alaskan petroleum and the leasing of the outer continental shelf, always within the context of preserving the fullest possible protection for the environment. We will reduce complexity and delays involved in siting, licensing and the regulatory procedures affecting power generation facilities and refineries.

Coal, America's most abundant energy resource, is of inestimable value to the American people. It can provide the energy needed to bridge the gap between oil and gas and nuclear and other sources of energy. The uncertainties of governmental regulation regarding the mining, transportation and use of coal must be removed and a policy established which will assure that governmental restraints, other than proper environmental controls, do not prevent the use of coal. Mined lands must be returned to beneficial use.

Uranium offers the best intermediate solution to America's energy crisis. We support accelerated use of nuclear energy through processes that have been proven safe. Government research on the use of nuclear energy will be expanded to include perfecting a long-term solution to the problems of nuclear waste.

Among alternative future energy sources, fusion, with its unique potential for supplying unlimited clean energy and the promise of new methods of natural resource recovery, warrants continued emphasis in our national energy research program, and we support measures to assure adequate capital investment in the development of new energy sources.

Carter answers energy questions. Jimmy Carter, who won the Presidential election, had been criticized during the primary campaign in January 1976 for refusing to answer questions on energy policy as posed by a citizens' group called the Energy Action Committee. The questions were published during the Iowa primary campaign, and Carter then answered them in an advertisement printed in the Des Moines Register.

These were the questions:

1. The major oil companies are structured so that they either own or are in control of most petroleum products from the time they are in the ground to final retail distribution. This is known as vertical integration of the oil industry. There have been proposals to restructure the industry so that each phase of production and sale would be spearated and more competition stimulated within the oil industry. This is known as "vertical divestiture."

Do you support or oppose legislation in Congress which would require the oil companies to "divest" themselves of all but one phase of the oil production-sale chain?

2. A second pattern in the energy industry is that the large oil companies own other energy sources such as natural gas companies, coal companies, uranium mines and processing plants. This development has become known as "horizontal integration."

Do you support or oppose legislative efforts in the U.S. Congress to prevent "horizontal integration" and require major energy companies to divest themselves of such competing energy sources?

3. Public doubts about the integrity of government have been intensified by the large gifts to presidential and congressional candidates by oil companies and their executives.

Will you accept or refuse to accept gifts and contributions from executives and officers of major oil companies and from political action groups controlled by them?

4. The people of the United States, through our government ,own large amounts of oil and gas located on the Outer Continental Shelf of the United States and other public lands.

Will you support or oppose legislation to assure that the earnings from these public resources go principally to the public rather than to increase oil company profits?

5. A major issue before the U.S. Congress is whether the price of natural gas should continue to be set under a system which insures gas producers a "reasonable return" or should be allowed to be set by the oil and gas companies themselves at whatever price (probably at least triple current prices) they think gives them the highest possible profits and allows them to take advantage of the high monopoly prices for oil set by the international oil cartel.

What is your position?

6. Another idea to increase the American People's control over energy costs is to create a Federal Oil and Gas Corporation which could, independent of oil companies, search for new sources of gas and oil, and then sell it to provide additional competition and lower prices.

Do you support or oppose the creation of a Federal Oil and Gas Corporation?

Carter answered the questions in an advertisement that said:

The questions could not be answered honestly and directly with the simple yes or no demanded by the Committee. Some of them dealt with more than one issue—others presented a choice between a promise that no candidate could keep or an answer that totally misrepresented Jimmy Carter's position. In several cases the propositions actually presented to the candidates, in fine print in the ad, were significantly different from those presented in large type.

Although Jimmy Carter is in sympathy with many of the goals of the Energy Action Committee, he felt this country could no longer afford oversimplified answers to complex problems. He decided it was time to draw the line.

Here are statements by Jimmy Carter on these important issues. Read and decide for yourself.

1. Divestiture to end vertical integration: "I support restrictions on the right of a single company to own all phases of production and distribution of oil. However, support of this proposition as worded in fine print would make it illegal for the same company to explore for oil and then extract that oil from the ground once discovered. This would clearly result in tremendous price increases to the consumer."

2. Divestiture to end horizontal integration: "I support legal prohibitions against ownership of competing types of energy, oil and coal for example. However, I cannot promise to oppose any joint responsibility for any phase of production of competing energy sources. Fuel oil and some propane, for example, are produced from crude oil. Their production clearly cannot be separated until after extradition and refining take place. It may not be beneficial to the consumer to separate control of these two competing energy sources until even further down the distribution line."

3. Contributions from officers, executives and politcial action committees of major oil companies: "I will not accept contributions from political action committees controlled by major oil companies. However, no candidate can honestly promise to accept no contributions from any officer or executive. When large direct mail lists were used to raise funds, and most of the candidates who responded to the questionnaire use those lists, it is impossible to determine the employer of a contributor. The only information provided under the new election law is name, home address, principal business address, and occupation. The name of the contributor's employer is not required."

4. Profits from oil and gas located on the publicly owned Outer Continental Shelf; "These natural resources are the property of

the people of this country. I oppose any crash program for massive leasing to the oil companies. I support the proposition of the Energy Action Committee to assure that the earnings from these public resources go principally to the public."

5. Deregulation of oil and natural gas: "This proposition began with a discussion of natural gas and ended talking about oil. There was no way to respond accurately with a single yes or no. I support legal restrictions to allow a 'reasonable profit' on oil and natural gas rather than allowing prices to be set without restriction. I oppose deregulation of old oil. Price for domestic oil should be kept below O.P.E.C. price levels. I support the overwhelming position of the National Governor's Conference to limit deregulation of natural gas to that small portion (less than 5%) of production not under existing contracts. This deregulation should not exceed five years."

6. Creation of a Federal oil and gas corporation: "I oppose the creation of this new federal agency. The Federal Government is presently unable to handle competently and equitably the responsibilities it already has in this area. Agencies designed to regulate and control special interests have almost invariably become the tools of those very interests. The Federal Energy Office is the best, most recent example. It is impossible to tell where the major oil companies stop and the Federal Eenergy Office starts.

"I would consider such a proposal only after a thorough reform and reorganization of the Federal Executive Branch, particularly the so-called Regulatory Agencies."

Carter urges nuclear plant curb. Jimmy Carter called May 13 for a voluntary international moratorium on the purchase or sale of nuclear fuel enrichment and reprocessing plants. He said the goal of his plan was to curb the spread of nuclear explosives.

Addressing a privately-sponsored conference at the United Nations, Carter said, "An alliance for survival is needed, transcending regions and ideologies, if we are to assure mankind a safe passage to the 21st Century."

Carter emphasized that supplier countries were "entitled to a fair share" of the reactor market. "What we must prevent, however," he said, "is the sale of small pilot reprocessing plants which sell for only a few million dollars, have no commercial use at present and can only spread nuclear explosives around the world."

Carter proposed the convening of a world energy conference, under UN auspices, to encourage alternative sources for developing countries that were making, he said, a "premature nuclear commitment." Out of this, he said, an agency for research and development of nonnuclear sources might be developed to work alongside the International Atomic Energy Agency. That agency, he said, could then concentrate on safeguards and international assistance projects.

For the U.S., Carter advocated an effort to keep dependence on nuclear power at a minimum. This effort, he said, should be combined with attempts to conserve energy and to develop alternative sources, such as solar energy.

Carter deplored as "wholly inadequate" the U.S.-Soviet treaty limiting the size of underground nuclear explosions for peaceful purposes. "We can and should do more," he said. He recommended an agreement to bar all nuclear tests for five years. The agreement would be subject to renewal, he said.

In a nuclear policy speech made in San Diego Sept. 25, Carter pledged to end sales of nuclear fuel and technology to countries engaging in nuclear weapons development or building a national plant for reprocessing reactor fuel.

In repeating his call for a voluntary moratorium on the sale or purchase of nuclear fuel reprocessing plants, Carter said that the moratorium "should apply retroactively" to agreements already concluded. "The contracts have been signed, but the deliveries need not be made," he said.

Carter also pledged "to insure that the U.S. is once again a reliable supplier" of enriched uranium, which, he pointed out, unlike plutonium, was unsuitable for weapons.

Carter criticized the Ford Administration for not providing leadership in this area. "We find only the faint footsteps of secret diplomacy," he said, "the constant yielding" to those who saw proliferation and increased production of dangerous nuclear fuels as inevitable.

Carter addresses Nader forum. Carter told a luncheon gathering Aug. 9 that he hoped to challenge consumer advocate Ralph Nader "for the title of top consumer advocate in the country."

Carter spoke in Washington at a Nader-sponsored Public Citizen Forum. During the luncheon, which was open to the public, Carter answered questions from a panel of reporters and from the audience on a wide range of subjects.

On energy matters, Carter said he would make atomic power the lowest energy priority of his administration, emphasizing instead efforts to conserve existing energy sources, substitute coal for oil and increase research on and development of solar power.

Carter promised to make drastic budget cutbacks in the nuclear breeder reactor that produced its own radioactive fuel. (The program had been at the top of the Ford Administration's energy priorities.) Instead, Carter said, the U.S. should examine the working breeders used by France and Great Britain.

Carter also said he opposed legislation pending in Congress that would allow private industry to get into the business of enriching uranium for fueling nuclear power. He refused, however, to endorse antitrust legislation backed by Nader that would "break up" the vertically integrated major oil companies whose operations ranged from exploration to the retail sale of gasoline.

Carter proposes energy reorganization. In a statement issued Sept. 21, 1976, Carter outlined a proposal to form a new Cabinet-level department to take over the functions of the Federal Energy Administration, the Federal Power Commission, the Energy Research and Development Administration and the Energy Resources Council, all of which would be abolished.

The new department also would assume the energy-related functions of the Departments of Commerce and Treasury, the authority of the Securities and Exchange Commission to enforce public-utilities legislation, the powers of the Interstate Commerce Commission to regulate pipelines and the "economic" regulatory functions of the Nuclear Regulatory Commission.

The paper referred to the 1973 oil embargo imposed by Arab states and charged that "our country still has no energy policy."

The debates. Ford and Carter had three televised debates during the campaign. In the first debate, Sept. 23, Carter charged that "the energy policy of our nation is one that has not yet been established under this Administration."

The government's energy structure needed reorganization, he said, citing his own reorganization proposal, the oil supply was being depleted—"we've got about 35 year's worth of oil left in the whole world"—and oil imports had risen. His proposals:

"Shift from oil to coal, emphasize research and development on coal use and also on solar power, strict conservation measures, not yield every time that special interest groups put pressure on the President, like this Administration has done, and use atomic energy only as a last resort with the strictest possible safety precautions. That's the best overall energy policy in the brief time we have to discuss it.

Ford said that Carter "skims over a very serious and a very broad subject." Ford referred to his own proposal in January 1975 for "the first comprehensive energy program recommended by any president."

It called for an increase in the production of energy, Ford said, and for conservation measures. Producers of oil and gas would have to be given an opportunity "to develop their land or their wells." Coal extraction would have to be improved, he said, as would utilization of coal—"make it more efficient, make it cleaner." Ford continued:

"In addition we have to expand our research and development. In my program for energy independence we have increased, for example, solar energy research from about $84 million a year to about $120 million a year. We're going as fast as the experts say we should. In nuclear power we have increased the research and development, under the Energy Research and Development Agency, very substantially to insure that our nuclear power plants are safer, that they are more efficient, and that we have adequate safeguards. I think you have to have greater oil and gas production, more coal production, more nuclear production and in addition you have to have energy conservation."

In the second debate, Ford said Oct. 6 that, on his orders, the Commerce Department the next day would release the names of companies that complied with the Arab requests to boycott Israel.

Carter had called the Arab boycott situation "a disgrace." On the oil-embargo issue, Carter said he would consider it "an economic declaration of war" and ship the Arab nations "nothing" if they repeated their oil embargo first imposed in 1973.

Eastern states urged to seek oil. In New Orleans Oct. 30, Carter said that he had told a New Jersey radio interviewer recently that Eastern states should "start trying to find oil and help states like Louisiana and Texas" in meeting the nation's energy needs.

Carter Policy

Jimmy Carter, victor in the 1976 election, was inaugurated as President of the U.S. Jan. 20, 1977

Carter cited his campaign promises and his initial action on several of them and plans to carry out the rest. He said:

"Some of these efforts will also require dedication—perhaps even some sacrifice—from you. I don't believe that any of us are afraid to learn that our national goals require cooperation and mutual effort."

Carter 'fireside chat' outlines plans. President Carter urged the American people to tackle their problems in a united spirit of "joint efforts and mutual sacrifices." He said this Feb. 2, 1977 in a nationally televised talk reminiscent of the "fireside chats" over radio by President Franklin Delano Roosevelt.

Sitting casually in the White House library, wearing a sweater and tie, Carter spoke at one point of having eliminated "expensive and unnecessary luxuries" of government, such as limousine service for many top officials.

One of the most urgent projects, he said, was development of a comprehensive energy policy. His proposals for such a program would be ready by April 20, he said, and he would ask Congress later in February to reorganize the various energy agencies into a new energy department.

His energy program would emphasize conservation, he said, adding:

"We must face the fact that the energy shortage is permanent. There is no way we can solve it quickly.

"But if we all cooperate and make modest sacrifices, if we learn to live thriftily and remember the importance of helping our neighbors, then we can find ways to adjust, and to make our society more efficient and our own lives more enjoyable and productive.

"Utility companies must promote conservation and not consumption. Oil and natural gas companies must be honest with all of us about their reserves and profits. We will find out the difference between real shortages and artificial ones. We will ask private companies to sacrifice, just as private citizens must do."

'Sacrifices' needed. At a news conference Feb. 8, 1977, President Carter said that his energy conservation program, then being drafted, would require "substantial sacrifices" by Americans. The recent trend of importing more than half the oil used in the U.S. "has got to stop," he asserted. While he was opposed to nationalization of the oil or natural gas industries, he said he would try to make sure that they did not derive "unwarranted profits" during the energy crisis.

Carter submits budget revisions. President Carter sent Congress Feb. 22 his proposed changes in the fiscal 1978 budget submitted by the Ford Administration in January. The President proposed to cut spending on some controversial long-range nuclear power projects and to focus more research on energy conservation.

The oil stockpile project was hiked to $3 billion from the Ford level of $1.69 billion for fiscal 1978. This would put 250 million barrels of oil in storage by December 1978 and 500 million barrels by December 1980. Ford's figures were 150 million barrels by December 1978 and 500 million by the end of 1982.

The budget for the liquid-metal, fast-breeder nuclear reactor was cut $199 million to $656 million for fiscal 1978. And the Energy Research and Development Administration was asked to study the feasibility of continuing the program.

Nuclear fusion research to find a technology to produce commercial electric power was reduced $80 million to the $433-million level for fiscal 1978.

Energy conservation research was boosted $160 million to the $318-million level. Increased spending was budgeted for solar-heating demonstration projects, electric-car research and ways to make synthetic gas from coal and to extract deep oil and gas.

On the issue of deregulation of natural gas, Carter said he would work with Congress on that "as a part of an overall energy policy."

The Administration was conducting studies "to see whether or not the reserve supplies are adequate and whether or not the oil companies are giving us accurate data," he said.

"I think it's obvious to all of us that there are some instances where natural gas is withheld from the market," he added. "That's understandable. If I was running an oil company, I would reserve the right to release or reserve some supplies of natural gas."

Carter suggested that the oil companies would be less likely to withhold supplies of oil and natural gas if the government could come up with a reliable long-range policy so the companies would know "in a predictable way" what the government's policy was going to be in the near and more distant future.

Energy Department bill submitted. President Carter sent Congress his proposal to establish a Department of Energy March 1, 1977. He proposed a department with 20,000 employes and a $10.6-billion budget.

The new department would absorb and replace three independent agencies—the Federal Energy Administration, the Energy Research and Development Administration and the Federal Power Commission. It also would include programs currently housed in other departments and agencies.

In an effort to insulate regulatory functions from policy-making and development programs, the President proposed establishment of a hearings and appeals procedure and an administrative law judge to handle the regulatory issues.

Text of the President's message on creating the new department:

To the Congress of the United States:

I hereby transmit to the Congress proposed legislation which will create a new Cabinet Department of Energy.

This legislation is a major step in my Administration's program for a comprehensive reorganization of the Executive Branch.

Nowhere is the need for reorganization and consolidation greater than in energy policy. All but two of the Executive Branch's Cabinet departments now have some responsibility for energy policy, but no agency, anywhere in the Federal Government, has the broad authority needed to deal with our energy problems in a comprehensive way.

The legislation I am submitting today will bring immediate order to this fragmented system:

—It will abolish the Federal Energy Administration, the Energy Research and Development Administration, and the Federal Power Commission, thereby eliminating three agencies whose missions overlap and sometimes conflict, and whose specialized perspectives have impeded progress toward a unified energy policy.

—It will allow us, for the first time, to match our research and development program to our overall energy policies and needs. This is particularly important if we are to make use of renewable energy sources such as solar power.

—It will enable us to move more quickly toward effective energy conservation by combining conservation programs which are now split between FEA and ERDA. And, to make certain that we will see results, the legislation creates an Assistant Secretary for Conservation, who will be personally responsible for seeing that the conservation program is carried out.

—It will place under one roof the powers to regulate fuels and fuel distribution systems, powers which are now shared by the FEA and the FPC along with the Securities and Exchange Commission and the Interstate Commerce Commission. An institutional structure built on the premise that fossil fuels are abundant and cheap no longer serves well in an era of fuel scarcity.

As this winter has shown us, uncoordinated regulatory policies can have serious impacts on our economic and social well-being. This reorganization can help us bring currently fragmented policies into a structure capable of both developing and implementing an overall national energy plan. At the same time, we must guard the quasi-judicial aspects of the regulatory process against improper influence. The legislation meets this concern by establishing a Board of Hearings and Appeals within the Department which is free from the control of the Secretary of Energy.

In addition to abolishing the FEA, ERDA, and the FPC, the legislation submitted today will transfer into the new Department several significant energy-related authorities and programs which now belong to other departments. These include the building thermal efficiency standards from Housing and Urban Development, the voluntary industrial compliance program from Commerce, and the Navy petroleum and oil shale reserves from Defense. The legislation provides for consultation between the Energy Department and the Department of Transportation on auto fuel efficiency standards, and establishes a role for the Energy Secretary in the REA loan program at Agriculture. Where it is appropriate, these departments will still carry out the program, but the new Energy Department will give them the policy guidance needed to bring unity and rational order to our energy program.

Finally, this legislation transfers certain parts of the Interior Department—those concerning fuels data collection and analysis, and coal mine research and development—into the new Department. Coal mine health and safety research will not be transferred. This will leave the Department of Interior still in charge of leasing energy resources under Federal control. We are leaving those functions in Interior because we believe that the responsibility for multiple-use of public lands, and for their environmental protection, belongs in one department—Interior—that can reflect a broad spectrum of concern. The Energy Department, however, will set long-term production goals and will have policy control over economic aspects of the leases. This will help us foster competition within the energy industries and encourage production of leased resources as expeditiously as possible.

This reorganization will also bring together our energy data gathering and analysis capabilities. More than twenty executive departments and agencies now operate more than 250 energy data programs. The FEA, ERDA, FPC and the Interior Department's Bureau of Mines together have more than 100 such programs. This fragmentation is not only uneconomic and frustrating: it can also have serious consequences. We have seen in recent weeks that, under our present system, we have no single source of information about where natural gas shortages were greatest and where supplies were still available to help make up those shortages. Consolidating these major data programs in an Energy Information Administration within the new department will now give us the ability to compile information which is complete, accurate and *believable.*

There are many things this legislation does not try to do.

I believe that health, safety and environmental regulation relating to energy—unlike economic regulation—should not be brought into the new Energy Department. Because public concerns about the safety of nuclear power are so serious, we must have a strong, independent voice to ensure that safety does not yield to energy supply pressures. Therefore, the Nuclear Regulatory Commission will remain as an independent body. For similar reasons, the Environmental Protection Agency should remain independent to voice environmental concern.

Even with a new Department of Energy, problems of interdepartmental coordination will remain, since virtually all government activity affects energy to some extent. Establishing this department, however, will give us one government body with sufficient scope and authority to do the massive job that remains to be done. Thus this legislation will abolish the Energy Resources Council. I intend to establish by Executive Order a non-statutory interdepartmental coordinating body, with the Secretary of Energy as its chairman to manage government-wide concerns involving energy.

This legislation contains no new substantive authorities. Instead, by eliminating three agencies and uniting a variety of existing energy authorities, the legislation I am submitting today will help reorganize the Executive Branch in a rational, orderly way. It is long overdue. I hope to work with the Congress to achieve our initial goal of a realistic and effective energy policy.

JIMMY CARTER.

THE WHITE HOUSE, *March 1, 1977.*

Carter at energy seminar. Carter flew March 17, 1977 to Charleston, W. Va., where he participated in a seminar on energy. Carter said a comprehensive energy plan for the country was "long overdue." He also said he was willing to sacrifice some of his "personal popularity" in order to make the nation face the "brutal facts" of the energy crisis.

Carter said his plan would stress conservation to reduce the annual growth rate in energy consumption to 2% or less (compared with a current growth rate of 3% to 4%). He said he would seek to increase use of coal and decrease reliance on imported oil. However, he said he saw no "prospect or need for our country to be self-sufficient" in energy.

James Schlesinger, White House energy adviser and also a member of the seminar, said the Carter plan envisioned use of the current generation of nuclear power plants as a major energy source.

CIA oil forecast pessimistic. The White House April 18 released a report by the Central Intelligence Agency forecasting "sizable" price increases for oil by 1982 or 1983 unless "large-scale conservation measures cut demand sharply."

"In the absence of greatly increased energy conservation," the report said, "projected world demand for oil will approach productive capacity by the early 1980s and substantially exceed capacity by 1985."

The report was prepared for President Carter, who referred to it at his news conference April 15 as "deeply disturbing." "Reserve estimates that had been used as a basis for decisions in the past were found to be quite excessive," he told reporters. "Reserves are not as great as we thought they were."

However, there were no estimates of world oil reserves in the released report, which projected demand on the basis of conservation measures currently in effect in major consuming countries.

A rising demand for oil imports was predicted for the U.S., Western Europe and Japan. The U.S. import demand was expected to rise to between 12 million and 15 million barrels a day in 1985, without new conservation measures. For 1976, the import figure was about seven million barrels a day.

The report predicted further competitive pressure on the export market as the Soviet bloc countries became substantial oil importers by 1985.

As for supply, the CIA found that most nations in the Organization of Petroleum Exporting Countries did not have the capacity for a significant expansion in production. Saudi Arabia did, the report said. But it predicted that the Saudis, wanting to conserve their resource "and having no immediate need for more money," would "have no economic incentives to expand production."

As for new oil producers, the report estimated that Mexico would be producing as much as five million to six million barrels a day in the mid-1980s. Its production was 900,000 barrels a day in 1976.

Carter Presents Program

Sacrifices required from every sector. President Carter presented a comprehensive energy policy to Congress April 20, 1977 with a warning that it would require "sacrifices" from every economic sector.

The policy called for higher prices and taxes for petroleum products and production. Conservation was the keynote. The goal was to switch the nation from dependence on oil to renewed use of coal and, eventually, to renewable energy sources, such as solar power.

The President had told the nation in a televised address April 18 that delay in coping with the energy problem could bring on "a national catastrophe."

Carter had conceded at a press conference April 15 that his energy program "will be inflationary in nature."

While "a termination of unnecessary waste of energy is anti-inflationary in itself," he said, "the inevitable increases in the price of energy as it becomes scarce are inflationary. So we're going to try to balance those two to minimize the impact of the inflation rate of any energy policy, or absence of energy policy."

He presented his proposals to Congress, assembled in joint session in the House chamber, in a second nationally televised address April 20.

Address to nation—Carter told the nation April 18 that his talk was an "unpleasant" one about a problem "unprecedented in our history." "With the exception of preventing war, this is the greatest challenge that our country will face during our lifetimes," he said. "The energy crisis has not yet overwhelmed us, but it will if we do not act quickly."

He said many of his proposals on energy policy "will be unpopular." But, he warned in drastic terms, "the alternative may be a national catastrophe. Further delay can affect our strength and our power as a nation. . . . This difficult effort will be the 'moral equivalent of war'—except that we will be uniting our efforts to build and not to destroy."

The stark fact was that "the oil and natural gas we rely on for 75% of our energy are simply running out," Carter said. Unless "profound" changes were made to lower oil consumption, he said, "we now believe that early in the 1980s the world will be demanding more oil than it can produce."

As for the United States,"Ours is the most wasteful nation on earth," Carter said. "We waste more energy than we import. . . . If we do not act, then by 1985 we will be using 33% more energy than we use today."

"We can't substantially increase our domestic production, so we would need to import twice as much oil as we do now," he continued. ". . .If we wait, we will constantly live in fear of embargoes. We could endanger our freedom as a sovereign nation to act in foreign affairs. Within 10 years we would not be able to import enough oil—from any country, at any acceptable price."

At home, failure to act would affect our factories and transportation system, he said. "We will feel mounting pressure to plunder the environment. . . . Intense competition for oil will build up among nations and also among the different regions within our own country, which has already started. . . . We will face an economic, social and political crisis that will threaten our free institutions."

But there was "another choice," Carter said. "We can begin to prepare right now. We can decide to act while there is still time."

The President then listed the 10 "fundamental principles" of his national energy plan. They were:

The first principle is that we can have an effective and comprehensive energy policy only if the government takes responsibility for it and if the people understand the seriousness of the challenge and are willing to make sacrifices.

The second principle is that healthy economic growth must continue. Only by saving energy can we maintain our standard of living and keep our people at work. An effective conservation program will create hundreds of thousands of new jobs.

The third principle is that we must protect the environment. Our energy problems have the same cause as our environmental problems—wasteful use of resources. Conservation helps us solve both problems at once.

The fourth principle is that we must reduce our vulnerability to potentially devastating embargoes. We can protect ourselves from uncertain supplies by reducing our demand for oil, by making the most of our abundant resources such as coal and by developing a strategic petroleum reserve.

The fifth principle is that we must be fair. Our solutions must ask equal sacrifices from every region, every class of people and every interest group. Industry will have to do its part to conserve, just as consumers will. The energy producers deserve fair treatment, but we will not let the oil companies profiteer.

The sixth principle, and the cornerstone of our policy, is to reduce demand through conservation. Our emphasis on conservation is a clear difference between this plan and others which merely encouraged crash production efforts. Conservation is the quickest, cheapest, most practical source of energy. Conservation is the only way that we can buy a barrel of oil for about $2. It costs about $13 to waste it.

The seventh principle is that prices should generally reflect the true replacement costs of energy. We are only cheating ourselves if we make energy artificially cheap and use more than we can really afford.

The eighth principle is that government policies must be predictable and certain. Both consumers and producers need policies they can count on so they can plan ahead. This is one reason I am working with the Congress to create a new Department of Energy, to replace more than 50 different agencies that now have some control over energy.

The ninth principle is that we must conserve the fuels that are scarcest and make the most of those that are plentiful. We can't continue to use oil and gas for 75% of our consumption as we do now when they make up only 7% of our domestic reserves. We need to shift to plentiful coal while taking care to protect the environment, and to apply stricter safety standards to nuclear energy.

The tenth and last principle is that we must start now to develop the new, unconventional sources of energy we will rely on in the next century.

Carter listed these "goals we set for 1985":

—To reduce the annual growth rate in our energy demand to less than 2%.

—To reduce gasoline consumption by 10% below its current level.

—To cut in half the portion of United States oil which is imported, from a potential level of 16 million barrels to 6 million barrels a day.

—To establish a strategic petroleum reserve of one billion barrels, more than six months' supply.

—To increase our coal production by about two-thirds to more than 1 billion tons a year.

—To insulate 90% of American homes and all new buildings.

—To use solar energy in more than two and one-half million houses.

Progress toward the goals would be monitored, Carter said, and strict conservation measures would be in order "if we fall behind."

The energy program "will demand that we make sacrifices and changes in every life," Carter said. But the sacrifices "can be gradual, realistic, and they are necessary," he said. "Above all, they will be fair."

As part of that effort, he said, "We will monitor the accuracy of data from the oil and natural gas companies for the first time."

"There should be only one test for this program," he said, "whether it will help our country."

The following is an abridged text of the Carter address to the nation:

Good evening. Tonight I want to have an unpleasant talk with you about a problem unprecedented in our history. With the exception of preventing war, this is the greatest challenge that our country will face during our lifetimes. . . .

It's a problem that we will not solve in the next few years, and it is likely to get progressively worse through the rest of this century.

We must not be selfish or timid if we hope to have a decent world for our children and our grandchildren.

We simply must balance our demand for energy with our rapidly shrinking resources. By acting now we can control our future instead of letting the future control us.

Two days from now, I will present to the Congress my energy proposals. Its members will be my partners and they have already given me a great deal of valuable advice. Many of these proposals will be unpopular. Some will cause you to put up with inconveniences and to make sacrifices.

The most important thing about these proposals is that the alternative may be a national catastrophe. Further delay can affect our strength and our power as a nation.

Our decision about energy will test the character of the American people and the ability of the President and the Congress to govern this nation. This difficult effort will be the moral equivalent of war—except that we will be uniting our efforts to build and not to destroy.

Now I know that some of you may doubt that we face real energy shortages. The 1973 gas lines are gone, and with this springtime weather, our homes are warm again.

But our energy problem is worse tonight than it was in 1973 or a few weeks ago in the dead of winter. It's worse because more waste has occurred, and more time has passed by without our planning for the future. And it will get worse every day until we act.

The oil and natural gas we rely on for 75% of our energy are simply running out. In spite of increased effort, domestic production has been dropping steadily at about 6% a year. Imports have doubled in the last five years. And our nation's economic and political independence is becoming increasingly vulnerable. Unless profound changes are made to lower oil consumption, we now believe that early in the 1980s the world will be demanding more oil than it can produce.

The world now uses about 60 million barrels of oil a day, and demand increases each year about 5%. This means that just to stay even we need the production of a new Texas every year, an Alaskan North Slope every nine months, or a new Saudi Arabia every three years. Obviously this cannot continue.

We must look back into history to understand our energy problem. Twice in the last several hundred years there's been a transition in the way people use energy.

The first was about 200 years ago, when we changed from wood—which had provided about 90% of all fuel—to coal, which was more efficient. This change became the basis of the Industrial Revolution.

The second change took place in this century, with the growing use of oil and natural gas. They were more convenient and cheaper than coal, and the supply seemed to be almost without limit. They made possible the age of automobile and airplane travel. Nearly everyone who is alive today grew up during this period and we have never known anything different.

Because we are now running out of gas and oil, we must prepare quickly for a third change, to strict conservation and to the renewed use of coal and permanent renewable energy sources, like solar power.

The world has not prepared for the future. During the 1950s, people used twice as much oil as during the 1940s. During the 1960s, we used twice as much as during the 1950s; And in each of those decades, more oil was consumed than in all of man's previous history combined.

World consumption of oil is still going up. If it were possible to keep it rising during the 1970s and 1980s by 5% a year as it has in the past, we could use up all the proven reserves of oil in the entire world by the end of the next decade.

I know that many of you have suspected that some supplies of oil and gas are being withheld from the market. You may be right, but suspicions about the oil companies cannot change the fact that we are running out of petroleum.

All of us have heard about the large oil fields on Alaska's North Slope. In a few years when the North Slope is producing fully, its total output will be just about equal to two years' increase in our own nation's energy demand.

Each new inventory of world oil reserves has been more disturbing than the last. World oil production can probably keep going up for another six or eight years. But some time in the 1980s it can't go up any more. Demand will overtake production. We have no choice about that.

But we do have a choice about how we will spend the next few years. Each American uses the energy equivalent of 60 barrels of oil per person each year. Ours is the most wasteful nation on earth. We waste more energy than we import. With about the same standard of living, we use twice as much energy per person as do other countries like Germany. Japan and Sweden.

One choice, of course, is to continue doing what we have been doing before. We can drift along for a few more years.

Our consumption of oil would keep going up every year. Our cars would continue to be too large and inefficient. Three-quarters of them would continue to carry only one person—the driver—while our public transportation system continues to decline. We can delay insulating our homes, and they will continue to lose about 50% of their heat in waste.

We can continue using scarce oil and natural gas to generate electricity, and continue wasting two-thirds of their fuel value in the process.

If we do not act, then by 1985 we will be using 33% more energy than we use today.

We can't substantially increase our domestic production, so we would need to import twice as much oil as we do now. Supplies will be uncertain. The cost will keep going up. Six years ago, we paid $3.7 billion for imported oil. Last year we spent $36 billion—nearly 10 times as much—and this year we may spend $45 billion.

Unless we act, we will spend more than $550 billion for imported oil by 1985—more than $2,500 for every man, woman, and child in America. Along with that money that we transport overseas, we will continue losing American jobs and becoming increasingly vulnerable to supply interruptions.

Now we have a choice. But if we wait, we will constantly live in fear of embargoes. We could endanger our freedom as a sovereign nation to act in foreign affairs. Within 10 years we would not be able to import enough oil—from any country, at any acceptable price.

If we wait, and do not act, then our factories will not be able to keep our people on the job with reduced supplies of fuel. Too few of our utility companies will have switched to coal, which is our most abundant energy source.

We will not be ready to keep our transportation system running with smaller, more efficient cars and a better network of buses, trains and public transportation.

We will feel mounting pressure to plunder the environment. We'll have a crash program to build more nuclear plants, strip-mine and burn more coal, and drill more offshore wells than if we begin to conserve right now. Inflation will soar, production will go down, people will lose their jobs.

Intense competition for oil will build up among nations and also among the different regions within our own country, which has already started.

If we fail to act soon, we will face an economic, social and political crisis that will threaten our free institutions.

But we still have another choice. We can begin to prepare right now. We can decide to act while there is still time. . . .

I can't tell you that these measures will be easy, nor will they be popular. But I think most of you realize that a policy which does not ask for changes or sacrifices would not be an effective policy at this late date.

This plan is essential to protect our jobs, our environment, our standard of living, and our future.

Whether this plan truly makes a difference will not be decided here in Washington, but in every town and every factory, in every home and on every highway and every farm.

I believe that this can be a positive challenge. There is something especially American in the kinds of changes that we have to make. We always have been proud, through our history, of being efficient people.

We always have been proud of our ingenuity, our skill at answering questions. Now we need efficiency and ingenuity more than ever.

We always have been proud of our leadership in the world. Now we have a chance again to give the world a positive example.

We've always been proud of our vision of the future. We've always wanted to give our children and grandchildren a world richer in possibilities than we've had. They are the ones we must provide for now. They are the ones who will suffer most if we don't act.

I've given you some of the principles of the plan.

I am sure each of you will find something you don't like about the specifics of our proposal. It will demand that we make sacrifices and changes in every life. To some degree, the sacrifices will be painful—but so is any meaningful sacrifice. It will lead to some higher costs, and to some greater inconvenience for everyone.

But the sacrifices can be gradual, realistic and they are necessary. Above all, they will be fair. No one will gain an unfair advantage through this plan. No one will be asked to bear an unfair burden. We will monitor the accuracy of data from the oil and natural gas companies for the first time, so that we will know their true production, supplies, reserves, and profits.

Those citizens who insist on driving large, unnecessarily powerful cars must expect to pay more for that luxury.

We can be sure that all the special interest groups in the country will attack the part of this plan that affects them directly. They will say that sacrifice is fine, as long as other people do it, but that their sacrifice is unreasonable, or unfair, or harmful to the country. If they succeed with this approach, then the burden on the ordinary citizen, who is not organized into an interest group, would be crushing.

There should be only one test for this program: whether it will help our country.

Other generations of Americans have faced and mastered great challenges. I have faith that meeting this challenge will make our own lives even richer. If you will join me so that we can work together with patriotism and courage, we will again prove that our great nation can lead the world into an age of peace, independence and freedom. Thank you very much and good night.

Address to Congress—Carter began his "sober" and "difficult" presentation to Congress April 20 with an appeal that "we must work together" and "we must act now."

He had come to realize very clearly why a comprehensive energy policy had not already been evolved, he said. "It's been a thankless job, but it's our job, and I believe that we have a fair, well balanced and effective plan to present to you. It can lead to an even better life for the people of America."

The heart of the problem was, he said, "that we have too much demand for fuel—it keeps going up too quickly while production goes down—and our primary means of solving this problem is to reduce waste and inefficiency."

Citing increasing demand for oil and dwindling supplies, Carter said, "The time has come to draw the line."

Carter recited the goals of his energy policy and asked Congress to adopt them by joint resolution "as a demonstration of our mutual commitment to achieve them."

The President said he preferred to reach the goals "through voluntary cooperation with a minimum of coercion." But he cautioned that "voluntary compliance will not be enough—the problem is too large and the time is too short. In a few cases, penalties and restrictions to reduce waste are essential."

"This is a carefully balanced program, depending for its fairness on all its major component parts," he told Congress. "It will be a test of our basic political strength and ability."

Among details of his program:

Gas guzzlers—Using fuel efficiency standards for new automobiles set by Congress, the President proposed "a graduated excise tax on new gas guzzlers that do not meet federal mileage standards." All of the money from the tax would be returned to consumers through rebates on automobiles that were more efficient than the mileage standard. The mileage standard would rise to 27.5 miles per gallon in 1985, when penalties would reach a maximum of $2,488 and rebates a high of $493.

Gasoline tax—Carter proposed a gasoline tax of at least 5¢ per gallon each year that national consumption exceeded stated targets, beginning Jan. 15, 1979. The tax would be reduced if goals were met later. A ceiling of 50¢ was set.

Any proceeds from the tax, Carter said, "should be returned to the general public in an equitable manner."

Homes and buildings—In the area of homes and buildings, Carter urged that new buildings be made as efficient as possible and that old buildings be equipped with insulation and heating systems that dramatically reduce the use of fuel."

He said he would issue an executive order establishing strict conservation goals for both new and old federal buildings.

For homes and businesses, Carter said, "Those who weatherize buildings to make them more efficient" would be eligible for a tax credit of 25% for the first $800 invested and 15% for the next $1,400.

Other proposals in this sector included: direct federal help for low-income residents; an additional 10% tax credit for business investments in conservation; federal matching grants to nonprofit schools and hospitals; public works money for weatherizing state and local government buildings.

Household appliances—The President proposed legislation that would impose "stringent" efficiency standards for household appliances by 1980.

Utility rates—To reform utility rate structures, Carter proposed legislation to eliminate "promotional" pricing that made natural gas and electricity cheaper per unit for major users. The legislation also would require utilities to offer "peak-load pricing techniques" that provided cheaper rates for electricity used during low-demand periods.

Oil taxes—Carter strongly opposed, as "disastrous for our economy," proposals for immediate and total decontrol of domestic oil and natural gas prices.

Instead, he proposed that the price of newly discovered oil be allowed to rise over a three-year period to the 1977 world market price, with allowances for inflation from then on. The price of previously discovered oil would be left alone, except for inflation adjustments.

The President proposed a wellhead tax on existing supplies of domestic oil, equal to the difference between the current controlled price of oil and the world price.

Natural gas—Carter called for an end to "the artificial distortions in natural gas prices in different parts of the country." The price limit for all new gas sold anywhere in the country, he said, should be set at the price of the equivalent energy value of domestic crude oil, beginning in 1978. The ceiling would apply, for the first time, to natural gas sold within a state in which it was produced.

Coal—The plan called for a 65% increase in the use of coal by industry and utilities. To encourage this, the President recommended a sliding-scale tax, starting in 1979, on large industrial users of oil and natural gas.

Nuclear energy—Even with the conversion effort, there would be a gap between need and availability of energy, Carter said. "Therefore, as a last resort we must continue to use increasing amounts of nuclear energy."

He said the U.S. had 63 nuclear power plants, producing about 3% of the country's total energy, and about 70 more plants were licensed for construction.

Carter said there was "no need to enter the plutonium age" by licensing or building a fast-breeder reactor such as the proposed demonstration plant on the Clinch River in Tennessee.

He called, however, for an increase in the capacity to produce enriched uranium fuel for light-water nuclear power plants using new centrifuge technology.

A reform of licensing procedures was urged to upgrade the safety of the plants and accelerate decisions.

Solar energy—To promote growth "of a large new market," Carter proposed a gradually decreasing tax credit, to run through 1984, for those who purchased approved solar heating equipment for their homes. At the start, the tax credit would be 40% of the first $1,000 invested and 25% of the next $6,400.

Fairness—"During this time of increasing scarcity, competition among energy producers and distributors must simply be guaranteed," Carter said. "I recommend that individual accounting be required from energy companies for production, refining, distribution and marketing, separately for domestic and foreign operations.

". . . Profiteering through tax shelters should be prevented, and independent drillers should have the same intangible tax credits as the major corporations.

"The energy industry should not reap large unearned profits. Increasing taxes, increasing prices on existing inventories of oil should not result in windfall gains but should be captured for the people of our country."

Strategy—The White House reported that the Carter energy program would be implemented under the following "strategy":

1. Implementation of an effective conservation program for all sectors of energy use.

2. Conversion of industry and utilities using oil and natural gas to coal and other more abundant fuels to reduce imports and make natural gas more widely available for household use.

(These two strategies would help achieve the short-term objectives of reducing dependence on foreign oil and limiting supply disruptions, and the medium-term objective of weathering the eventual natural decline in the availability of world oil supplies.)

3. A vigorous research and development program (to help achieve the long-term objective of providing renewable and essentially inexhaustible sources of energy for sustained economic growth in the next century).

'No significant effect' on growth—Bert Lance, director of the Office of Management and Budget, said April 21 that the Administration's energy program would have "no significant effect" on the country's economic growth.

"There may be a temporary adverse impact, but it would be strictly that," he said. "Over the next four years, if the impact is not favorable, it's certainly not unfavorable."

The position, stated at a news conference, was a retraction of the prediction contained in the White House "fact sheet" that the program would have "small but generally positive impacts on the economy." The early prediction was said to have been included by mistake on the "fact sheet."

Lance also gave the Administration's first cost estimates for the energy program. He said it would increase the federal deficit by a range of $1.2 billion to $2.6 billion in fiscal 1978 and by $7 billion to $15 billion by 1985. He pointed out that some of the cost would go toward establishment of the strategic petroleum reserve.

"This doesn't change our determination to balance the budget," he added.

There also would be an inflationary impact because of the energy program, he said. His estimate was an increase by one-quarter of one percent without the gasoline tax and one-half of one percent with it.

Program Summarized*

CONSERVATION: ***Transportation***—Imposition of a graduated excise tax on new automobiles and light-duty trucks whose fuel efficiencies did not meet applicable standards under existing law. (By 1985, the tax on a new car that went 15 miles on a gallon of gasoline and one that went 11 miles per gallon would be $1,600 and $2,500, respectively.) Graduated rebates would be available on purchases of cars and light-duty trucks whose fuel efficiencies exceeded the standards. Electric vehicles would be eligible for the maximum rebate. Rebates would be available automatically for vehicles made in the U.S. and Canada. For vehicles manufactured in other countries, rebates would be available on the basis of bilateral treaty or executive agreement. (legislative)

*Source: The White House. The designation in parentheses following each item in the energy program indicates whether the proposal would have to be implemented administratively, by congressional action, through the regular federal budget process, or through a combination of means.

—Review and possible raising of vehicle fuel-efficiency standards by the transportation secretary after 1985. (administrative)

—Promulgation of fuel-efficiency standards by the transportation secretary for all trucks that weighed less than 10,000 pounds. (administrative)

—Request that states strictly enforce the national speed limit of 55 miles per hour or face possible loss of U.S. highway trust fund grants. (administrative)

—Imposition of a standby tax on gasoline that would take effect following the first year in which gasoline consumption exceeded the predetermined target level for that year by 1% or more. The yearly targets would increase slightly until 1980 and decrease between 1983 and 1987 to a level of 6.9 million barrels a day, 200 million barrels below actual 1976 consumption. The tax would work as follows: If gasoline consumption in 1978 exceeded the 1978 target by 1% or more, a tax of 5¢ a gallon would be imposed in 1979. In subsequent years, the tax would be adjusted toward the goal of a 5¢-per-gallon tax for each full percentage point by which consumption in the previous year had exceeded the target. The tax could only be raised or reduced by 5¢ per year. (It would not be permitted to exceed 50¢ a gallon.) Any revenue from the tax would be rebated to the public progressively through the federal income tax system, with direct payments to people who paid no taxes. (legislative)

—Use of the federal highway trust fund to compensate states for loss of gasoline tax revenue due to decreased gasoline consumption, thus insuring adequate funds for highway maintenance. (administrative)

—Removal of the 10% excise tax on intercity buses. (legislative)

—Elimination of existing federal excise tax preferences for general aviation and motorboat fuel. Commercial airlines and commercial fishermen would not be affected. Motorboat-fuel tax revenue would go to the U.S. Land and Water Conservation Fund. (legislative)

—Implementation of energy-conscious purchasing practices for new federal government vehicles (administrative) and implementation of a demonstration commuter van-pooling system for federal employes. (legislative)

(The programs described would not be sufficient to achieve a 10% reduction in gasoline consumption. If voluntary conservation efforts were not sufficient to achieve the goal, a commuter tax and other mandatory measures would be considered.)

Buildings—Approval of a tax credit for homeowners amounting to 25% of the first $800 and 15% of the next $1,400 spent on approved measures to conserve energy through home insulation and use of efficient heating systems. (legislative)

—Requirement that utilities offer their customers a residential energy conservation service (such as insulation) performed by the utility and financed by loans repaid through monthly utility bills. (legislative)

—Removal of federal barriers to residential conservation loans through the Federal Home Loan Mortgage Corp. and the Federal National Mortgage Association. (legislative)

—Increase in funding for the existing residential conservation (weatherization) program for low-income homes. (budget)

—Action by the labor secretary to insure that recipients of Comprehensive Employment Training Act funds would supply labor for the residential conservation program. (administrative/budget)

—Approval of a 10% tax credit, in addition to the existing investment tax credit, for businesses investing in approved conservation measures. (legislative)

—Use of $300 million a year for three years for matching grants for nonprofit schools and hospitals to take energy conservation measures. (legislative/budget)

—Encouragement by the commerce secretary to state and local governments to include energy conservation items in their proposals for Commerce Department public-works funds. (administrative)

(The programs described above were voluntary, except in the case of the role of electric utilities. If they were not sufficient, mandatory measures would be considered, such as a requirement that all homes be insulated before they were sold.)

—Advancement from 1981 to 1980 of the effective date of the mandatory standards for new residential and commercial buildings under the 1976 Energy Conservation and Production Act, with funds to be made available to states to aid their compliance. (administrative/budget)

—Adoption of procedures to reduce energy use in federal buildings by 1985 by 20% of 1975 levels for existing buildings and by 45% for new buildings. (administrative/budget)

—Spending of up to $100 million over the next three years to add solar hot water and space heaters to suitable federal structures to demonstrate the commercial potential of solar energy use. (administrative/budget)

Home Appliance Efficiency—Replacement of voluntary targets with mandatory standards on certain home appliances, such as air conditioners, furnaces, water heaters and refrigerators. (legislative)

Industrial conservation—Legislation for a five-year, 10% tax credit for investment in approved energy-saving equipment, including solar energy devices, that could be incorporated into existing plants. (legislative)

Public Utilities—Legislation to encourage "cogeneration"—the production of electric power and other forms of energy through the recapturing of heat that would otherwise be wasted in the production of electricity. Exemptions from federal and state public-utility regulations and possibly from the energy plan's requirement for conversion from oil and natural-gas use would be available to industrial cogenerators. Also, an additional 10% tax credit would be provided for the purchase of cogeneration equipment. (legislative)

—Encouragement of state public-utility commissions to site new power generating facilities in such a way as to allow waste heat to be used for heating in the surrounding areas. (administrative/budget)

—Phasing out of promotional and other electricity rate systems that did not reflect the true cost of energy. (legislative)

—Use of peak-load pricing and other incentives to shift power consumption from peak to off-peak periods. (legislative)

—Installation of individual electricity meters in each apartment or office of new buildings. (legislative)

—Authorization of the Federal Power Commission to require interconnection and power pooling among utilities and to require "wheeling"—the transmission of power between two noncontiguous utilities across a third utility's system. (legislative)

PRODUCTION: ***Oil and Natural Gas***—Retention of domestic oil price controls to prevent windfall profits for oil producers, which otherwise would be able to charge the price determined by the Organization of Petroleum Exporting Countries.

—Revision and extension of the 1975 Energy Policy and Conservation Act to create a new long-range pricing system. Under the new system, current price ceilings for previously discovered oil would remain in effect, with adjustments for inflation, and prices of newly discovered oil would be allowed to rise over a three-year period to the 1977 world price, adjusted for domestic inflation. (Newly discovered oil would be defined as oil from a well drilled more than 2½ miles from an onshore well in existence as of April 20, 1977, or more than 1,000 feet deeper than any well within a 2½-mile radius. New offshore oil would be that from tracts leased after April 20, 1977.) (legislative)

—Phased imposition of a wellhead tax on domestic crude oil that by 1980 would bring the price to refiners up to the 1980 world price for oil. Thereafter, the tax would increase as the world price increased. The net funds collected as a result of this tax would be returned to the public on a per capita basis through tax credits or, for those who owed no taxes, direct payments. (legislative)

—Limitation of the price of all new natural gas sold anywhere in the U.S. to the average price refiners would have to pay for domestic crude oil (before the new tax) that yielded the same amount of energy, as measured in BTUs (British Thermal Units). (New gas would be defined by the same criteria as newly discovered oil.) The price ceiling would be approximately $1.75 per thousand cubic feet of natural gas. It would apply to gas sold under new intrastate contracts as well as to new gas. It would not affect existing contracts. The discrepancy between prices of gas sold within the producing state and gas sold to other states thus would gradually disappear. (legislative)

—Guarantee that prices of currently flowing gas would remain at current levels, with adjustments for inflation. (legislative)

—Allocation of the cost of the more expensive new gas to industrial users, not to residential and commercial users. (legislative)

—Expansion of the national strategic petroleum reserve to one billion barrels, or a 10-month supply, from the currently projected 500 million barrels. (administrative/budget)

—Replacement of the Ford Administration's limitations on imports of liquefied natural gas with a policy of case-by-case consideration. (administrative)

—Establishment of a federal task force to identify areas where synthetic natural-gas plants should be constructed. (administrative)

—Elimination of gasoline price controls and allocation regulations, allowing greater market competition. Controls would be reimposed if prices rose above a predetermined level. (administrative)

—Amendment of the 1976 Tax Reform Act to grant independent oil and natural gas producers the same tax breaks for intangible drilling costs as the major oil companies received. Investors who financed oil and gas exploration with income earned in other occupations would not receive a tax break. (legislative)

Information monitoring—Takeover by the proposed energy department of the audit and verification roles of the American Gas Association and the American Petroleum Institute. The system of estimating oil and natural gas reserves would be consolidated and supervised by federal officials. (administrative/budget)

—Establishment of a comprehensive financial reporting system under which oil and natural gas producers would submit to the government detailed information on capital expenditures and operating results by geographic region and type of fuel. Companies would have to conform to a uniform system of accounts and report information relating to every phase of their operations—domestic and foreign—allowing the government to assess the performance of individual firms and the industry as a whole. (administrative/budget)

—Establishment of an emergency management information system to provide the government with information on local energy supplies and demand. (administrative)

Industry competition—Promotion and maintenance of competitive practices within the industry, through government programs, information monitoring and, if necessary, passage of new laws.

CONVERSION: ***Tax disincentives***—Imposition of a tax, beginning in 1979, on industrial use of natural gas. The tax would amount to the difference between the user's average cost of natural gas and a price target that by 1985 would rise to the BTU-equivalent price of distillate oil. Fertilizer manufacturers and certain agricultural users who had to use only natural gas would be exempt from the tax. (legislative)

—Imposition of a similar tax on utility use of natural gas, but to begin in 1983 and peak in 1988, since utilities required a longer lead time to convert to coal use. (legislative)

—Taxation of industrial use of petroleum at a flat rate that would rise to $3.00 a barrel between 1979 and 1985. A flat tax of $1.50 a barrel would be imposed on utility use of petroleum beginning in 1983. (legislative)

—Legislation to make industry eligible for either an additional 10% investment tax credit for conversion expenditures or a rebate on any of the above natural gas or petroleum taxes paid (up to the amount of expenditures incurred for conversion to coal or other fuels). (legislative)

(Oil and gas taxes collected from utilities would be set aside to help utilities accelerate the conversion to coal and other fuels.)

Conversion regulatory policy—Prohibition of the use of natural gas or petroleum in new boilers by industry or utilities, with few exceptions. Existing facilities with coal-burning capacity would be prohibited from burning gas or oil. With only limited temporary exceptions, no utility would be permitted to burn natural gas after 1990. (legislative)

—Requirement that government approval be obtained for conversions from coal to petroleum or natural gas and, in the case of utilities, from natural gas to petroleum. (legislative)

—Permission for any industrial firm or utility prohibited from using natural gas to sell its existing contract to purchase gas at a price that would provide adequate compensation. (legislative)

Coal environmental policy—Requirement that the best available pollution-control equipment and procedures be used in all new coal-fired plants. (legislative)

—Maintenance of air quality in clean-air areas. (legislative)

—Appointment of a special committee to study health and environmental effects of increased coal production and use. (administrative)

—Support for tough, uniform national strip-mine legislation.

Nuclear power—Indefinite deferral of commercial reprocessing and recycling of spent fuels produced in U.S. civilian nuclear power plants. (administrative)

—Indefinite deferral of construction of the Clinch River liquid-metal fast-breeder reactor demonstration project in Tennessee. The federal breeder-reactor program would be re-oriented toward research and development of non-plutonium fuels, with emphasis on safety and nonproliferation of plutonium technology. (administrative)

—Expansion of the U.S. uranium-enrichment capacity, with use of energy-saving centrifuge plants rather than the gaseous diffusion plants currently in use. (administrative/budget)

—Legislation to guarantee sales of enriched uranium to any country that agreed to comply with U.S. nonproliferation objectives and certain other conditions. (legislative)

—Expansion of the Nuclear Regulatory Commission (NRC) staff; increase in the number of nuclear plant inspections by the NRC; development by the NRC of siting criteria for new nuclear plants; mandatory rather than voluntary reporting to the NRC of all minor mishaps and component failures at nuclear plants, and review of the NRC licensing process for nuclear plants to insure its objectivity and efficiency. (administrative)

—Review of the federal nuclear waste-disposal program. (administrative)

Hydroelectric power—Study by the Army Corps of Engineers of the potential for additional hydroelectric power installations at existing dams throughout the country, especially at small sites. (administrative)

DEVELOPMENT: ***Coal***— Major expansion of the federal government's coal research and development program, focusing primarily on meeting environmental requirements more effectively and economically and seeking to expand the uses of coal as a substitute for petroleum products and natural gas. (budget)

Solar energy—Approval of a tax credit of 40% of the first $1,000 and 25% of the next $6,400 spent on installation of qualifying solar energy equipment. The credit would decline by 1984 to 25% of the first $1,000 and 15% of the next $6,400. (legislative)

—Inclusion of investments in solar equipment for industrial and commercial purposes among the approved conservation measures eligible for the proposed 10% tax credit for energy-saving investments. (legislative)

—Encouragement of states to exempt solar energy installations from property tax assessments and to enact legislation to protect access to the sun.

Geothermal energy—Approval of a tax deduction for intangible costs of drilling for geothermal energy sources, similar to the deduction available for oil and gas drilling costs. (legislative)

—Removal of administrative barriers to development of geothermal resources. (administrative)

Other energy development measures—Creation of an office of small-scale technology within the proposed energy department to tap more fully the potential of individual inventors and small business firms. (administrative/budget)

—Increased funding for development of alternative energy sources and new uses for them. Included were photovoltaic systems; solar space cooling and other solar building technologies; gas-fired heat pumps and small fuel cells for residential and commercial heating and cooling; identification of new hydrothermal energy sources, and uses of geothermal energy other than for generation of electricity. (budget)

—Establishment of a commission to study and make recommendations concerning the national energy transportation system. (administrative)

Reaction to Program

Carter energy plan generally praised. President Carter's efforts toward the establishment of a national energy policy were applauded by industrialists and environmentalists, businesses and foreign governments and others throughout the country and the world; however, almost all found one or more specific proposals to criticize.

As could be expected, the nature of the criticism varied according to the particular concern of the critic:

Congressional reaction—Nearly everyone on Capitol Hill applauded the President for tackling a major national problem in presenting his proposal for a national energy policy—but several elements of the program immediately came under fire from lawmakers with particular interests.

About the only proposal to win universal support was the suggestion for tax credits for home insulation—a plan similar to the one approved by the House in June 1975.

A recurring criticism, which knowledgeable observers said could lead to some reshaping of the energy package, was that the Carter plan overplayed conservation and underplayed efforts to increase energy production. This was the main theme of Republican lawmakers and some important Democrats from energy-producing states.

Louisiana Sen. Russell Long (D), chairman of the Senate Finance Committee, said, "The President's diagnosis is essentially correct. He's right about our need for conservation. I wish he had placed more stress on production."

House Majority Leader Jim Wright (D, Tex.), a strong spokesman for oil and natural gas producers, said, "Conservation is important but it is only half the job."

Sen. Jennings Randolph (D) of West Virginia, a major coal-producing state, lauded the plan in general, but complained about "unnecessary restrictions" he said might impede coal production.

Sen. Edward Kennedy (D) of Massachusetts, a large energy consuming state, criticized the prospective increases in natural gas prices.

But the natural gas-price proposals were criticized from the opposite side by Republican Sen. John G. Tower of Texas, who asked for "deregulation and decontrol—and let people make a profit out of it." Tower's colleague, Sen. Lloyd Bentsen (D, Tex.), specifically assailed the plan to put a price ceiling on gas sold within the producing state—so-called intrastate gas.

Among the least popular of the President's proposals was the one calling for a standby tax on gasoline. It drew heavy fire from rural lawmakers who said, as did Senate Minority Leader Howard Baker (R, Tenn.), that the plan would discriminate against states in which people relied on their cars for their livelihood and had no other means of transportation.

Rep. Morris Udall (D, Ariz.), who said he would support the program in its entirety, summed up the congressional response by saying, "It's like it was the day after Pearl Harbor, and you interviewed the congressman from Detroit and he said, 'The Japanese attack was outrageous, but before we rush into war, let's see how it would affect the automobile industry,' and then somebody else said, 'It was dastardly, but consider the effect on oil,' and another congressman said, 'War could be very serious for recreation and tourism.'"

For all the criticism of specific proposals, one of the prevailing sentiments in Congress was that expressed by Sen. Bentsen: "The President is doing what has to be done."

Energy producers, utilities—John E. Swearingen, chairman of the board of Standard Oil Co. (Indiana), expressed the prevailing sentiment among oil and natural gas producers April 24 when he deplored what he said was the plan's lack of incentive for development of new fossil fuel supplies. Appearing on the NBC television program "Meet the Press," Swearingen said, "If you really want to accelerate domestic exploration of oil and gas production . . . , the industry has to be given the money to do it."

(Presidential energy adviser James Schlesinger, appearing on CBS-TV's "Face the Nation" show the same day, called such statements "absolutely invalid" and "misleading." He said the provision to allow the price of newly discovered oil to rise to the world market price would be sufficient to spur exploration. He added that the oil industry was dissatisfied with the plan because it wanted higher incentives, bigger profits and a larger share of the nation's gross national product.)

The oil industry's public criticism was in general somewhat restrained, however, reportedly because the companies were wary of setting themselves up as a collective "villain" that Carter could attack in an effort to win public support for his plan.

A spokesman for the Independent Petroleum Association of America, a group of relatively small entrepreneurs, said April 21 that the plan "retreats to the worn, discredited policy of the past—increased government meddling with domestic energy production—the very real policy which has created the situation we now face."

According to a report April 21, oil officials also were unhappy about the requirement for division-by-division reporting of their profits.

A Natural Gas Supply Committee executive warned April 21 that the program represented "a serious backward step" from fostering new supplies. He said reducing prices for gas sold in the state in which it was produced was "the only bright spot in the picture."

The president of the National Coal Association (NCA), Carl E. Bagge, said on "Meet the Press" April 24 that current environmental restrictions were the major factors constraining the production and use of coal. Yet, he noted, the President had endorsed stringent strip-mining controls and continued clean-air efforts. "Something has to give," Bagge said.

The NCA April 21 had said that to achieve the Carter plan's goals, the government would have to relax air-quality standards, lift current prohibitions on surface mining and prevent the use of investment capital on expensive conversions of plants to coal from oil and natural gas when the money was needed for construction of new coal-fired plants.

Another guest April 24 on "Meet the Press" was W. Donham Crawford, president of the Edison Electric Institute. Crawford contended that the proposal to require exhaust stack gas-cleansing devices, known as scrubbers, would "add hundreds of millions of dollars to electric power bills." Crawford April 20 had said the plan would place "a particularly severe burden on electrical utilities" and warned against cutting back "so deeply that we cut into the economic muscle of the country."

Auto industry, business—Auto industry officials told the energy and power subcommittee of the House Interstate and Foreign Commerce Committee April 22 that the proposed tax and rebate scheme to encourage fuel efficiency in cars would be "redundant and disruptive," and would reduce car sales and create unemployment in the industry.

Executives of the four auto manufacturers said the fuel economy standards enacted into law in 1975 would achieve the Carter Administration's goal of a 10% reduction in gasoline use by 1985.

Henry Ford II, chairman of Ford Motor Co., said April 21 that the proposal "seems to challenge the industry's intent and ability to meet the law." Similarly, Chrysler Corp. president John Riccardo, on the ABC-TV program "Issues and Answers" April 24, said the plan to tax gas-guzzling cars was "like changing the rules in the middle of the game." General Motors Corp. chairman Thomas A. Murphy said April 21 that the penalty plan was simplistic and unnecessarily risky.

An American Motors Corp. executive said April 24 that his company's new car sales had fallen off sharply since April 18, when Carter first outlined the basis of his energy plan. The official recommended that the rebates and taxes be made retroactive to the day after the plan actually was proposed, on April 21, so potential buyers would not wait for congressional action.

Other businessmen were left in a state of increased uncertainty by the Carter plan as well, the Wall Street Journal reported April 21. Businessmen were saying they were unsure how to plan ahead, since they did not know how much of the plan would be enacted into law, when it might be passed or what its effects might be, the Journal said.

State officials, others—Louisiana Gov. Edwin Edwards (D) expressed the typical reaction of officials from oil- and gas-producing states April 20 when he called the plan "a duster," or dry hole, "which won't produce a bucket of coal or a barrel of oil."

Gov. Dolph Briscoe (D) of Texas called the President's plan "a cocked gun aimed at Texas" and promised a court fight against the proposal for federal control of prices in the intrastate natural gas market. That promise was immediately seconded by Sen. John Tower (R, Tex.).

Gov. Michael Dukakis (D) of Massachusetts, which was heavily dependent on foreign oil supplies, as were other Northeastern states, said the Carter "strategy was the cheapest, most practical and most attainable." Gov. James Thompson (R) of Illinois called the plan "basically fair."

Newark, N.J. Mayor Kenneth Gibson (D) told the President in a letter April 15, after receiving a briefing on the details of the plan, that the plan lacked sufficient emphasis on mass transit and help for city governments.

Vernon Jordan Jr., executive director of the National Urban League, April 21 criticized the plan's "casual vagueness" with regard to low-income persons. Renters would suffer when fuel prices went up, Jordan said.

Clarence Mitchell Jr., director of the Washington Bureau of the National Association for the Advancement of Colored People, said he feared the program would put people out of work and that the campaign to insulate homes "would open the field for crooks of all kinds operating in the neighborhoods of the poor," according to a report April 23.

Frank Fitzsimmons, president of the International Brotherhood of Teamsters, said April 21 that the proposed standby tax on gasoline was a tax on working people. It would be "awfully hard on the people who must commute some 20 and 30 miles to work," Fitzsimmons said.

Environmentalists—Nine national environmental groups issued a joint statement April 20 praising the plan as "fundamentally fair and farsighted" in its emphasis on conservation and use of renewable energy sources, such as solar energy. The statement also commended Carter for his stress on "local initiatives and decentralized energy sources, and on avoiding nuclear proliferation." Brock Evans, director of the Sierra Club, which joined in the statement, said April 21, "The energy message sounded like we wrote it ourselves."

However, a spokesman for Environmental Action, which also signed the statement, noted some "obvious omissions," such as mention of mass transit and proposed legislation to require payment of deposits on beverage bottles.

The 'People'—President Carter's energy publicity blitz during the week of

April 18–22 appeared to have had the desired effect. A special survey conducted for Newsweek magazine by the Gallup organization April 20–21 indicated that 54% of those sampled believed the energy situation in the U.S. was "very serious" and 36% thought it "fairly serious." An earlier Gallup poll, reported April 15, had shown that 45% believed it to be "very serious" and 37% "fairly serious."

A poll taken by ABC News and Louis Harris and reported April 23 indicated public support for new taxes on gas-guzzling cars (60% in favor, 33% opposed); tax credits for home insulation (85% in favor, 14% opposed); "increasing strip mining of coal but also protecting the environment" (82% in favor, 8% opposed); tax credits for installation of solar heating in homes (68% in favor, 22% opposed), and "speeding up construction of conventional nuclear power plants" (63% in favor, 21% opposed). The poll indicated that 54% of those surveyed were opposed to the proposed standby gasoline tax, while 39% thought it a good idea.

Foreign—Despite concern in many quarters over President Carter's proposals on nuclear energy, foreign governments and international agencies reacted generally favorably to Carter's energy policy statement.

Officials of the European Community (EC) Commission hailed the program as "important and courageous" April 21 and expressed the hope that Carter's message would inspire the EC to take similar action.

In Paris, an International Energy Agency spokesman April 21 called the program "well balanced."

Officials of the Organization of Petroleum Exporting Countries (OPEC) would not speak for the record but were reported April 22 to regard the Carter decision to raise the price of U.S.-produced oil to $13.50 a barrel as a vindication of OPEC's recent price hikes.

British energy officials said April 21, "The U.S.A. is easily the biggest consumer of energy and if it is able to achieve the energy conservation program and development of indigenous resources outlined, the energy outlook for the rest of the world will be brighter." Similar sentiments were expressed the same day by the West German and Japanese governments.

Pressure For & Vs. Program

Carter affirms tax commitment. President Carter expressed a strong commitment April 22, 1977 to his proposal for a standby gasoline tax as part of his comprehensive energy program.

The President defended the tax and other aspects of his energy policy at a news conference, his sixth since assuming office. The session was nationally televised.

The question on the gasoline tax referred to a suggestion that the proposal was "a bargaining chip to be traded later for something else that you really want." Carter replied that he was "deeply committed" to the tax and would "fight for it until the last vote in the Congress."

As for his stated intention to return to the public revenue from the gas tax in the form of rebates and refunds, Carter said "we still have to have some flexibility about exactly what we do."

"I can't certify today that every nickel of the taxes collected will be refunded to consumers," he said. But "my present inclination is to see that the gasoline tax to a substantial degree, and the fuel tax increase to a substantial degree, are refunded directly to the people of the country in the form of tax credits," he said.

Asked about the possibility of gasoline rationing, Carter said it would be "a viable alternative" in case of national emergency, such as an oil embargo. Was he saying "that rationing would be a fallback position if milder measures" proved insufficient? "That's correct," he replied. He added that if the entire energy package were put into effect, as he hoped and expected, "then I see no reason for gasoline rationing."

The President said he had not included any emphasis on mass transit in his energy policy because "this is a separate item that will be handled under the Transportation Department."

He said the energy policy did provide for one transportation problem, maintenance of highways. "As we reduce the consumption of gasoline, we'll have to make that up to the states so they can continue an adequate maintenance program," he said, "because they'll sell less gas in those states and therefore collect less gas tax."

Carter said he did not think it was possible in the immediate future to have complete deregulation of oil and natural gas prices. "I think the adverse impact on consumers and on our economy would just be too severe," he said.

As far as the oil companies were concerned, Carter said that his program contained a "prohibition against their deriving additional income as they produce oil from the presently discovered supplies compared to what the world market price would bear." He predicted that "there will be a move throughout the industrial world, in our country, away from oil and gas toward coal."

In referring to the reporting procedures being required of the oil companies, Carter said, "I think, when this information is analyzed it will be almost instantly obvious that unfair competitive procedures are in effect within the energy-producing area."

"And the antitrust laws can take care of it," he continued. "If I ever feel convinced that there is still an absence of competition within the energy field, after this proposal is put into effect, I would not hesitate to recommend divestiture."

On the economic impact of his energy plan, the President said "the most conservative and unfavorable analysis shows that it will have no adverse impact."

"Some computer model studies show that it will increase the number of jobs several hundred thousand and have a beneficial effect on our economy," he said.

The total net outlay from the federal government for his energy plan, through 1985, would be $4 billion, he said. He said the figure was "conjectural" but based on computer analysis.

Carter said the U.S. had "now taken the leadership in moving toward a comprehensive energy policy" and he hoped that other nations "would do a similar thing."

Republicans attack energy taxes. Republican national leaders suggested April 25 that President Carter's proposed energy conservation taxes were a disguised general tax increase to finance welfare reform and help the President fulfill his commitment to balance the federal budget.

House Minority Leader John Rhodes (Ariz.), Senate Minority Leader Howard Baker (Tenn.) and Bill Brock, chairman of the Republican National Committee, raised the tax issue while acknowledging that the Republicans had no broad energy program of their own.

Baker said the proposed taxes might generate as much as $70 billion a year in federal revenue by 1985, an estimate derided by Administration officials. Baker noted that White House energy adviser James R. Schlesinger had said April 24 that Carter wanted flexibility in the use of the tax receipts after "the early years." He also cited the President's own comment at an April 22 news conference that he could not guarantee that "every nickel" collected in energy taxes would be returned to the public.

House forms energy panel. The House of Representatives April 21, by voice vote, set up a 40-member Ad Hoc Select Committee on Energy to coordinate House action on a comprehensive energy program.

The new energy committee would try to "reconcile and harmonize" the different elements of an energy bill that would be written by the different committees with jurisdiction over the various parts of the bill. (They were, principally, the Ways and Means Committee, the Commerce Committee and the Interior Committee.)

The energy committee could not rewrite the proposals it received from the other committees before submitting the legislation to the full House, but it could offer amendments on the House floor.

House Speaker Thomas P. O'Neill (D, Mass.) appointed Rep. Thomas L. Ashley (D, Ohio) chairman of the energy committee. The panel had 27 Democratic and 13 Republican members.

GOP rebuts Carter energy plan. The Republican Party gave a televised rebuttal June 2 to President Carter's energy program.

The half-hour program was carried by NBC free of charge in fulfillment of the equal-time provision of the federal law governing political broadcasts. It was a reply to Carter's presentation of his energy program in April.

Republican national chairman Bill Brock presented the party's alternative in general terms. The party put emphasis on mass transportation as a major way to achieve conservation. It urged more efficient cars and favored tax credits for insulation.

It also called for more production of oil and gas through tax incentives and for more research into new energy sources.

One of the participants, in a four-minute appearance filmed in California, was Ronald Reagan, former governor of California and candidate for the 1976 Republican presidential nomination. He derided the argument that "we're running out of everything these days: energy, food, space, even love." "We're not running out of anything except confidence in ourselves," he said. "Our problem isn't a shortage of fuel; it's a surplus of government."

Other participants included former Transportation Secretary William T. Coleman Jr., Sens. John C. Danforth (Mo.) and H. John Heinz 3rd (Pa.) and Reps. Clarence J. Brown (Ohio), Jack F. Kemp (N.Y.) and Jack Cunningham (Wash.).

Much criticism was directed at Carter's proposal to increase the gasoline tax.

An energy resolution adopted by the Republican National Committee April 30, at the end of a two-day meeting in Chicago, had affirmed the party's opposition to "any attempt by the government to deny people their family cars."

Other critics abound—President Carter's energy program was under fire from many sectors, such as business, labor, congressional and consumer groups. Even the Senate Democratic leadership was suggesting changes.

At a White House meeting with congressional leaders May 19, Senate Democratic Leader Robert C. Byrd (W. Va.) reportedly urged the President to return the proposed tax on gas-guzzling cars to transportation needs, especially mass transit. The President had proposed a rebate of the tax to buyers of cars with good fuel efficiency.

Senate Democratic Whip Alan Cranston (Calif.) was doubtful that Congress would accept either that rebate or the proposed rebate of standby gasoline taxes to the general public.

President Richard L. Lesher of the U.S. Chamber of Commerce told the House Ways and Means Committee May 18 the gas-guzzler tax was "punitive" and the gas tax rebate "little more than a new means of redistributing the wealth."

At the committee's hearing May 19, AFL-CIO spokesman Andrew J. Biemiller deplored the gas tax as falling hardest on low-income people and the gas-guzzler rebate as a subsidy for the already popular small foreign cars.

The Consumer Federation of America denounced Carter's energy program May 18 as a "bonanza" for the oil industry. The technique of raising prices and taxes to attain conservation was "anticonsumer and regressive," it said. "The President has chosen the most reactionary form of rationing for the American people—high prices."

The spokesman was Lee C. White, former chairman of the Federal Power Commission and White House aide to President Kennedy. White was heading an energy-policy task force for the federation. The task force represented labor and farm groups, electric cooperatives, the Consumer's Union and the U.S. Conference of Mayors.

The Congressional Budget Office May 31 disagreed with the Administration's estimates. The Carter program, it said, would save 20% less fuel than claimed, would result in only a "small and gradual" rise in prices and would not bring about "dramatic" changes in the American style of living.

The General Accounting Office in a report to Congress June 8 said the Carter energy program "should be redesigned to provide a reasonable opportunity of achieving the stated goals." The agency said the Carter plan would not bring much reduction in energy demand nor an increase in the supply of oil, natural gas or nuclear power beyond the growth expected without the energy program.

Schlesinger on administered pricing—Carter's energy adviser James Schlesinger told the Joint Economic Committee of Congress May 25 that the oil industry was "not an industry that has embraced the market mechanism" and therefore a federal factor was necessary.

"When supply exceeds demand, the industry asks for regulation, and when demand exceeds supply, they want to use the market mechanism," he said. Since the price of oil was administered by oil-producing countries and did not reflect the market value, he said, "the industry should not be the beneficiary" of the sharp price increases expected in the near future.

Struggle over legislation—Several of President Carter's most important proposals were rejected in House committee and subcommittee action June 9. A House Commerce subcommittee approved decontrol of new natural gas, contrary to the President's plan. The House Ways and Means Committee voted to reject Carter's proposed standby gasoline tax and tax rebates for purchase of fuel-efficient cars, and it reduced and delayed until 1979 the proposed tax on "gas-guzzler" cars.

White House Press Secretary Jody Powell met with reporters June 10 to tell them the President was "deeply concerned" about the decisions. Carter felt, he said, that "yesterday the oil companies, the auto companies and their lobbies won significant preliminary victories" in the fight over energy policy. Powell said the vote to deregulate the price of natural gas was "particularly serious." Deregulation, he said, was a "ripoff of the American consumer."

Senate Democratic leader Robert C. Byrd (W. Va.) June 11 rejected the White House criticism as "overreacting." House committee action was "only the first pitch in an energy ballgame that may go 10 innings or more," he said.

At a news conference June 13, Carter again decried "the inordinate influence" of the oil and automobile industries. He warned that failure to adopt a strong energy policy would have "catastrophic" economic and political consequences.

But the President pulled back from criticism of Congress. "We don't consider ourselves infallible," he said, and "I don't say that everything we've proposed has got to be passed just as . . . we put it forward."

The Ways and Means Committee June 14 approved a key component of the Carter energy program, an oil tax to raise the price of domestic crude oil to the world price level, as a means to discourage consumption. The committee rejected an industry-oriented proposal to return some of the tax revenue to producers for searching out and drilling more wells. Afterward, Powell said the President was "extremely pleased" by the "courageous action" of the committee "in the face of an intense lobbying campaign by the oil industry."

New Energy Department created. Legislation establishing a new Energy Department was signed by President Carter Aug. 4 at a White House Rose Garden ceremony.

The President announced at the same time his nomination of White House energy adviser James Schlesinger, 48, as energy secretary. The new department would be the 12th Cabinet-level department, the first to be established since the Transportation Department in 1966.

Senate confirmation of the Schlesinger nomination came later Aug. 4, on a voice vote. A hearing on the anticipated nomination had been held Aug. 3 by the Senate Energy and Natural Resources Committee, which voted unanimous consent to the nomination less than an hour after it was made.

The Energy Department legislation had been cleared by Congress Aug. 2 by a 76–14 Senate vote and 353–57 House vote.

The new department's size was projected at the 20,000-employe level and an annual budget for fiscal 1978 of $10.6 billion. It would acquire the functions and employes of the Federal Energy Administration, the Energy Research and Development Administration and the Federal Power Commission. It would absorb portions of other departments and agencies as well.

A new five-member federal energy regulatory commission, its members appointed to four-year terms by the president and subject to Senate confirmation, would be established within the new department but independent of its secretary. The commission would set rates for the transportation and sale of natural gas and electricity and for the transportation of oil by pipeline.

The Energy Department also would have regional marketing authority over electric power, control over the rate of energy production on public lands and jurisdiction over the naval petroleum and oil shale reserves.

It would have authority to set energy conservation standards for buildings and oversee voluntary industrial conservation programs. Coal development and energy data programs also would be housed in the new department.

A "sunset" provision, requiring the department to go out of existence after five years unless Congress extended its life, was deleted by the Senate-House conference committee that prepared the final bill. A Republican move in the House to have the provision reattached failed prior to final passage of the legislation.

In place of the "sunset" provision, the conferees inserted a requirement for a comprehensive presidential review of the department's programs within five years.

Carter warns of 'impending crisis'—At the Rose Garden ceremony Aug. 4, President Carter warned of an "impending crisis of energy shortages."

The President had sounded the same warning as the House faced a week of debate and decision on the Administration's package of energy proposals. The public "is not paying attention" to energy problems and "has not responded well" to calls for voluntary conservation, Carter told 26 editors and radio and television news directors in an interview session July 29 (a transcript of the interview was released July 30).

The President expressed concern in the interview that it would take "a series of crises" to make the American people "quit wasting so much fuel."

"I think voluntary compliance is probably not adequate at all," he said.

House passes Carter energy package. The House passed a massive energy bill Aug. 5 patterned in large part after President Carter's proposals in April.

The prompt House action on the complicated legislation, which had more than 100 separate provisions, was credited to House Speaker Thomas P. O'Neill (D, Mass.), who took personal charge of the package.

The Carter program remained substantially intact in the House version, except for his major proposal for a 50¢ standby gasoline tax increase, which the House rejected.

The package was passed by a 244-177 vote. Prior to passage, the House rejected, by a surprisingly thin margin, 219-203, a Republican attempt to dilute the bill and delay passage.

On another tight vote Aug. 5, a GOP move to drop a key component from the bill—a crude oil tax—was defeated, 219-203.

By a vote of 221-198, the House strengthened the bill further Aug. 5 by removing a tax exemption on industrial use of oil and gas for an industry that was unable to convert to coal.

Crude oil tax—The crude oil tax was designed to bring the price of domestically produced oil up to the world price. It would be imposed at the point of sale to refineries. (The price of domestic crude, which was subject to price controls, averaged $8.71 a barrel, compared with $13.30 on the uncontrolled world market.) The revenue from the tax would be rebated to the public through income tax credits or special payments to persons who did not pay income taxes. Oil used to heat homes, schools, hospitals and churches would be exempted from the tax.

Natural gas prices—Reflecting a major victory for the Administration, in support of Carter's stand against deregulation at this time, the House agreed to an increase in the price of newly discovered natural gas to $1.75 a thousand cubic feet from $1.45. The ceiling would apply for the first time to gas produced and sold within a state. The figure of $1.75 was pegged to the average price of the energy equivalent in domestic crude oil.

The legislation specified that the gas in question would be that produced from a well drilled 2½ miles or more from an existing well or at least 1,000 feet deeper than an existing well. State regulatory agencies would certify the designation as new gas, subject to change by the Federal Energy Regulatory Commission, the successor to the Federal Power Commission in the new Energy Department.

The brunt of the price increase was to be put initially upon industrial users rather than homeowners.

Conversion to coal—The bill would provide authority for the government to require utilities and industrial plants to burn coal instead of natural gas or oil. The authority was backed up by stiff user taxes. The tax could be offset by conversion to coal or fuels other than gas or oil.

Gasoline taxes—The federal income tax deduction for state and local gasoline taxes would be repealed. The existing 4¢-a-gallon federal tax on gasoline, scheduled to drop to 1½¢ on Oct. 1, 1979, would be retained at 4¢ for six more years. The 2¢-a-gallon tax break for noncommercial motor boats would be removed.

Tax on gas guzzlers—New cars with poor gas mileage, the so-called gas guzzlers, would be taxed, beginning in 1979. Buyers of new cars would face a tax that would range from $339 to $553 for cars getting fewer than 15 miles a gallon. For 1980 cars, the tax would be $249–$666 for less than 17 miles a gallon. By 1983, the range would be $397–$3,856 for less than 23.5 miles a gallon.

The House authorized the proceeds from the tax to be put into a trust fund for payments on the national debt.

Utility rates—The legislation would establish national standards for determining utility rates. Utilities would be required, through state regulatory agencies, to relate fees to the cost of providing service.

Insulation—A tax credit would be available to homeowners who installed insulation and other energy-saving devices on existing housing. The maximum would be $400, or 20% of outlays up to $2,000. Persons in the middle-income brackets or below could apply for grants and low-interest loans for installing the insulation. Federal funds also would be distributed, on a matching basis, for schools, hospitals and nursing homes to install insulation.

Energy conservation plans, with special attention paid to solar energy, would be established for all federal buildings.

Solar and wind power—Tax breaks would be provided homeowners for installing solar-energy equipment or windmills. The maximum credit would be $2,150, or 30% of the first $1,500 plus 20% of the next $8,500 invested.

The new tax break would apply to existing or new housing.

Buses—All federal taxes on bus operators would be repealed.

Appliances—Mandatory minimum efficiency standards would be set for major appliances, from furnaces to television sets.

Energy studies—A study would be conducted to determine the potential fuel economies from off-highway vehicles, such as aircraft and motor boats. Another study would be made of the savings potential from increased bicycle use.

Resources & Tactics

Tactics Proposed

Government, business and professional leaders have achieved some consensus on tactics to be followed in the U.S. attack on the energy crisis.

There seems little argument with the thesis that conservation of energy must be accorded high priority in the program. There is also wide agreement with the calls for intelligent and vigorous actions to develop energy sources, although specific source-development proposals have produced strong disagreement.

Coal is often spoken of as the nation's most abundant fossil fuel, but there is much opposition to strip-mining operations and frequent warnings of the pollution hazards posed by the use of coal. Nuclear power is favored by some industry spokesmen but is also attacked on environmental grounds. Other energy sources—ranging from the sun to the undersea oil off the nation's coasts—also have their proponents and detractors.

Engineers' proposal is typical. An energy policy statement published by the National Society of Professional Engineers in the December 1975 issue of the Professional Engineer is fairly typical of some of the proposals made by responsible U.S. leaders

The statement held that "a comprehensive, coordinated, and priority-charged energy program, involving the full breadth of political and social decisions by government, industry, the technical professions, and the public, must be developed without further delay; and, that this program should include, as a minimum: (1) effective conservation efforts; (2) increase in production of all existing domestic forms of energy; (3) rapid expansion of sources with demonstrated reserves and productive potential; (4) provision for adequate research, development, funding and other needed stimuli of new energy sources."

The statement also said:

While gas, oil, coal and nuclear fuel are the real workhorses of energy supply, much interest has been aroused in energy sources such as Geothermal, Solar, Wind, Tidal, and Ocean Currents. Whether these sources of energy can make a significant impact before 2000 must be further researched. Even with the most intensive development, question has been raised about how much they could actually supply. While some think they could supply no more than 4 percent of our energy needs by 1990, others are more optimistic. The potential of solar energy must also be kept in perspective. With a well-managed, well-funded, aggressive program, we may be able to provide 2 percent of our energy by solar heating, cooling and other uses. Other energy sources, such as the incineration or conversion of refuse, cannot contribute significantly to the overall energy picture. The potential capacity of these energy sources must be considered before massive amounts of capital funds are poured into these projects.

When we search for domestic energy sources to substitute for imported oil, we must look at the complete energy picture. The energy sources such as gas, oil, coal, and nuclear power, are capable of meeting our energy needs for many years, provided they are used effectively.

Natural gas supplies presently available to the United States are not sufficient to meet all the demands for this premium fuel. This supply can be extended by gas manufactured from coal, oil shale, organic and inorganic waste, and increased exploration and drilling, including offshore, secondary recovery, and deep well.

Oil, which continues to be the leading source of primary energy, is not expected to grow significantly in supply. As in the case of gas, the lag in discovery rates has begun to darken the domestic supply picture. If efforts fail to reverse the disappointing discovery trends of recent years, the vast reserves of tar sands and shale will have to be developed to make significant contribution to the Nation's energy supply by the end of the present decade.

Of all fossil fuels, coal is by far the most abundant. United States reserves amount to an estimated 1.6 trillion tons—3,000 times the 500 million tons burned in 1971. With this abundance of energy in coal, we must further develop its use by producing synthetic oil and pipeline quality gas. To attain independence from energy imports, we must accelerate all programs aimed at the production of oil and gas from coal. The greater use of coal is an immediate option. Environmental considerations and governmental regulation must also be tempered with the needs for industrial growth, increased employment, and popular demand for comfort and convenience.

The best hope for reduction in the spiraling costs of electricity can be and has been realized by the increased use of nuclear power. For example, in one area of Pennsylvania alone, 900 Megawatts of nuclear capacity has significantly reduced the need to burn high-priced, low-sulfur oil, and the resulting savings have reduced customers' fuel adjustment charges by about 40 percent in the first five months of 1975. For a typical residential consumer using 500 kilowatt-hours the monthly bill reduction was about $2.50.

Nuclear generating capacity can be expanded and brought on line to accommodate an increased electric use. Barring limitations on and regulation of the supply of economically recoverable uranium ore, there are no technological or resource limitations blocking a rapid expansion of nuclear generating capacity. However, effective licensing and siting procedures must be developed to permit prompt processing of applications for nuclear power plants consistent with the need for adequate safety and environmental protection. Environmental considerations, of course, cannot be ignored. And with respect to all energy sources, consideration should be given to practicalities, such as the siting of nuclear projects in areas with no water supplies. Availability of necessary water sources for some energy recovery processes may be a critical consideration. But a reasonable balance among all interest must be maintained. Nuclear energy is an essential power source for the immediate future. Moratoriums and other delaying actions in the construction of nuclear power plants are non-productive. With close regulation, nuclear power is the best approach for supplementing fossil fuel sources in the near term. The increased use of nuclear power will, furthermore, free use of coal for conversion to liquid and gaseous hydrocarbons. . . .

Mathias on coal & the future. Sen. Charles M. Mathias Jr. (R, Md.), during Senate debate May 20, 1977 on a bill to control strip-mining of coal, discussed the "delicate balance [that must be struck] between our nation's need for energy sources and our need to protect our environment." According to his statement, as it appeared in the Congressional Record:

"We, as a people, have become much more sophisticated in recognizing the tradeoffs which are inherent as our energy needs confront the environment where we all exist. We have had to meet and resolve so many difficult questions in these areas over the last few years that public awareness is greater now than it has ever been.

"One thing we have very definitely learned is that you cannot separate energy development from environmental quality. In short, you cannot say, let us have all the energy development that this Nation can ever need, and leave those areas of the Nation that are unsuitable for energy development as 'environmental areas.' The relationship between our supplies of clean air, pure water, soil, and minerals is simply too complex for such a view.

"Coal, once referred to as 'the homeliest hydrocarbon,' is enjoying a rebirth as a potential solution to this Nation's energy problems. Coal constituted 70 percent of the Nation's energy consumption in 1900. By the fall of 1973, however, coal's share was down to 18 percent. But the recent limitations on natural gas hookups, and the whopping increase in foreign oil prices has redirected the spotlight to coal.

"The demise of the deep mining industry is proceeding at an everincreasing pace. Since 1966, over 2,400 deep mines have shut down and production has fallen by more than 84 million tons in Appalachia alone. At the same time, strip mining operations have increased by well over 700 and production by 59 million tons. But everyone in this Chamber knows, or should know, that strip mined coal is a finite resource. We only have 45 billion tons of strippable coal left. As we shoot for coal production of over a billion tons a year, and 2 to 3 billion in the years ahead, as we get into coal gasification and liquefaction, it is easy to see that by the end of this century, all strippable coal may be exhausted.

"The Department of the Interior study entitled, 'Energy Research Program,' predicted that Western strippable coal will all be gone by 1996 and most of the Eastern strippable coal will be exhausted as well. But while we have 45 billion tons of strippable coal, we have 30 times that much deep minable coal. Even taking the most conservative statistics supplied by the U.S. Bureau of Mines, we find 356 billion tons of deep minable coal as against the 45 billion tons I mentioned earlier. Now that is an extremely conservative analysis of the amount of underground coal this Nation possesses. But even using those most conservative figures, we see a ratio of 8 to 1.

"Given these statistics, what would be a rational coal policy? Would you encourage stripping wherever you can physically get the coal out? Would you encourage strip mining on steep slopes where you know that the final result will include environmental degradation and a possible threat to the safety of the people who live in the valleys below? Would you, in essence, encourage strip mining to grow by leaps and bounds, and thereby guarantee the mass exodus of deep miners from the mines and the closing of those deep mines? Would you do all this when you know by looking at statistics—the very ones I have quoted; the most conservative statistics available—that massive strip mining is a short-term operation and that eventually the deep mines must be opened and expanded and the deep mine labor force rebuilt and expanded? Since 1966 there has been a loss of 19,000 jobs in the mines. In the final analysis when we look at all the figures, we have to recognize the need to guarantee the continued existence of deep mining in Appalachia. The Nation desperately needs the 67.6 billion-ton reserve of deep minable coal in Appalachia, but the recovery of those vast reserves is seriously jeopardized by a rush headlong to strip mining of that region of the country. Not only is it going to spell the economic death of the deep mine industry, but the blasting and other activities on the surface will make deep coal seams technically impossible to mine.

"I have briefly alluded to the exodus of deep miners to other types of employment, but this bears some further analysis. I am convinced that unless we halt this steady exodus of skilled labor, other Senators will meet in this Chamber years hence to discuss ways of retraining a massive labor force to meet this Nation's by then urgent requirement for coal. That retraining will be a very expensive undertaking, but if we act today to ensure a continuing stimulus to deep mining operations, the expense will be avoided.

"In the final analysis, I would hope that one lesson would be learned as a result of the current energy shortage: We must plan for the future in our handling of energy. To some the future is tomorrow. To others, it is measured in years. But with regard to energy, the public interest requires that it be measured in decades. We must start with this bill to establish a policy for coal which will respect the land, the air, and the water—the elements which sustain us—a policy which will safeguard the deep mine coal industry so that when the strip mines are exhausted, there will be a viable industry to serve this country.

"Because coal has once again become a competitive energy source, its price is bound to go up. Already this upward trend in price is apparent. And, of course, the profits to be realized by the coal extraction industry are increasing substantially. . . .

"We must remember in our search for energy independence and self-sufficiency that coal is but one of the elements and that it is a finite resource. It should not be viewed as a panacea for our energy crisis.

"And in the long-range perspective we must remember some other equally important goals: to improve the air we breathe; to tread gently on our mountains, valleys, streams, and forests; to protect our marine animals and wildlife.

"These things together form our world and our view of it. To despoil our environment in a race to find energy sources is to desecrate a God-given gift as well as to do violence to our psyches. We must guard against this happening.

"We must pursue all avenues which hold potential for new energy sources: solar; wind; geothermal; the oceans. And we must pursue them with a view toward costs and benefits, weighing carefully those tradeoffs I mentioned earlier between energy independence and environmental—and human—protection."

Conservation

Zarb presses conservation. Sen. Edward M. Kennedy (D, Mass.), chairman of the Energy Subcommittee of the Congressional Joint Economic Committee, told the Senate Feb. 19, 1976 that Federal Energy Administrator Frank G. Zarb had estimated before his subcommittee that "over $200 billion of energy conservation investments will be required over the next 10 years." Kennedy cited Zarb as estimating that "energy demand by 1985 can be reduced by 14%, or the equivalent of seven million barrels of oil per day, from levels anticipated before the oil embargo." Zarb's testimony included these statements:

During the past two years there has been much discussion of the need for energy conservation. However, despite the fact that conservation has been the subject of considerable public debate, several widely held misconceptions somehow still remain. These have not only delayed the enactment of important legislation but have also engendered confusion among the general public. I would like, therefore, to begin my testimony today by identifying, and hopefully dispelling, some of these myths.

First, and perhaps most widespread, is the myth that intelligent conservation of energy will hinder economic growth, increase unemployment or lower our high standard of living. There is no question that the dramatic increases in the price of imported oil instituted by the OPEC Nations during the past two years pose a threat to our economy. Because of this threat, it is absolutely necessary that individuals and businesses take steps to use energy more efficiently. Contrary to the myth, conservation is vital to our efforts to sustain our high standard of living and rekindle economic growth. Moreover, several recent analyses have shown that reducing the inefficient use of energy would not result in an employment penalty and may, in fact, create more jobs.

A second is the myth that energy conservation is only an environmental concern and that conserving energy is not an economic proposition. While energy conservation would result in a cleaner environment, the key motivation behind virtually all efforts to conserve energy is and should be economics. Saving energy is synonymous with saving dollars and can, in fact, be considered as one of the least expensive energy supplies this Nation has.

A third myth is that higher energy prices will not induce greater energy conservation. Since the dramatic rise in oil prices at the end of 1973, petroleum demand has declined markedly. In comparison to pre-embargo forecasts, 1975 petroleum demand declined by approximately 14 percent or 2.7 million barrels per day. Of that 2.7 million barrels per day, over a third or about one million barrels per day is attributable to increased awareness and response to higher prices. Thus, as energy prices climb higher, saving energy becomes more attractive—for both businesses and individuals.

A fourth myth is that conservation is only a stop-gap measure and can't really make a significant contribution to the resolution of our longer term energy needs. We estimate that anticipated increases in energy prices along with Government initiatives will result in the adoption of conservation measures that will reduce energy demand, including oil, gas, coal, nuclear power and other energy sources, by about 14 percent from levels anticiapted before the embargo—or the equivalent of more than 7 million barrels per day of oil—by 1985. This reduction is just slightly less than our current rate of production of domestic crude oil. Although a large part of thees savings are likely to occur in response to higher energy prices alone, the full amount would not be achieved without Government involvement to accelerate the adoption of conservation measures. A good example of a desirable conservation measure that would result in long-term savings, if adopted, is the updating of standards for new residential and commercial buildings.

I should like also to try to dispel several myths which are often engendered by many advocates of energy conservation. This is necessary to obtain a balanced understanding of the conservation issue.

One is the myth that energy conservation alone—or in combination with the development of solar and other inexhaustible energy resources—can solve our energy problems. Even when we achieve our estimate of reduced energy demand, which I cited earlier, we would still require the energy equivalent of approximately 44 million barrels per day of oil to meet the needs of our economy in 1985. This is 24 percent more than what we use today. Even the most optimistic projection of the contribution to our national energy needs that could be made by solar and other inexhaustible energy resources is far below this figure. Obviously, unless we reverse the trend of rapidly declining domestic oil and gas production, we will be forced to rely even more heavily on imported energy.

A second myth is that the Federal Government, by enacting a law or issuing regulations can swiftly and painlessly ensure that energy is conserved. As this past year has clearly indicated, there are no such simple solutions. In fact, encouraging greater energy conservation is, in many respects, a more complex and difficult task than encouraging increased domestic energy production. While only several thousand companies produce and/or distribute our domestic energy supplies, literally millions of diverse businesses, institutions and individuals consume energy. While increased energy prices have stimulated conservation actions, in a few circumstances the President and the Congress have taken a mandatory approach. It will be no simple task to manage these complex programs; great care must be exercised to avoid the large bureaucracies and economic distortions that often are the result of Government regulation.

Finally, there is myth that energy conservation is free—or nearly free. While it is true that significant energy savings can be realized for little or no cost, it is also true that many measures that could result in large energy savings require significant investment. The installation of storm windows, heat pumps, heat recovery systems, and power recovery turbines has a cost, just like measures to increase energy supplies. The choice between whether or not to adopt any specific conservation measure must be made by the individual or firm concerned, on the basis of hardnosed economic analysis. . . .

Energy is only a means to economic well being, not the end product. If a fuel becomes overly expensive, or unavailable, then common business sense dictates that the thing to do is replace it with the lowest cost substitute. Let's consider a simple example of a consumer faced with rising fuel oil prices. The consumer has available several alternative responses. First, he could simply continue to pay the higher heating bills. Second, he might consider switching from fuel oil to some other source of heat—such as natural gas, electricity, or possibly solar energy. Another alternative, however, would be to reduce his use of fuel oil by installing insulation. How does he choose among the alternatives?

Continuing to pay the higher bills for fuel oil would cost more than $17 for each barrel of oil used. If the homeowner were able to switch to natural gas, then he would be paying only around $9 for the energy equivalent of that barrel of fuel oil. (Natural gas is still a "bargain" because it is regulated at unrealistically low prices. However, many areas, including Washington, D.C., have moratoria on new gas hook-ups. Consequently, the natural gas alternative is increasingly unlikely to be available.) If the homeowner lived in Massachusetts and chose to heat his home electrically by installing baseboard or some other form of resistance heating, he would be paying more than $30 for the equivalent of that barrel of oil. However, if this homeowner chose to install ceiling insulation to improve his home's thermal efficiency, he could effectively save a barrel of oil or $17 for every $5 that he spent on insulation. Thus, conservation turns out to be, in this case, the most economically attractive alternative. . . .

Industrial examples include the installation of an air pre-heater on a boiler for $11 per barrel and addition of power recovery turbine, also costing $11 for each barrel saved. Naturally, in every sector conservation measures range in cost from virtually zero for "housekeeping" actions to more than the cost of simply purchasing more fuel.

The point of such examples is that conservation measures are not only viable alternatives, but generally they represent some of the most cost-effective ways we have of dealing with energy problems. . . .

ERDA focuses on energy conservation. The Energy Research and Development Administration (ERDA) said April 19, 1976 that it was redirecting its long-range energy policy to put conservation efforts on a par with development of new energy supplies. In its initial policy plan submitted to Congress in 1975, the agency had stressed increased production from all sources—nuclear, coal, solar and geothermal.

In its latest report to Congress, Robert C. Seamans Jr., ERDA administrator, said it was "impossible to exaggerate the need to make more efficient use of energy."

"Ordinarily it costs less to save a barrel of oil than to buy it," he told Congress. "Each barrel saved means one not imported."

Details of the conservation effort had not been worked out yet, but Seamans told reporters at a press conference April 19 that "some very significant projects" were planned. In his report to Congress, he cited ERDA's goals of increasing the efficiency of energy use in factories, buildings, and transportation and curbing energy waste in the energy-production process. The report noted a "widening" gap between demand and domestic production of oil.

Wolfcreek Statement. An energy analysis called the Wolfcreek Statement, produced by participants in a meeting held in October 1976 at Wolfcreek Wilderness in the North Georgia Mountains, stressed the priority of energy conservation. Among the statement's assertions:

The keystone of a rational energy policy must be a national commitment to energy conservation. Such a commitment will provide the time we need to develop and deploy energy technologies that are safe and sustainable, appropriately scaled, and economically sound. To achieve these objectives we recommend: (1) removing all subsidies for non-renewable fuels; (2) placing a royalty on non-renewable fuels in a manner that is equitable to all and beneficial to the economy; and (3) eliminating all institutional barriers to efficient energy use.

The advantages of this course are overwhelming: employment can be increased; inflation can be more easily controlled; pollution can be reduced; scarce resources can be conserved; social equity and the quality of life can be improved; dependence on foreign energy sources can be reduced; and the pressure toward nuclear proliferation can be relieved. But time is short. More than half of our recoverable domestic supply of our two choicest fuels has been exhausted. . . .

The largest source of energy available to the United States is the energy we currently waste. Our country ranks only fourteenth among the eighteen members of the International Energy Agency in terms of effective energy conservation efforts (10). Sweden, West Germany, and Switzerland, at comparable standards of living, consume only 60 percent as much energy per capita. In 1976 Americans wasted more fossil fuel than was used by two-thirds of the world's population. This excess, which currently comprises one-half of our energy budget, represents our nation's largest, cheapest, cleanest, and safest near-term source of new energy.

It is feasible for the United States to sharply increase its energy efficiency by the year 2000 and in the process greatly extend its energy supply. Energy consumption can be made more efficient by carefully matching the quality of fuel with the quality of work desired. Such matching would eliminate, for example, the burning of high quality natural gas and oil at temperatures in excess of 1000° to obtain room temperatures of 70°. Energy consumption can also be reduced by changing the patterns of consumption to favor efficiency through the use of mass transit, bikeways, cluster housing, district heating, and recycling (11).

The list of good ways to conserve energy is long. For example, ceiling insulation in a typical home costs about $300 but will save about seven barrels of oil each year for the lifetime of the house. These seven barrels, which are as valuable as new oil pumped out of the ground, can then be employed more just 10 years would amount to 70 barrels, productively elsewhere. The energy saved in which means that we are "producing" oil at $4.30/barrel. When heating oil costs $3 per barrel, insulation was no bargain. But today, heating oil often costs $16 per barrel, and insulation has become a cheap source of new energy.

Industry currently consumes 40 percent of this country's fuel. Recent studies by the Conference Board and the Ford Foundation's Energy Policy Project suggest that enormous energy savings can be made with existing technology without denting industrial productivity (2,4). The primary metals industries use about one-fifth of all industrial fuel. By adopting technologies now widely employed in other countries, the steel industry can reduce its inordinate fuel demands by about 50 percent by 1995. Using the new chloride process instead of the traditional Hall method to refine aluminum yields energy savings of about one-third. Recycled scrap aluminum requires only 5 percent as much energy as aluminum refined from virgin ore.

Forty-five percent of all industrial fuel is used to generate process steam. If this steam were first used to generate electricity and then used as process steam, considerably more electricity would be produced than the entire industrial sector now purchases from utilities. The additional fuel and capital needed per kilowatt-hour of industrial "co-generation" is only half that required by the most efficient new centralized powerplants.

From 30 to 50 percent of the operating energy in most buildings can be economically conserved now. Moreover, the American Institute of Architects estimates that the use of advanced conservation technologies including solar devices, heat pumps, and total energy systems can conserve up to 80 percent of energy consumption in new buildings (1). A national commitment to upgrading the energy efficiency of buildings would, by 1990, save the equivalent of 12.5 million barrels of oil a day.

Equally dramatic energy savings are within practical reach in the transportation sector. The use of manual transmissions would save one-tenth of all automotive fuel consumed. . . .

Energy conservation will give us the time we need to make a transition from an energy base consisting of geologic capital (oil, coal, natural gas, and uranium) to one based on renewable, safe, direct and indirect solar income. This time provides a reprieve in which we can examine alternative energy technologies. If our present knowledge advances even modestly over the nevt several decades, we may confidently expect a rising level of energy derived from the renewable income sources (9). . . .

It is apparent that there are two mutually exclusive energy paths before the nation. One option is based upon continued drift down the path of "hard" technology. It is fraught with technical risks and with economic and social uncertainities. The second option, which begins with a national commitment to energy conservation, represents a more cautious, sustainable approach. But while the choice of either tends to preclude the other, only the hard path is irreversible. At issue is our willingness to gamble against great odds that we can achieve a high technology solution to our energy needs and that we would be wise enough to manage the result. The option represented by a sustainable energy approach is more modest in what it promises, but it is also much less likely to imperil our future. The time for decision is now—continued drift could soon remove any possibility of moving in sustainable directions. Continued depletion of fossil fuels and growing investment of our capital and our reputation in hard technology will at some unknown date effectively preclude flexibility.

REFERENCED SOURCES

(1) American Institute of Architects, "Energy Conservation in Building Design" (1974); "Energy and the Built Environment" (1974); "A Nation of Energy Efficient Buildings by 1990" (1975).

(2) *The Conference Board Record*, (February, 1975)

(3) Energy Research and Development Administration, *A National Plan For Energy Research, Development and Demonstration: Creating Energy Choices for the Future*, Vol. 1 (June, 1975).

(4) Ford Foundation, Energy Policy Project, *A Time to Choose* (Cambridge: Ballinger 1974).

(9) Amory Lovins, "Energy Strategy: The Road Not Taken?" *Foreign Affairs* (October, 1975).

(10) Organization for Economic Co-operation and Development, *Energy Conservation in the International Energy Agency* (Paris, 1976).

(11) Real Estate Research Corporation, *The Cost of Sprawl* (U.S. Government Printing Office, 1974).

Alliance to Save Energy. A private, non-profit Alliance to Save Energy (ASE) was formed in February 1977 by Sens. Hubert H. Humphrey (D, Minn.) and Charles H. Percy (R, Ill.) in what Percy described in the University of Chicago Journal of Energy & Development as an effort "to develop a broadly based constituency—and to set quantitative goals—for energy conservation in every sector of our society. . . ."

President Carter welcomed the creation of the ASE and announced Feb. 10 his appointment of ex-President Gerald Ford and Vice President Walter Mondale as honorary chairmen of the alliance.

Rand study urges peak-load pricing. Vast quantities of energy could be conserved and electricity bills could be lowered if U.S. utilities adopted the European practice of charging more for electricity during hours and seasons of peak use, according to a study by California economists at the Rand Corp., reported March 11, 1977.

After an eight-month study for the California Energy Resources and Development Commission, the Rand team concluded that the fuel costs of supplying power to 18 industries in the state could be cut by as much as $1.3 million a month with such a system. They said the utilities supplying those industries could achieve a 3% reduction in energy consumption during peak-load hours, for savings of $2.3 million a month in operating and capacity costs and corresponding reductions in customers' electricity bills. Were the incentive-pricing system applied to all California industry, the savings in energy costs would be nearly twice as large, according to the study.

For several reasons, utilities produced electricity more efficiently at off-peak hours than during high-use periods. The European system was designed to encourage the use of cheaper energy during the off-hours. U.S. utilities sold power at the same price all year round and at all hours of the day, in effect subsidizing the use of electricity at peak hours and penalizing its use at other times.

Bridger M. Mitchell, leader of the study team, said that if major industries reduced their use of electricity at peak hours, fewer new generating plants would be needed and construction of those could be postponed.

California utilities to test voltage cuts—Nearly all the major utilities in California had agreed to begin testing the effects of a four-volt, or 3%, reduction in electrical voltage, it was reported March 18. The circuit-by-circuit test to determine where voltage could be cut without adverse effect was to provide the basis for a long-term conservation effort to save fuel and, at the same time, reduce the need for new generating facilities.

The program, under consideration for years, was not directly tied to the hydroelectric power problems caused by the current drought. However, informed observers said the drought helped the state's Public Utility Commission finally to convince Pacific Gas & Electric—one of the utilities hardest hit by the lack of rain—to be the first to try the experiment.

Accent on Coal

Carter stresses coal. Sen. Jennings Randolph (D, W. Va.) told the Eastern

Coal Forum in Charleston, W. Va. Nov. 15, 1976 that newly elected Jimmy Carter planned increased use of coal as a key element of his energy program. Randolph said:

It was in Charleston on Aug. 14 that candidate Jimmy Carter stated his commitment to coal. He said: "The future of West Virginia Appalachian coal is indeed bright if we can have a government policy that recognizes this valuable resource."

In an interview a month later, Governor Carter said "a shift from oil to coal" is a key part of his proposed energy program for our nation.

President-elect Carter gave me several opportunities to discuss this subject. So, I have reason to anticipate that we will have an administration aware of the importance of coal and determined to use this vast resource for the benefit and security of the United States. This is essential to creation of an orderly energy policy and the reduction of our perilous dependence on foreign sources of energy. . . .

The stark reality is that until at least 1985 the United States cannot satisfy its tremendous appetite for energy without increasing oil imports. The production of domestic alternatives, such as coal, cannot grow fast enough. The alternatives to oil imports are mechanisms such as rationing or severe and mandatory energy conservation.

Because of our frustrating inaction, many possibilities earlier considered viable are no longer realistic. For example, 1970 projections for synthetic oil and gas supplies from coal by 1985 have been reduced drastically to approximately one-seventh the earlier estimates. The resultant shortage must be filled by oil imports. Similar shortfalls for nuclear power also will have to be filled by oil imports—assuming they are available. . . .

Energy conservation can minimize the economic impact of energy independence, but its influence on growth of energy demand is yet to be realized in any adequate degree. For the most part, the energy savings achieved since the 1973 embargo have resulted mainly from our economic recession, higher energy prices and the 55 miles-per-hour speed limit. . . .

But even should major conservation efforts reduce the rate of growth of energy consumption, our country's energy future would depend on greater coal utilization for the next 80 to 100 years.

Some persons question the wisdom of expanded coal production and utilization. These critics advocate the use of renewable energy sources. But such sources as solar, geothermal, wind and tidal power have inherent disadvantages. Long research and development times are required before their commercial application. Solar heating is an exception.

Renewable energy resources offer significant potential, but it may take 50 years to harness them. . . .

Illinois conversion plant. The Wall Street Journal Nov. 18, 1975 reported an Energy Research & Development Administration announcement that the country's first large plant to convert coal into gas and oil would be built near Belleville, Ill. The $237.2 million facility was to be the joint venture of Union Carbide Corp. and the General Tire & Rubber Co.

Coal situation. Sen. Bob Packwood (R, Ore.) Feb. 24, 1976 gave the Senate these facts on the U.S. coal situation:

"The U.S. Coal Reserve Base—known to exist—as of January 1974 is about 435 billion tons of coal, which constitutes more than half of the known coal deposits on earth. Assuming an average heating value of 10,000 Btu per pound, this U.S. coal reserve has an energy value equivalent to about 120 times the total U.S. energy consumption in 1974. This is a somewhat conservative estimate since some authorities estimate the world's known coal deposits to be in the order of 6 trillion tons with the United States holding more than half of that amount or about 3.2 trillion tons. Nevertheless, taking the more conservative estimate, about 298 billion tons of the U.S. reserves coal exists underground and the remaining is surface coal. Most of the surface coal exists in the Western United States.

"Approximately 60 percent of the Nation's coal reserves contain 1 percent or less sulfur by weight and most of this is found in the coal in the Western United States. Only about 50 percent of underground and about 90 percent of surface coal can be adequately recovered by present mining methods. The reason being that about half of the underground coal must remain in the mines in order to provide support for the mines, and some surface coal is inaccessible because of natural or people made surface obstructions.

"The present annual rate of coal consumption in the United States is about 600 million tons; however, 60 million tons of this amount is exported. The remaining 540 million tons constitutes about 17 percent of the Nation's annual energy consumption.

"At this present rate of consumption, our known reserves are capable of supplying coal for approximately five centuries. However, if we were to convert to 100 percent coal energy economy, these supplies could be reduced to less than 100 years. . . ."

Oil savings through coal use. FEA (Federal Energy Administration) Administrator Frank Zarb reported April 23, 1976 that conversion of MFBIs (major fuel-burning installations) to coal would ultimately save about 287 million barrels of oil a year (at a cost of about 79 million tons of coal) and free about 104 billion cubic feet of natural gas for industries facing curtailment, residential use and other purposes.

The conversions, under orders issued by authority of the ESECA (Energy Supply & Environmental Act of 1974), as amended by the EPCA (Energy Policy & Conservation Act), were expected to reduce electric utility fuel costs by about $1.74 billion annually by 1980. The report said:

Converting Existing Powerplants From Oil and Gas to Coal. The potential savings from converting existing powerplants from oil and gas to coal is significant.

Through a lengthy review of data relating to the 725 generating stations which responded to a Federal Power Commission's Emergency Fuel Convertibility Questionnaire, the FEA identified units located at eighty (80) stations which were currently burning oil or gas but capable of burning coal as a primary energy source.

On June 30, 1975, FEA issued Prohibition Orders affecting seventy-four (74) powerplant units located at thirty-two (32) generating stations owned by twenty-five (25) utility companies. . . .

Ensuring that Existing Powerplants Do Not Convert From Coal to Oil. During the late 1960's and early 1970's, many powerplants converted from coal to oil, particularly on the East Coast. Since EPA cannot issue compliance date extensions under ESECA to facilities currently burning coal, ESECA is ineffective to prevent such coal to oil switches before they occur.

Requiring New Powerplants To Have The Capability To Burn Coal. During those late 1960's and early 1970's, while coal-fired powerplants were being converted to oil, utilities were also building new plants to burn oil. In 1970, only 40 percent of new boiler orders provided for coal firing capability. In 1974, however, in response to the natural gas shortages and increased price of oil, 97 percent of new boiler orders provided for coal firing capability.

To avoid further deterioration in coal firing capability, FEA, on June 30, 1975, issued Construction Orders affecting 74 powerplants at 32 generating stations requiring that upon commencement of operations the units be fully capable of utilizing coal as a primary energy source.

Converting MFBI's From Oil and Gas to Coal. Unlike the utility powerplant sector, data indicating the location of MFBI's of certain size categories and their convertibility to coal firing was not readily available to FEA. Therefore, FEA began its MFBI program from ground zero. The following is indicative of the activity to date:

For the purposes of ESECA, FEA has determined an MFBI to be a unit or facility other than an electric utility powerplant where the design firing rate is equal to or in excess of 100,000,000 Btu/hr.

The far reaching data system necessary to support the selection and analysis of MFBI candidates has been established through the development and promulgation of an industrial coap capability questionnaire. The data base presently contains information pertaining to the convertibility of 6300 individual MFBI units and 3500 facilities exceeding the established firing rate threshold. . . .

As a natural spin-off of the non-utility universe definition, FEA has compiled an extensive inventory of federal facilities and is proceeding with analysis of these sites. Further, the FEA is working closely with representatives of various federal agencies to expedite increased coal consumption at existing federal facilities and to ensure that future installations are designed with an ability to utilize coal as a primary energy source. . . .

Coal is the most frequent substitute for oil and gas in industry and, by and large, served as the focus of the MFBI survey. Nevertheless, the primary objective of an ESECA-styled program is the prohibition of the use of two scarce fuels and not the dictated use of any one alternative energy source. Almost 15% of all MFBI combustors currently utilize something other than fossil fuels for their primary fuel. The use of municipal wastes, wood products, waste heat recovery systems, industrial by-products, etc. should be considered as alternatives to oil and gas where it is in the best interests of the installation. . .

The requirement to comply with federal and state environmental regulations constitutes an important consideration in the conversion of industrial sites away from coal. FEA has identified 35 combustors undergoing such a conversion and has initiated steps to categorize them as potential candidates for Prohibition Orders. There are undoubtedly many more such units, however, and FEA is currently establishing means by which to identify them.

A year later, after Jimmy Carter became President and John O'Leary had become FEA administrator, Sen. Floyd K. Haskell (D, Colo.) summarized for the Senate May 11, 1977 some of the findings of Senate Energy Production & Supply Subcommittee on coal conversion:

"Experience with the Energy Supply and Environmental Coordination Act has shown that, in many cases, significant consumer cost savings can be achieved through the conversion of existing facilities from petroleum to coal.

"On March 21, the Administrator of the Federal Energy Administration, John O'Leary, outlined the present Federal coal conversion program and reviewed the extensive delays that are occurring between the issuance of coal conversion orders and the actual use of coal. Mr. O'Leary stated:

". . . from the beginning of the current process until coal is actually burned, it could take from two to eight years. Nearly three years of operation with this program have led us to conclude that there are many problems. . . .

"In order to expedite implementation of this program the administration supported, in the words of Mr. O'Leary:

"The strengthening and expansion of current authorities and the establishment of clear guidelines and ground rules to all current and potential new users of oil and gas.

"Administrator O'Leary also stated that the conversion concept must be harmonized with clean air legislation and strip-mining bills, but noted that technological advances are needed.

"Representatives of the electric utility industry testified that substantial economic and technical problems must be overcome prior to conversion. They advocated trade-offs between industry and environmental interests, noting that proposed new, more stringent environmental restraints will seriously hinder any coal conversion program.

"According to the Federal Energy Administration there are potential savings of $3.6 billion per year in operating costs by existing and new electric powerplants. The accompanying reductions in oil consumption would be 613 million barrels per year. In addition, natural gas consumption would be reduced by 115 billion cubic feet per year. The resultant demand for coal would be increased by approximately 163 million tons per year.

"In the industrial sector, annual operating cost savings for existing and new installations potentially are $4.3 billion. The accompanying increase in coal demand would be about 199 million tons per year. The resultant oil savings would be 703 million barrels per year. The natural gas savings would be 355 billion cubic feet per year.

"Achievement of these cost savings and reduced levels of oil and natural gas consumption will depend upon the ability of the affected installations to obtain adequate coal supplies. In addition the affected facilities must install the necessary equipment to meet applicable environmental requirements.

"However, if coal-conversion is to be a national policy then the costs of such a program should be shared evenly by all Americans. . . ."

The FEA May 9 had proposed ordering 56 industrial plants to switch to using coal instead of oil or natural gas. The proposal was the first of its kind for industrial plants. Previously, the FEA had taken steps to force electric utilities to convert some of their power plants to coal.

The proposal was in the form of so-called notices of intent issued to the companies. The notices covered 24 existing plants and 32 planned or under construction in 25 states.

The conversions would result in the burning of 9.3 tons of coal annually.

Strip-mining rules adopted. The Interior Department May 11, 1976 issued rules for reclamation of federal lands strip-mined for coal.

The rules required restoration of leased federal land to about its original contour and vegetation level. The leasing procedure required that companies obtain approval of their exploration and development plans for a specific site before a lease is granted. A provision for variances was contained in the new rules.

The rules, which paved the way for vast new strip-mining operations, required mining companies to apply the "best practicable commercially available technology" to protect ground and surface water supplies.

State regulation of the environmental provisions would be permitted if the state regulations were at least as strict as the federal provisions.

The new federal rules were endorsed by both the Environmental Protection Agency and the Council on Environmental Quality, but they were criticized by Rep. Morris K. Udall (D, Ariz.). Udall said May 11 that the rules were "woefully inadequate" and had a "loophole" that would allow the federal government to permit exceptions to the contour requirement. It would also let the government "opt for mining" in cases where a conflict arose between coal production and possible water pollution, Udall said.

In announcing the rules, Interior Secretary Thomas Kleppe said that they would make possible an annual coal production from strip-mining on federal lands of 300 million tons by 1985. The production in 1975 was 32 million tons.

A moratorium on federal leasing of land for strip-mining coal had been in effect for about five years because of environmental issues and criticism of leasing procedures.

Until the moratorium, 537 federal coal leases had been granted covering about 799,000 acres and involving 16.4 billion tons of recoverable coal, mostly in the West. New leasing was not anticipated until adoption of further rules and completion of environmental studies.

Coal-leasing reforms enacted over veto. By votes of 75–18 in the Senate Aug. 3, 1976 and 316–85 in the House Aug. 4, Congress overrode a veto by President Ford and enacted legislation that set new regulations on the leasing of federal lands with coal reserves. The bill also increased from 37.5% to 50% the share of the leasing revenues going to the state in which the federal land was located.

Ford had objected that the bill's regulations would create an administrative thicket that would retard production and increase prices. (Another feature of the bill which Administration supporters said would lead to a price rise was an increase in royalties from 5¢ per ton to 12.5% of the value of the coal.)

The bill's backers—including the governors of nine western states in which most of the coal lands were located—contended, however, that the legislation would furnish a basis for the orderly development of federal coal reserves and might actually hasten production. Rep. Morris K. Udall (D, Ariz.) Aug. 4 termed the bill "vital to the West"; the Administration, Udall added, "time after time has been a wholly-owned subsidiary of the coal interests."

Other provisions were aimed at ensuring that land leased would be mined and not left unproductive, that mining would be conducted efficiently (through consolidation of tracts into "logical mining units") and that environmental values would be protected. The new legislation limited to 100,000 acres the amount of federal coal lands that any individual or corporate entity could control at one time.

The White House did not announce the veto—Ford's 51st—until after 11 p.m. July 3, less than an hour before the deadline for presidential action on the bill. In his veto message, Ford said he favored the provision increasing the states' share of lease revenues. That was the feature of the bill that had made it attractive to many western states.

Coal leases revoked. The outgoing Secretary of the Interior, Thomas S. Kleppe, Jan. 13, 1977 revoked several corporate leases and lease options on coal lands on the Crow Indian reservation in southern Montana. The strip-mining agreements had come under attack from members of the Crow tribe, who objected to the royalty payment—17.5¢ a ton on removed coal—specified in the lease held by Shell Oil Co.

Kleppe's action applied to the leases and options involving tracts larger than the regular limit of 2,560 acres.

The Indians had filed suit against the Interior Department, charging that the Bureau of Indian Affairs (an agency of the department) had not protected tribal interests in counselling acceptance of Shell's 17.5¢ offer. Although Shell subsequently upgraded its offer to 40¢ a ton, tribal government nearly had come to a standstill because of the controversy stemming from the tribal chairman's acceptance of the initial Shell offer, the New York Times reported Jan. 14.

Besides the Shell lease of 30,000 acres, AMAX Inc. had leased 14,000 acres and Peabody and Gulf had taken lease options for, respectively, 11,000 and 73,000 acres. Shell and AMAX had jointly paid over $1 million for their rights, but no coal actually had been mined. Although the controversy originated with the Shell lease, the Indians' suit sought to have all four agreements voided.

Kleppe urged the tribe and the companies to negotiate new agreements. He noted, however, that the negotiations would have to be conducted under department regulations spelling out the government's trust responsibilities to the tribe and that new environmental impact studies would be required.

Utah power project dropped. A consortium of utility companies announced April 14, 1976 the cancellation of a coal-fired electric power plant project on the Kaiparowits Plateau in southern Utah. The project, according to William R. Gould, executive vice president of Southern California Edison Co., "was beaten to death by the environmental interests."

Southern California Edison was the principal partner in the project with 40% ownership. The other partners were San Diego Gas and Electric Co. 23.4% ownership and Arizona Public Service Co. with 18%. The remaining 18.6% of the ownership was uncommitted.

The $3.5 billion plant would have been on federal land at Four-Mile Bench, 30 miles north of the Glen Canyon on the Colorado River. The site was within 200 miles of eight national parks and three national recreation areas. The plant would have burned more than 1,000 tons of coal an hour; this was expected to produce about 300 tons of air pollutants per day.

Environmental interests fighting the project included the Environmental Defense Fund and the Sierra Club. The project, which would have been the largest coal-fired plant in the country, was planned to supply electricity for Arizona and Southern California.

Gould attributed the cancellation to "a series of uncertainties" relating to costs, regulatory approvals and anticipated environmental lawsuits.

Gould said members of the consortium were retaining their interests in coal and other rights relating to the project and would determine later "how the project might be continued either in the present form or reconstructed in another form." There was a possibility the coal at the site might be used for gasification, he said.

A letter signed by 31 members of Congress had been sent to Interior Secretary Thomas S. Kleppe April 9 seeking a delay in a U.S. decision to permit the Kaiparowits project. The petitioners wanted an independent evaluation "of the need for the project" and an opportunity for Congress to establish new national air quality standards.

Western strip-mining. During Senate debate on strip-mining regulation, Sen. Orrin G. Hatch (R, Utah) May 20, 1977 inserted into the record these details on Western coal mining in opposition to a bill (S. 7) curbing strip mining:

"Surface mining in the West is in its infancy. Before the passage of the Clean Air Act there was no market for low-sulfur, low-Btu western coal, yet 57 percent of all U.S. coal reserves are in the West. In addition, there was no transportation system—railroad lines or coal slurry pipelines—that could get the coal to market. The Federal Government owns 80 percent of all the coal in the West, and yet in 1974 Federal coal production—that from Federal leases—was only 23 million tons, or 3.8 percent of our annual production of 600 million tons. Those advocating a ban on strip mining have focused on methods of preventing growth in western coal production. Drafted into S. 7 are provisions such as the 'alluvial valley floor ban.' Section 410(b)(5), page 59, is the permit approval or denial section, which prohibits surface mining on "alluvial valleys" west of the 100th meridian if such mining would adversely affect existing or potential farming or ranching operations. This again is a land use decision setting a priority for present or future farming or ranching opposed to being surface mined. This decision is made regardless of whether the land can be reclaimed.

"It is essential to note that the bulk of all strippable coal reserves in the West which are economic to mine today are located in alluvial valleys. This is why the ban proponents selected the alluvial valley technique to accomplish their purpose.

"The bill also bans surface mining, regardless of whether the land can be reclaimed, in all areas set out in section 422(e), page 114. Those are the lands within the National Park System, the National Wildlife Refuge System, the National System of Trails, the Wilderness Areas, the Wild and Scenic Rivers, the national recreation areas, and finally, the lands within national forests. It also bans surface mining on any lands included in the National Register of Historic Sites or lands within 100 feet of any public road, dwelling, public school, church, community, institutional building, public park, or even cemeteries. These are some of the major reasons President Ford vetoed this strip mining bill.

"The Democrats, President Carter included, have urged a doubling of coal production by 1985—to 1.2 billion tons.

Yet they have demanded this national strip mining bill. You cannot have increased coal production with this Federal law. This is the essential difference between the Republican approach, which would continue to allow State law to regulate surface mining. This Democratic initiative would penalize the 32 States which have surface mining laws by supplanting State codes which meet peculiar and specific climatic, geologic conditions. This bill is really prompted by past crimes committed in the 1930's, 1940's, and 1950's by the coal companies. There was little environmental ethic then, but those circumstances are now gone. Coal companies know the public will no longer stand those earlier outrages and the companies spend the money necessary to reclaim.

"Even without this bill there is a predicament which must be overcome if Federal lands are going to produce more coal. Since 1971 there have been a moratorium on leasing Federal coal. Leasing is the necessary first step to increasing production. In 1976 the Secretary of Interior finally lifted the 5-year leasing ban and adopted stringent new reclamation standards for Federal surface mining. Now the new administration refuses to go forward with any leasing.

"Approval of Federal strip mining necessitates the preparation of environmental impact statements under the National Environmental Policy Act. These impact statements have taken years to prepare and have taken years to resolve the lawsuits brought to challenge the adequacy of the statements. If S. 7 becomes law, then the National Environmental Policy Act will require hundreds and perhaps thousands of new impact statements to be written on private and State lands heretofore not required. Section 420, page 101 of S. 7, encourages citizen suits to be brought challenging the regulatory authorities' decisions to grant mining permits. This provision will further exacerbate increased coal production by dragging into the courts even those mining premits that were granted after all hurdles had been successfully overcome by meeting all the provisions of the bill.

"One could not conjure a more disastrous mechanism to frustrate coal production, unless it were to adopt a total ban on all strip mining."

Boom in Gillette, Wyo.—The energy crisis, precipitating a turn to coal, resulted in a boom for Gillette, Wyo., which was built atop the world's largest coal deposit. According to an article by Gillette Mayor Michael B. Enzi in the Los Angeles Times May 15, if the U.S. "had no coal except the coal in Campbell County, Wyo. —where Gillette is the only incorporated town—it would ... be the fourth-largest coal-producing country (instead of the first)."

Campbell County, with 12% of the world's known coal reserves, is exceeded in the size of its coal deposits only by Russia, China and West Germany.

Strip-mining bill enacted. Congress July 21, 1977 cleared legislation regulating coal strip-mining. President Carter signed the measure Aug. 3. This was the first legislation on the issue to get through Congress and escape a veto.

The final form of the bill was approved July 20 by the Senate, 85-8. The House cleared the bill the following day by a 325-68 vote.

Environmentalists for the most part were happy with the bill, although some expressed reservations about areas of the bill where the original proposals had been softened. The coal industry, on the other hand, did not view the bill so positively. Carter Manasco, a lobbyist with the National Coal Association, cited in the July 23 Congressional Quarterly, said "Every day they want more and more coal production, then they put more and more road blocks."

Manasco continued, "It's a difficult bill. We knew one was coming—we tried to make it work as well as possible. It will be hardest on small operators, and there's no question that it will cost more."

In its major provisions, the bill:

- Required strip-miners to restore stripped land to its approximate original contours. Before receiving a permit to strip-mine coal from farmland, a strip-miner would have to show the "technological capability" to restore the land to its original productivity.
- Limited mining on alluvial valley floors in the West to operations that would "not interrupt, discontinue or preclude farming."
- Permitted a mining technique known as 'mountaintop removal,' provided environmental values were not sacrificed.

- Set up a tax on coal production, the proceeds of which would be used to reclaim land previously damaged by coal mining. Most of that land was in the East, where only about half of 1.3 million acres of strip-mined land had been reclaimed. The tax was 35¢ per ton for strip-mined coal and 15¢ per ton for underground-mined coal (or 10% of the value of the coal at the mine, whichever was less). For lignite, or brown coal, the tax was 10¢ per ton or 2% of the value of the coal, whichever was less.

The bill set up an Office of Surface Mining Reclamation and Enforcement in the Interior Department to implement the bill's provisions. Money was authorized for state mining institutes to conduct research and train professionals in mining.

Under the bill, the individual states would have the primary responsibility for formulating strip-mining regulations, but the state regulations would have to meet the federal standards spelled out in the legislation.

The bill was essentially directed only at the strip-mining of coal. Earlier forms of the bill had applied to other minerals as well, but backers of the bill feared the broader coverage would increase opposition to the bill, making passage unlikely.

Coal slurry bill deferred. Two subcommitties of the House Interior Committee June 27, 1977 voted, 13-12, to put off until January 1978 any action on a bill to remove legal obstacles to coal slurry pipelines.

In the slurry process, coal was pulverized to about the consistency of sugar and then mixed with water so that it could be transported through pipes. Advocates of the pipelines maintained they would provide a cheaper way of transporting coal than carriage by railroads.

The railroads, worried about competition from the pipelines, had in many cases refused to allow the pipelines to cross under the railroad tracks. The bill shelved by the Interior Committee would have given the pipelines the right of eminent domain: the power to use privately owned land (in this case, land owned by the railroads) for a purpose deemed in the public interest.

The bill would have put certain restrictions on the pipelines' exercise of eminent domain, however. The pipelines would have been required to obtain certificates from the Interior Department. And the Interior Department would have been required to study the ecological, economic and other effects of each pipeline before granting the certificates.

Besides the opposition by railroads, some environmentalists feared coal slurrying would threaten the water reserves of western states. Slurrying required large quantities of water, and the western states were often short of water.

Backers of the move to delay action on the coal slurry bill argued the delay would give the congressional Office of Technology Assessment time to complete its study of coal slurrying. They also pointed out that the Administration had not yet formally presented its position on coal slurrying.

Supporters of the coal slurry legislation argued the Administration had backed the general concept of coal slurrying. Carter energy advisor James Schlesinger had made a statement to that effect May 5, when he was discussing the President's energy program. The Carter program emphasized reliance on coal, and Schlesinger had expressed doubt that the railroads could carry all the coal needed.

A letter formally expressing the Administration's support for the coal slurry legislation, with some modifications, was supposed to have been sent to the House in time for the June 27 meeting of the Interior Committee subcommittees. However, backers of the bill had not received the letter by the time of the meeting, and the motion to delay action on the legislation won by a 13-12 vote. The two panels that met for the vote were the mines and mining subcommittee and the public lands subcommittee.

Kennecott Copper sells Peabody. Kennecott Copper Corp. June 30, 1977 completed the sale of its Peabody Coal Co. subsidiary. Kennecott was the nation's largest copper producer, and Peabody was the largest U.S. producer of coal.

Peabody's U.S. assets were sold for $1.1 billion in cash and securities to Peabody Holding Co. The owners of the new company and their share in the firm were Newmont Mining Corp., a major metals and minerals extractor, 27.5%;

The Williams Cos., which was in the fertilizer, energy and metals field, 27.5%; Bechtel Corp., a construction and engineering firm, 15%; The Boeing Co., a large aerospace firm, 15%; Fluor Corp., another large construction and engineering company, 10%, and The Equitable Life Assurance Society of the U.S., the nation's third largest life insurance firm, 5%.

Broken-Hill Proprietary Co. of Australia paid $100 million in cash for Peabody's Australian holdings.

Kennecott's 1968 acquisition of Peabody for $475 million had been challenged by the Federal Trade Commission on antitrust grounds. The agency, which contended the purchase would reduce competition in the coal industry, ordered Kennecott in 1971 to divest itself of the subsidiary. Kennecott appealed the ruling through the courts until 1976, when the purchase agreement was arranged with the U.S. and Australian companies. The FTC June 7 had approved the sale. The divestiture was the largest ever ordered by the government.

De-Controlling Oil Prices

Two-tier oil pricing ruled lawful. The Temporary Emergency Court of Appeals July 7, 1975 voided an earlier ruling and ruled that the Federal Energy Administration's two-tier system for regulating crude oil prices was lawful.

The court, in a 4–3 decision, rejected a complaint brought by Consumers' Union that the FEA had acted illegally in failing to set ceiling prices for "new" and released oil. (Oil production from a given well for each month of 1972 formed the base level. "New" oil was the volume of domestic crude that exceeded the base level for any given month. "Old" oil was the base level minus an amount of released oil equal to new production from the well. Released oil, which was not subject to federal price controls, provided an incentive that allowed producers to double the amount of oil produced at decontrolled prices.)

Oil prices rolled back, then decontrolled. President Ford Dec. 22, 1975 signed into law an energy bill providing for an immediate roll-back in oil prices but then an end to oil price controls in 40 months.

The President also removed immediately the $2-a-barrel tariff on imported oil.

In signing the bill, despite pressure for a veto from the oil industry and conservative Republicans, Ford said "the time had come to end the long debate over national energy policy."

Deploring the legislation Dec. 22, Frank Ikard, president of the American Petroleum Institute, said it would "lead to higher energy costs in the future" and would "increase the nation's vulnerability to a potential embargo."

At a press briefing Dec. 22, Federal Energy Administrator Frank Zarb said the price rollback at the consumer level might not be as much as estimated by Congressional Democrats supporting passage of the bill. They had estimated a reduction of three cents to four cents a gallon at gasoline pumps for motorists as a result of enactment of the bill. Zarb said the maximum reduction possible there was perhaps a penny a gallon and no reduction at all also was a possibility. Despite the mandated price rollback for domestic crude oil, he said, the industry had a backlog of $1.2 billion in cost increases it was legally entitled to pass through to consumers but had not as yet because of competitive considerations. Furthermore, Zarb said, the recent 10% price hike by foreign oil-producing countries had not fully emerged in U.S. retail pricing as yet.

Congress had cleared the bill by 236–160 House vote Dec. 15 and 58–40 Senate Vote Dec. 17. The immediate price reduction made by the bill cut the average price of U.S.-produced crude oil from the current level of $8.75 a barrel to $7.66.

After that, the average price would be permitted to rise up to 10% a year to account for inflation and to spur production. If the price rose at the 10% rate annually, the current level of fuel prices would not be regained until mid-1977, well after the 1976 presidential election, observers noted.

Congress retained for itself the power of veto, by a majority vote of either house, of any increase in crude oil prices, after February 1977, that would be in excess of the national rate of inflation.

An exemption at the President's discretion was provided for Alaskan crude oil once it was in production, again subject to Congressional veto.

The bill also contained conservation provisions and others designed to cope with any new Arab oil embargo. The President would be given stand-by power to order rationing or other conservation measures in the event of an embargo, subject to Congressional approval. And a national oil reserve at the billion-barrel level was authorized.

A "maximum efficient rate" of production could be required by the President for oil and natural-gas producers from federally leased onshore and offshore fields.

Energy-conservation targets could be set for voluntary compliance by the 10 industries consuming the most energy.

Grants totaling $150 million would be provided to states drafting conservation plans, covering such things as lighting efficiency and building insulation.

The car industry would be put under statutory requirements for improved gasoline mileage—an average of 18 miles a gallon by 1978, 20 miles a gallon by 1980 and 27.5 miles a gallon by 1985.

Energy consumption labeling would be required for home appliances.

Finally, the oil industry's statistics would be opened to inspection by the General Accounting Office.

An early provision of the legislation for federal loan guarantees for reopening coal mines had been dropped from the bill.

2 vetoes preceded final bill—Earlier bills involved in the oil-price issue had been vetoed by President Ford July 31 and Sept. 9, 1975.

Ford July 14 had announced a compromise plan to decontrol oil prices gradually over a 30-month period. His plan, strongly opposed in Congress, would have raised the price of gasoline and other refined fuels by 7¢ a gallon by 1978—by 1¢ a gallon by the end of 1975, 4¢ by the end of 1976 and 7¢ by the end of 1977. The President threatened, if his plan was not enacted, to veto any extension of current controls on domestic oil prices, which were due to expire Aug. 31.

A domestic price rise to discourage consumption was one of the President's goals. He also wanted to stimulate production and lessen dependence on imported oil.

As part of his plan, Ford proposed, for the first time, a ceiling on "new" domestic oil of $13.50 a barrel, the current price of imported oil from the Mideast. New oil currently was selling at $12.75 a barrel. "Old" oil would be freed of price curbs by a rate of 3.3% of production each month, starting in August, under the Ford compromise plan. The price currently was $5.25 a barrel.

The Administration also wanted Congress to pass an excise tax of $2 a barrel on new oil and a windfall-profits tax on old oil to recoup some profits from the industry for return to the consumer. The President's plan was expected to cost the consumer, according to the Administration figures, $200 a year for energy and energy-related costs at the end of the 30-month period.

(Consumer advocate Ralph Nader put the cost of decontrol to a family at $900 a year. He made the estimate while appearing on the ABC-TV "Issues and Answers" program July 13 with John E. Swearingen, chairman of Standard Oil Co. [Indiana], who supported Ford's plan.)

With Congress moving in a direction opposite to decontrol, however, a bill requiring a rollback in the price of uncontrolled domestic oil was cleared by Congress July 17 and sent to the White House. The threat of a presidential veto, previously expressed, was reinforced by White House press secretary Ron Nessen, who said the bill was "absolutely the wrong course." The votes on it were 57-40 in the Senate July 16 and 239-172 in the House July 17.

Under it, the currently uncontrolled price of "new" oil, selling at the time for about $13 a barrel, would be rolled back to $11.28, the price prevailing as of Jan. 31. "Old" oil would continue under controls at $5.25 a barrel.

In other provisions of the legislation, price controls on oil would be extended four months beyond their scheduled Aug. 31 expiration date, and the period in which either house could veto a presidential price decontrol plan was increased to 20 days from its current five-day period.

The approach by Congress, to roll back the price of uncontrolled domestic oil, was vetoed by the President July 21. Lower prices, he said, would spur more consumption and oil imports. He said the bill "would increase petroleum imports by about 350,000 barrels per day in 1977 compared to my phased decontrol plan," and "it would even increase imports by about 70,000 barrels per day over continuation of the current system of mandatory controls through 1977."

The President conditioned his acceptance of continuation of the current system of controls, due to expire Aug. 31, upon acceptance by Congress of his plan to decontrol domestic oil prices gradually over a 30-month period. Under the 1973 Emergency Petroleum Allocation Act, either house of Congress could veto the President's proposals on petroleum allocation and price control by simple majority vote within five days after receipt of the proposals.

If Congress accepted his phased decontrol plan, Ford said in his veto message, he would accept extension of the oil-price controls. "If decontrol is not accepted," he warned, he would veto the six-month extension bill being processed by Congress. The simple extension bill was considered stop-gap legislation; without it, all price controls on oil would end Aug. 31 and the cost of petroleum products was expected to rise sharply immediately.

Alan Greenspan, chairman of the President's Council of Economic Advisers, made clear July 21 that the Administration preferred not to let the controls lapse. "The President's first preference is phased decontrol," he told reporters. But he gave assurance that the economy could sustain the fuel price hikes following instant decontrol "without severe economic consequences."

Congress's decision was made the next day. The House rejected Ford's plan for phased decontrol by a 262–167 vote July 22. The majority contended that the plan would augment inflation, unemployment and profits for the big oil companies and would not stimulate U.S. oil production. A study by the staff of the House Commerce Committee concluded that Ford's proposals would result by the end of 1977 in 800,000 more persons unemployed, a 2% rise in the consumer price index and an additional cost to the average family of $700–$900 a year.

(The consumer affairs advisory committee of the Federal Energy Administration adopted a resolution July 18 opposing Ford's decontrol plan and making the same points as the House Democrats—gradual decontrol would "eventually result in unjustifiably high fuel and other product prices to consumers, windfall profits to the major energy industries and further disruptions to the economy in the form of continued inflation, increased unemployment, etc.")

A compromise oil-price decontrol plan submitted by the President was rejected by the House July 30. The vote, which killed the proposal, was 228–189 (207 D & 21 R vs. 121 R & 68 D). The President's proposal, sent to Congress July 25, was a more gradual form of his earlier plan, which Congress also had rejected.

The compromise plan was to decontrol the price of "old" oil over 39 months, during which it would rise from the current ceiling of $5.25 a barrel to $11.50. The price of "new" production would have been allowed to rise, under the compromise proposal, 5¢ a month beginning Oct. 1 for 39 months, after an initial rollback to $11.50 a barrel from the current price of $12.50 a barrel.

The President announced on television July 25 that he would approve a 90-day extension of oil-price controls if Congress accepted his latest plan.

As before, the President urged along with his decontrol plan a windfall profits tax with the revenue being returned to the consumer in recompense for the higher prices resulting from his plan.

One feature of the new plan was a delay in any significant price increase from the plan beyond the November 1976 Congressional and presidential elections.

Earlier July 30, the House had voted 218–207 to attach a three-tier pricing amendment to an energy bill—$5.25 a barrel for "old oil," $7.50 for "new oil," $10 a barrel for a new category, termed high-cost oil.

Congress July 31 cleared a six-month extension of authority for price controls on domestic oil. The clearance was obtained by a 303–117 vote by the House. The Senate had approved the bill by a 62–29 vote July 15.

The bill was vetoed by President Ford Sept. 9, and the veto was sustained in the Senate Sept. 10. The vote was 61 to override the veto, 39 against, or six votes short of the two-thirds majority necessary to override.

In vetoing the legislation, the President contended that the cost of oil imports had risen 700% to $25 billion a year since controls were imposed in 1971. "I am talking about American dollars—your dollars," he said, "to pay for foreign oil and for foreign jobs. This $25 billion could provide more than one million jobs for Americans here at home." The Administration argument was that the price controls had increased reliance on imported oil and the only way to reverse this was to let the prices increase, which would discourage consumption and stimulate more domestic production.

Another facet of the argument was that an abrupt end to oil price control would cost consumers only six or seven cents a gallon more for fuels and the Administration would act, as pledged by Federal Energy Administrator Frank G. Zarb Sept. 9, to reduce that increase to three cents a gallon over the next nine months. President Ford had promised, in a speech at Vail, Colo. Aug. 15, to remove the $2-a-barrel tariff on imported oil if Congress upheld his veto of the bill to extend controls six months, which he said at the time he definitely intended to veto.

The Democrats, advocates of retaining controls, had different statistics. Sen. Edmund S. Muskie (D, Me.) Sept. 9 cited a Congressional Budget Office report stating that decontrol would cost 600,000 U.S. jobs by the end of 1977 and lead to a 4% increase in overall consumer prices and a 20% reduction in the anticipated growth of national production. Senate and House Democratic leaders issued a joint statement Sept. 9 warning that the nation would face a "deepening economic crisis" without controls.

Federal controls on all oil prices actually had expired Sept. 1 during the impasse between the President and Congress on extension. There had not been an immediate increase posted by the companies after the expiration of controls; Congress was expected to make any new controls retroactive to Sept. 1.

Congress and the President had been negotiating for a compromise on phasing out controls, and both sides were seeking, also without avail, interim legislation for a short-term extension of controls pending agreement on a program for the long-run. The President had agreed to sign, for example, a 45-day extension bill if there were progress on phase-out legislation. Ford also delayed his veto of the extension bill to allow more time for a compromise.

Largely to insure there would be no sudden price rise, the House approved, unanimously, Sept. 11 stopgap legislation to extend the controls, retroactive to Sept. 1, through October. The President's reaction was sought, and he said he would sign such a bill.

But the Senate Democrats, in caucus Sept. 11, approved a bill to extend price controls for 60 days and to prohibit the President from submitting a decontrol plan within the first 45 days of that period. Such a plan would become law if neither house rejected it within five days.

The Democrats wanted more time to develop a long-term plan and felt that a presidential plan interjected into their deliberations at that point would be disruptive.

Ford's reaction to the Democrats' plan, as relayed by Republican leaders: he could not accept it. The President also made a promise not to send Congress any decontrol plan before Oct. 20.

The Ford Administration and Congress Sept. 25 agreed to temporarily restore the recently expired oil price curbs.

The agreement, between President Ford and Congressional leaders of both parties, was ratified by voice vote later Sept. 25 by a caucus of Senate Democrats. Under it, the previous price controls would be made retroactive to Sept. 1 and be extended until Nov. 15.

The agreement also allowed the President to propose to Congress, after Nov. 1, a plan to raise oil prices. Congress was allowed to reject the plan by majority vote of either house within five working days.

Another aspect of the accord was an assurance that the Senate would have the opportunity to vote on any such presidential proposal within the five-day period even if a filibuster was attempted.

Legislation embodying the agreement was quickly drafted and was passed by both houses of Congress and sent to the President Sept. 26. The votes were 75-5 in the Senate and 342-16 in the House.

The purpose of the legislation was to give the President and Congress more time to try to reach agreement on a long-range energy policy. The White House made clear that the President would not agree to a further extension of controls legislation if the energy impasse persisted through Nov. 15. Press Secretary Ron

Nessen told reporters Sept. 26 "the President wants you to know that this is the last extension."

In signing the bill Sept. 29, Ford appealed for "an aroused citizenry" to press Congress to adopt a long-range energy program, one that "encourages Americans to produce our own energy with our own workers from our own resources and at our own prices." "When the price of gasoline goes up at the service station," he said, "I want the American people to know exactly where the blame lies."

In rebuttal, Senate Democratic leader Mike Mansfield (Mont.) said that "it wasn't Congress that put the $2 import tax on oil." Sen. Henry M. Jackson (D, Wash.) said the President "has done nothing but search for scapegoats, whether Congress or the Arabs."

60¢ oil products tariff lifted. Federal Energy Administrator Zarb Sept. 23, 1975 announced the lifting of the Admintration's 60¢-a-barrel import fee on refined petroleum products, retroactive to Sept. 1. The tariff had been imposed June 1 at the same time that President Ford had raised the import fee for crude oil to $2 a barrel.

Removal of the tariff on refined products would result in a 1.5¢-a-gallon drop in the price of heating oil, residual fuel, and other refined products, Zarb said. Most of the 2 million–2.5 million barrels of petroleum products imported daily to the U.S. were consumed in the New England and Middle Atlantic states and Florida.

Zarb said the White House acted to lift the tariff because of reports that importers were withholding their products from the market pending a resolution of the price-control debate between the White House and Congress.

"Failure to clarify the situation until Congress acted," Zarb said, "might result in inadequate heating oil stocks through an unusually cold winter."

The oil tariff increases imposed by President Ford in January and June were ruled illegal by a federal appeals court in August. While the issue was being appealed to the Supreme Court, the court-ordered ban on collection of the fees had been stayed.

In a speech at Vail, Colo. shortly after the court decision, Ford offered to lift the tariff voluntarily if Congress agreed to end federal price controls on domestic oil. Lifting the import fees, Ford said, would soften the impact of decontrol.

Fuel-price decontrol opponents lose. In separate votes June 30, 1976, both chambers of Congress decided to refuse to exercise their right to veto proposals by the Federal Energy Administration to lift price and allocation controls on diesel fuel, home-heating oil and similar petroleum products. Decontrol critics came closest to stopping the Ford Administration move in the House, where the vote was 194–208; the Senate thwarted an attempt to veto decontrol by 52–32.

The controls accordingly ended July 1. (Congress already had failed to veto two earlier FEA price actions. One, that came before Congress in April, lifted price and allocation controls on residual-fuel oil. The second, in May, ended an exemption from the entitlements program for small refiners. Congress derived its power to veto FEA "energy actions" by a simple majority vote in either house from language in the 1975 omnibus energy act.)

Carter rescinds gasoline decontrol order. Shortly after becoming President, Jimmy Carter Jan. 24, 1977 ordered rescinded a last-minute move by the Ford Administration to eliminate federal controls on the price of gasoline. Carter, however, did not rule out the possibility that he would propose his own decontrol plan at a later date.

Ford, in one of his last official actions, Jan. 19 had ordered an end to the controls March 1. Ford's outgoing Federal Energy Administration administrator, Frank Zarb, had said in a statement Nov. 23, 1976 that "preliminary findings indicate that gasoline decontrol would create no adverse price or supply situations." He said decontrol would increase price competition and act to limit price increases to consumers.

Oil Leasing & Production

California oil leases sold. In keeping with its accelerated program for energy independence, the Ford Administration Dec. 11, 1975 held a sale of oil leases off Southern California, the first in areas previously unexplored. The action followed by several days the announcement of test drills off the Atlantic coast and the invitation to oil firms to designate sites for exploration in the Gulf of Alaska.

The Interior Department accepted 56 high bids totaling $417.3 million, the Wall Street Journal Reported Dec. 22. Fourteen bids totaling $21 million were rejected as too low.

The sale, conducted by the Interior Department's Bureau of Land Management, involved some 1.3 million acres of ocean floor in 231 tracts located from three to 130 miles off the coast between Point Conception and San Clemente. All tracts were outside the Santa Barbara Channel, where an oil spill occurred in 1969. Interior Secretary Thomas S. Kleppe had announced Dec. 9 that four tracts earlier marked for bids were being withdrawn from the sale because of potential geological hazards related to steep undersea slopes and uncertain bottom conditions.

A total of $902 million in bids was received on 70 tracts, considerably less than the volume some analysts had predicted. A total of $438.2 million in high bids was offered as part of a procedure by which the Land Management Bureau later determined which tracts would be awarded to oil companies for exploration and which ones would remain unsold because bidding on them had not been high enough. The single highest bid, made by a group that included the Standard Oil Co. of California and two firms owned by J. Paul Getty, offered $105.2 million for a tract off San Pedro and Long Beach. The New York Times Dec. 12 quoted an unidentified oilman in the audience as saying, "They just left $70 million on the table. I'd hate to be the guy that has to fly to London and explain why they bid so much to J. Paul Getty." Under apparently ad hoc directives, the government was to receive as royalties one-third of the income after production from this tract and the ones that had drawn the two next highest bids at the sale. On all other tracts, 16⅔% of the income was to be charged as royalties. There was substantial variety in the estimates for how much oil was off the California coast and how soon production could begin. (The New York Times Dec. 14 reported that lease sales in the Gulf of Mexico in 1974 and 1975 had shown a pattern of diminishing bids which confirmed the belief that oil resources there were growing thinner.)

Interior Secretary Kleppe's Oct. 31 approval of the sale, deferred since July, had come amid persistent criticism of the oil lease program. Kleppe declared that day: "Our decision is based on the urgent need to develop available domestic energy resources and upon the legitimate environmental concerns raised by public comments and government officials."

Kleppe said the sale area was being reduced from 297 tracts (about 1.6 million acres) to 235 and that none of these would be in Santa Monica Bay or off San Miguel Island, a nesting place for sea lions. There was to be a three-quarter-mile buffer zone between state and federal tracts where the prospect of improper oil drainage was likely; there would be oil containment provisions in areas of "critical environmental concern."

A request for a preliminary injunction to halt the sale, brought by attorneys for California and several municipalities, was denied Dec. 5 in Washington by Federal District Judge Aubrey Robinson Jr. The suit charged the Interior Department with having failed to thoroughly examine the environmental impact of drilling, to insure a fair return on the sale and to keep the right to terminate leases if drilling caused environmental damage. Robinson said most of the issues had been litigated in a case won by the department in federal district court in Los Angeles Nov. 17.

Among other issues raised by opponents of the program were the lack of coordination between state and federal policies, required under the Federal Coastal Zone Management Act of 1972, and the lack of separation between exploration and production, which oil companies claimed had to be interrelated in order for the second to pay for the cost of the first.

In related developments, the New York Times reported Oct. 18 that the California Coastal Zone Conservation Commission had rejected an application by the Atlantic Richfield Co. to drill 17 new wells within the three-mile limit at Goleta, just west of Santa Barbara. By a vote of 10–0, with two absentees, the commission withheld the permit on the grounds that the new wells would mar the beauty of the coastline and that Arco should consolidate its existing 13 installations. It noted that the application covered only drilling, and not also support facilities, and said that such a "piecemeal approach to energy production is not consistent with Coastal Act requirements for orderly, balanced use of coastal resources and not consistent with sound conservation principles." The Times also said that the California legislature had passed in August a law banning until 1978, or until the state adopted a long-term coastal development plan, all construction of pipelines from federal waters outside the three-mile limit to onshore facilities.

Atlantic tests authorized—The Interior Department's Geological Survey Oct. 31, 1975 approved a request of a consortium of 20 oil firms to drill two deep test wells off the East Coast in order to determine the potential oil and gas deposits under the outer continental shelf.

Both wells would be drilled in about 16,000 feet of water, one of them in the Baltimore Canyon off New Jersey, the other in the Georges Bank area off Cape Cod, Mass. A spokesman for the Geological Survey called the tests "the continuation of an accelerated program to gather fundamental scientific data on the continental margin before decisions are made about leasing." According to the Washington Post Nov. 1, the drilling would cost $18 million and would be borne by the consortium, which included Amoco, Atlantic Richfield and Getty.

The department Nov. 7 said that nine oil firms had chosen 778 tracts in South Carolina, Georgia and Florida as likely places to find oil. A spokesman said the tracts ranged from 21 to 83 miles offshore and covered about 4.5 million acres of ocean floor.

A study of the offshore region of seven East Coast states, carried out by the New Jersey consulting firm of Woodward-Clyde, found that less than four square miles of onshore land would be required for facilities to support production of oil and natural gas, according to the Journal of Commerce Oct. 28. The study, sponsored by the American Petroleum Institute, forecast that 13,000 workers directly connected with the industry would earn as much as $177 million in annual wages in the 16th year of production, the peak year, and that 15,000 other jobs would be created indirectly.

A draft environmental statement for the Baltimore Canyon area off New Jersey and Delaware, released by the Interior Department Dec. 10, said offshore oil and gas production would be an "unavoidable threat" to the environment. Although the industry would create over 15,000 jobs, the statement emphasized that "oil spills are considered statistically probable" and "some disturbance to fishery and wildlife values will occur." Dr. Wilson Laird, exploration director for the American Petroleum Institute, said that day he hadn't read the report but that "no one knows if there is a drop of oil out there."

Alaska sites set—The Interior Department Oct. 31 set aside an area in Alaska's Western Gulf from which it would receive nominations from oil companies to lease tracts. The area was about 430 miles long and 90 miles wide, stretching from Montaque Island and the Kenai Peninsula south to about a dozen miles past Chirikof Island. There were 2,915 tracts covering some 16 million acres of ocean floor.

The department April 26, 1976 accepted high bonus bids totaling $459.8 million for leases in the northern Gulf of Alaska.

93 Atlantic bids accepted—The Interior Department announced Aug. 25, 1976 that it had accepted 93 of the 101 high bonus bids submitted by oil and natural-gas companies seeking leasing rights to drill in Atlantic Ocean tracts.

Proceeds from the first U.S. sale of oil and gas leases off the East Coast totaled $1.13 billion. That was the amount the government drew in bonuses from the bidding companies. A bonus bid was the front money offered the government solely for the right to drill.

Eight high bids, worth $7.9 million, were rejected as "insufficient," Interior Department officials said. The government accepted or rejected the highest bid for a tract submitted by competing companies after comparing the high bid with the Interior Department's own estimate of the tract's potential value.

Two of the rejected bids were for high-royalty tracts, considered to have an unusually high potential for oil and gas deposits. On these tracts, companies whose bonus bids were accepted agreed to pay the government 33.3% of the value of each barrel of oil produced. (The government usually received a 16.6% royalty in addition to the bonus payment once production had begun.)

Four of the eight rejected bids were submitted by Exxon Corp., which bid more money for tracts than any other company and which also was the only company to bid alone for the tracts. (Of the more than 60 companies that competed in the lease sale, all but Exxon bid in partnership for the tracts.)

Despite having four bids turned down by the government, Exxon emerged as the sale's biggest spender and the biggest winner in terms of tracts. The government accepted Exxon's bids totaling $342.75 million for 30 tracts.

The sale began in New York City at 6 p.m. Aug. 17, eight hours after its scheduled start. The sale was delayed by court challenges filed by New York State, local governments in the state and environmental groups.

U.S. District Judge Jack B. Weinstein had issued a preliminary injunction in New York Aug. 13, temporarily blocking the sale until critics had a hearing on their request for a permanent injunction.

Weinstein acted on a petition for a stay filed by New York, Nassau and Suffolk counties in Long Island, five Long Island municipalities, private citizens' groups and the Natural Resources Defense Council.

Weinstein ruled that the Interior Department's environmental impact report on the offshore-drilling proposal was deficient in its assessment of the impact on coastal states of decisions about the transporting of offshore oil and gas and its onshore storage and processing. The Second Circuit Court of Appeals in New York Aug. 16 overturned Weinstein's ruling. "Nothing in the case satisfies us that the sale . . . will cause . . . irreparable injury. On the other hand, the national interest, looking toward relief of this country's energy crisis, will be clearly damaged if the proposed sale is aborted," the court stated.

In a last-ditch appeal, opponents of the offshore-drilling program took their case to U.S. Supreme Court Justice Thurgood Marshall, who had jurisdiction over the Second Circuit. Marshall heard arguments from opposing lawyers at 1 p.m. Aug. 17, but refused to block the sale. When Marshall's ruling was received, bids were opened that evening.

Offered at the sale were 154 tracts of about nine square miles each lying 50-90 miles off the coast of Atlantic City, N.J. in a geological basin known as the Baltimore Canyon trough.

The Interior Department believed that total oil deposits in the Baltimore Canyon area ranged from 400 million to 1.4 billion barrels. Natural gas deposits were believed to total up to 9.4 trillion cubic feet. (The total domestic production of natural gas in 1975 was 20 trillion cubic feet.)

Bonus bids worth $3.5 billion were received for 101 of the 154 tracts. The $1.13 billion in high bids accepted by the Interior Department was about double what the government had expected to receive from the lease sale.

The sale of Atlantic Coast leases was twice as profitable as earlier lease sales for tracts off Alaska and California.

Oil-lease renewal tied to production. Interior Secretary Cecil D. Andrus was ordering oil and natural gas companies to agree to strict production schedules or face the unprecedented penalty of losing their federal leases on offshore drilling sites, according to the Washington Post March 15, 1977. (Andrus had become Interior Department head after the Carter Administration assumed office in January.)

Andrus already had reached an agreement with Aminoil USA Inc. under which the company, a subsidiary of R.J. Reynolds Industries Inc., would begin pumping natural gas by the fall of 1978 or forfeit its lease. The secretary had said he personally would review renewal applications for 60 other leases in the Gulf of Mexico.

Previously, drilling leases had been extended virtually automatically by regional supervisors of the U.S. Geological Survey, a branch of the Interior Department.

East Coast drilling ban lifted. The 2nd Circuit U.S. Court of Appeals Aug. 25, 1977 overturned a lower court ruling that effectively had blocked oil companies from drilling in East Coast offshore tracts leased from the federal government.

Federal Judge Jack B. Weinstein earlier had declared the sale of the leases to 93 mid-Atlantic tracts invalid because the Interior Department had neglected environmental considerations.

In approving the lease sales, the appeals court ruled unanimously that it was satisfied the department would be able to control any environmental hazards posed by the offshore drilling and would "deal with them thoroughly."

Oil companies that had purchased drilling rights to the area off New Jersey, Delaware and New York said they expected to begin exploratory drilling before the end of 1977.

Interior Secretary Cecil D. Andrus welcomed the appeals court's ruling, saying the department would prepare additional environmental-impact statements before approving development plans to transport any oil found in the offshore tracts.

The government's sale of the 93 leases had yielded $1.13 billion in high bids. The government also would collect royalties from the oil companies once production was underway. Thirteen of the 93 tracts would pay a royalty rate of 33.3% on all oil discovered there, while the other sites would pay the usual royalty of 16.66%.

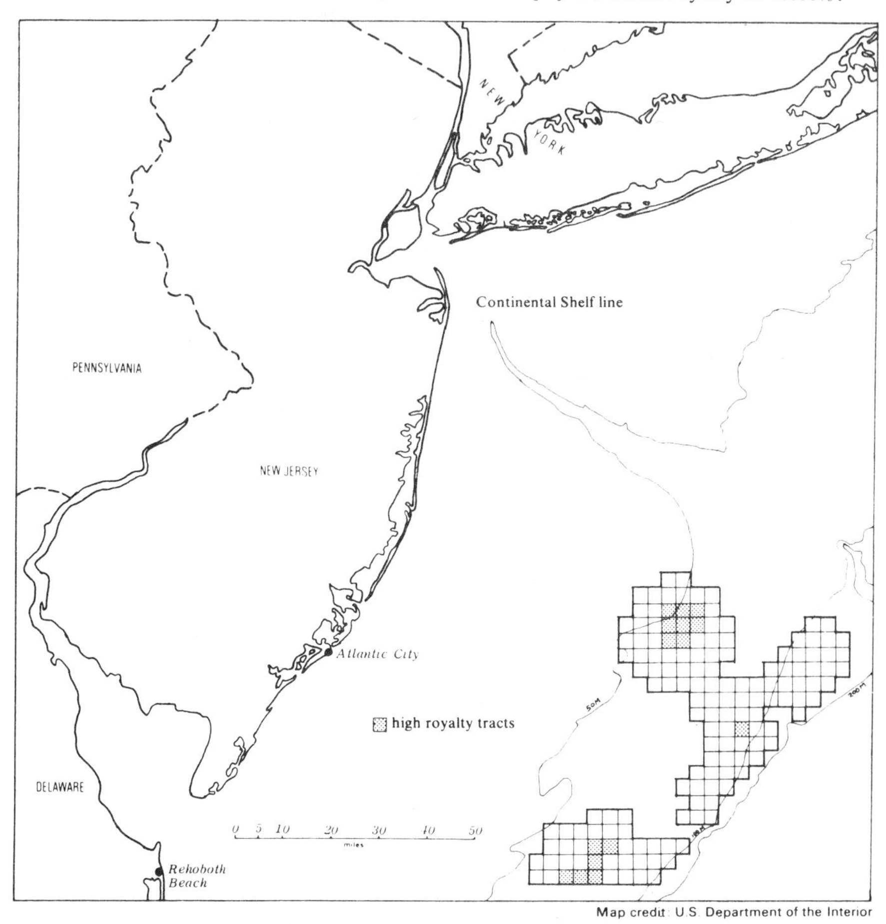

Map credit: U.S. Department of the Interior

Squares in lower right mark sites in the Mid-Atlantic Outer Continental Shelf that the Interior Department had leased in 1976 to oil companies for exploration.

Amoco's Wyoming reserves opened to independents. The Federal Trade Commission had announced Oct. 7, 1975 that it had signed a consent decree with the Amoco Production unit of Standard Oil Co. (Indiana) by which Amoco agreed to make available its Wyoming oil reserves to independent refiners over a period of twenty years.

Also signing the agreement were Pasco, Inc. and Studebaker-Worthington, Inc., which owned more than 55% of Pasco. The decree required Pasco to obtain FTC approval before selling any of its remaining assets—including refining, marketing and pipeline facilities—to any buyer with refining capacity in the Rocky Mountain area or to any buyer with refining capacity in excess of 30,000 barrels a day anywhere in the U.S.

Record U.S. oil search budgets set. The oil and natural gas industry had budgeted a record total of $28.4 billion for capital and exploration spending in the U.S. in 1977, according to the Oil and Gas Journal (reported March 8, 1977). This was a 6.7% increase over the 1976 outlay of $26.6 billion. Some 69% of current spending was earmarked for drilling and exploration, production operations and outer continental shelf lease bonuses.

According to a related report March 10, 41,209 exploratory wells were drilled in the U.S. in 1976, compared with 39,097 the year before.

Oil well drillings increase. The oil industry had drilled 13% more wells in the second quarter of 1977 than it had the same period of the previous year, according to American Petroleum Institute figures reported July 28.

The API said 9,514 oil and gas wells were completed in the second quarter of 1976. This figure rose to 10,633 in the same quarter in 1977. Of those wells drilled in 1977, 2,376 were exploratory wells and 8,257 were development wells. While 74% of the wells drilled in the 1976 quarter were found to be dry holes, that figure decreased to 73% in the 1977 quarter.

API president Frank Ikard cited a steady increase in the number of completed wells since the post World War II low year of 1971. Ikard was not optimistic that the increase would continue. He claimed political uncertainties and the U.S. economic climate made it difficult for producers to attract capital essential to continue the upward trend in U.S. drilling.

Oil imports up. As U.S. production of petroleum declined, imports of petroleum and petroleum products rose to record levels during the first half of 1977. The increase in oil imports was at least partly spurred by a cold winter and a consequent rise in the demand for heating fuels.

The reserves of petroleum distillates, which included home heating fuels, had dropped to 132.8 million barrels nationwide during the week that ended Feb. 11, the American Petroleum Institute (API) reported Feb. 17. Stocks were down 8.3 million barrels from the week before and 26.7 million barrels from the level of the comparable week in 1976.

Imports of crude oil and oil products had risen to 8.9 million barrels a day, from 8.7 million barrels a day the week before, the institute said. Distillate imports also surged during the week, reaching 626,000 barrels a day. The average of the previous week had been 398,000 barrels a day; that of a year earlier, 207,000 barrels a day. (U.S. refinery output of distillates had reached an all-time high during the week, averaging 3.7 million barrels a day.)

The FEA said the sharp increase in imports of distillates was due mainly to increased demand in the New England states, which depended heavily upon imported heating oil. In order to encourage imports into New England, the FEA Feb. 10 had adopted a previously proposed rule under which refiners with access to U.S. oil would in effect subsidize importers of the more expensive foreign oil. Domestic refiners would pay the importers $2.10 per barrel of No. 1 and No. 2 heating oil imported.

The decline in nationwide distillate stocks was halted during the week that ended Feb. 18 when reserves rose for the first time since the week that ended Nov. 27, 1976, according to an API report Feb. 24. The reserves rose slightly to 133.53 million barrels, still substantially below the comparable figure for the previous year, 157.57 million barrels.

Oil imports hit a new record during the week that ended Feb. 18, increasing to 9.92 million barrels. The comparable figure for 1976 was 7.61 million barrels. Observers noted that the U.S. appeared to be relying more than ever on imported oil, since estimated total production from U.S. wells was only 8.02 million barrels per day and consumption was running at more than 17 million barrels a day.

Imports of petroleum and petroleum products continued to climb to new record levels during the week ended March 4, according to API figures released March 9. Total imports rose from the 10.1 million-barrel-a-day record of the previous week to a 10.3 million-barrel daily mark.

While imports of crude oil dipped in the March 4 week to 6.6 million barrels a day from 6.95 million barrels, product imports rose to a record 3.69 million barrels a day from 3.13 million barrels a day during the week that ended Feb. 25.

A record 1.17 million barrels a day of middle distillates, including home heating oil, was imported during the week that ended March 4. The sharp increase from the previous week's figure of 790,000 barrels a day was attributed to orders placed in early February when the weather was extremely cold.

Oil imports rose 22% during March to a seasonally adjusted $4.06 billion. Most of the increase "consisted of greater fuel oil entries at sharply higher prices, reflecting heavy U.S. energy consumption during the exceptionally cold winter," the Commerce Department reported April 27. The cost of oil imports then declined to $3.69 billion in April and $3.1 billion in May, rose to $3.9 billion in June and declined to $3.78 billion in July, according to Commerce Department reports.

Oil & Gas From Alaska

Alaska pipeline problems assessed. Congressional investigators looking into construction problems of the 789-mile Alaska pipeline had "serious doubts" that it would be operational by mid-1977, when oil was scheduled to begin flowing, according to a staff report prepared for the House subcommittee on energy and power. The report's conclusions were printed in the Sept. 5, 1976 editions of the Washington Post and New York Times.

William Darch, president of Alyeska Pipeline Service Co., a consortium of eight oil companies that was building the pipeline, said the target would be met.

Subcommittee investigators in July conducted a two-week on-site probe of the construction. A team of Transportation Department investigators, headed by Deputy Secretary John W. Barnum, spent three days in July doing the same thing.

The separate government investigations were prompted by an Interior Department disclosure May 21 that it was reviewing 3,955 "problem welds" that joined the pipeline's sections. Officials said Alyeska had found 28 welds with cracks that needed replacement to prevent a possible rupture in the pipeline. Repair of the 28 defective welds would cost $5–$10 million and would not delay completion of construction, company spokesmen said. If 1,700 welds that did not meet Transportation Department specifications were corrected, the cost could run to $55 million and cause a delay in completing the project, according to Alyeska.

Government officials also said that Alyeska had confirmed that five radiographs—X-rays used to maintain quality control over the welding process—had been "falsified" and that another 70 were questionable. X-rays for 360 welds were missing, government officials said.

In disclosing the problem welds May 21, the Interior Department said that it would ask the Transportation Department to review each one. Officials also said that Arthur Andersen & Co., a leading accounting firm, had been hired to validate Alyeska's own audit that had uncovered the 3,955 problem welds.

Results of the Andersen audit were announced July 6 by Ron Nessen, White House press secretary. Nessen said that "there could be" more flaws in the pipeline than Alyeska's internal investigation had uncovered in its examination of the 3,955 welds. Andersen auditors had found it difficult to verify through Alyeska records the soundness of the remaining 27,000 welds.

For this reason, Nessen said, President Ford had ordered Deputy Transportation Secretary Barnum and others to make an on-site inspection of the pipeline and determine if re-testing of all 31,000 completed welds was necessary.

At the same time, Rep. John Dingell (D, Mich.), chairman of the House Commerce Committee's subcommittee on energy and power, said that his staff also would investigate these problems.

While the government probes were under way, Alyeska announced July 15 that it had made repairs, most of them minor, to 2,300 of the 3,955 problem welds. The construction company also said that the pipeline project was two-thirds completed, with 543.7 miles of mainline pipe installed.

Alyeska said that a seven-foot, underground section of the pipeline had burst during testing. Alyeska's explanation for the rupture—"excessive pressure on the pipe"—contradicted the assessment of Charles Champion, Alaska's pipeline coordinator. Champion said that the failure occured at "very low pressure" (187 pounds per square inch). The pipe was designed to operate at 1,180 pounds per square inch.

Champion said that the pipe, rather than the weld, was the cause of the failure. Alyeska spokesmen blamed the break on "human error" in applying too much pressure.

On his return from the fact-finding trip to Alaska, Barnum said that Alyeska would have to make additional checks on the 31,000 welds. He said this would delay completion of construction for at least two months beyond the September target date, but he insisted that the pipeline would meet its start-up date of mid-1977.

On the basis of the Transportation Department review, the Interior Department July 21 ordered Alyeska to improve its record-keeping system for verifying the soundness of the welds.

Interior Secretary Thomas S. Kleppe said Aug. 12 that the department was "dissatisfied" with Alyeska's efforts to upgrade its quality-control system and had warned the company that "not one drop of oil will flow through the Alaska pipeline until it has been thoroughly tested and we are assured of [its] integrity. . . ."

Kleppe said that about 2,600 of the 3,955 problem welds had been repaired or brought into compliance with federal standards, but that 45–109 radiographs remained missing. "It still is not possible to relate a specific X-ray picture to every welded joint put in place in 1975," Kleppe said.

Of the more than 1,000 welds that remained in need of correction, he said, 600–900 were in critical areas, such as underneath either rivers or the permafrost. If these welds were to rupture after the pipeline started operation, it was feared that the oil, flowing at 140 degrees Farenheit, would cause thawing, heaving and settling of the Arctic tundra and lead to widespread pollution of the environment.

When released in September, the conclusions of the two investigating teams examining the pipeline's construction problems differed widely. The subcommittee investigators said that despite months of publicity about deficiencies in construction, the pipeline still showed sloppy workmanship, inadequate quality control and insufficient on-site inspection by government officials.

The investigators charged that Interior and Transportation Departments were failing to find and correct the causes of a "collapse" in the quality-control system, and that welding problems involved not only pipe installed in 1975 but also pipe still being laid.

(In subcommittee hearings June 21, a pipeline technician had testified that he was ordered by superiors to falsify weld X-rays. He also said that during six months of work on a 70-mile segment of the pipeline, he had never seen a government inspector.)

The subcommittee investigators said that they were "regaled with horror stories detailing . . . threats of physical harm and abuse" against quality-control inspectors by construction crews.

The investigators also said they were told that once the pipeline became operational, "it would be possible for leaks up to 500 barrels a day to occur without detection indefinitely."

The Transportation Department's report on its three-day evaluation trip was released Sept. 10. In releasing the study, Barnum said he believed that welding flaws were confined to 1975 work. Barnum also said he was "satisfied" that welding and pipeline quality-control "is being done adequately this year."

Barnum added that if welding problems were not corrected until spring 1977, it "might not impose substantial delay on the completion of the entire system," including terminal facilities in the south. He said that added the repair work would be "enormously expensive."

(The last section of the 800-mile Alaskan pipeline laid Dec. 5. Welding was not completed, however.)

Background. The oil field had been discovered in 1968 by Atlantic Richfield Co. (Arco) and Exxon Corp. It was believed to contain about 9.7 billion barrels of recoverable oil (out of an estimated total of 20 or 30 billion barrels) and some 25.4 trillion cubic feet of natural gas. (The gas was to be injected back into the ground for storage until a transport system for it could be devised.) The primary owner of the North Slope oil was The Standard Oil Co. (Ohio). Exxon and Arco were the other major owners; 13 other companies held minor shares.

Plans to build a pipeline to carry the oil from the North Slope had been blocked by environmentalists' lawsuits until November 1973, when Congress, at the urging of the Nixon Administration, had passed the Trans-Alaska Pipeline Authorization Act.

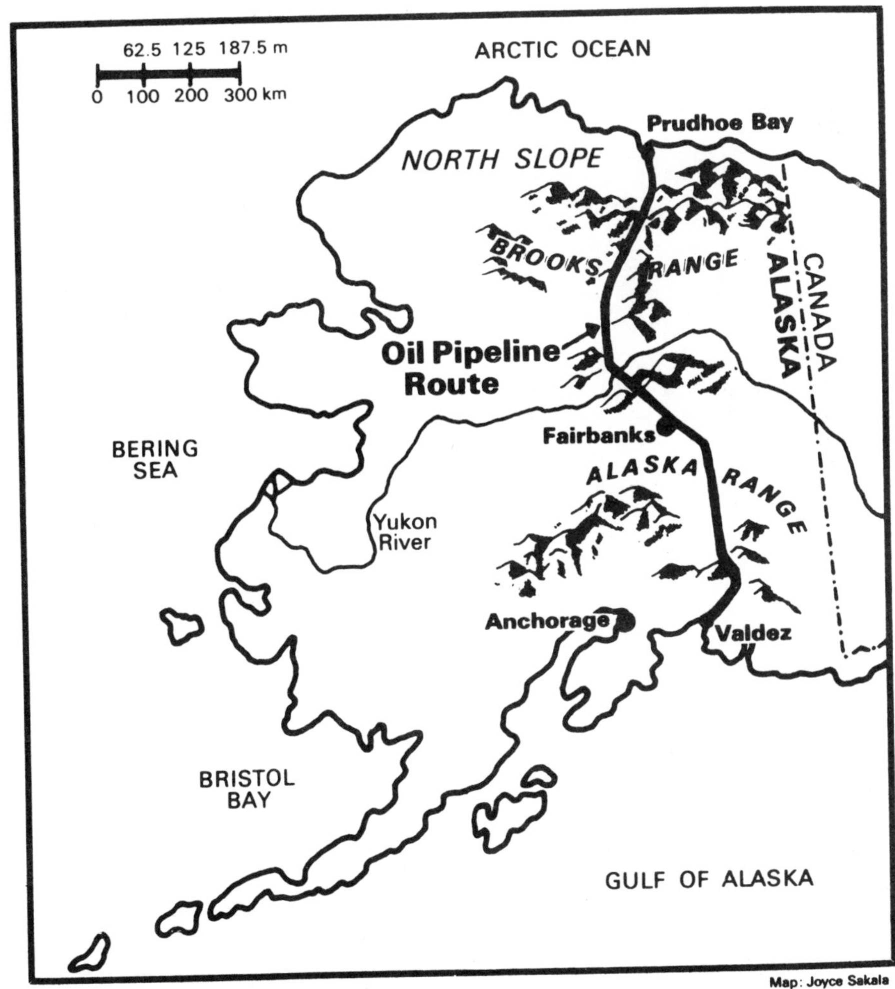

Map: Joyce Sakala

Crude oil began flowing through the 789-mile-long Trans-Alaska pipeline June 20.

Alyeska Pipeline Service Co., a consortium of eight oil companies, was responsible for building and operating the line. Alyeska's major owners were Standard Oil (Ohio), Arco, Exxon and British Petroleum Co. Ltd. The other owners were Mobil Oil Corp., Union Oil Co. of California, Phillips Petroleum Co. and Amerada Hess Corp.

Alyeska estimated the cost of the 48-inch diameter pipeline at $7.7 billion (not including some $3.6 billion for equipment, construction of work camps and drilling of the oil wells). In 1970, the cost of a pipeline with a smaller diameter had been estimated at $900 million. Alyeska's first estimate, made in 1974, had been $6 billion. Company officials said the largest components in the increase in cost were $3.2 billion caused by inflation during the four-year construction delay and $1.1 billion for an above-ground support system that the company had not expected to have to build. (About half the pipeline ultimately was built above ground in regions where the oil temperature—about 130 degrees Fahrenheit—might have thawed the permanently frozen ground and damaged the line and the environment. Original plans had called for burying the entire line.)

The entire issue of the pipeline's cost and the reasons for it was at the heart of a major rate-setting battle among the oil companies, the State of Alaska and the federal government.

Illegal environmental damage done—Alyeska was able to complete the Trans-Alaska pipeline almost precisely on schedule in part because it cut corners on environmental safeguards, according to the Wall Street Journal June 20, 1977.

The Journal's sources said Alyeska had "frequently violated state and federal environmental rules . . . [it] agreed to follow when it signed the pipeline right-of-way lease in 1974." The result, according to the newspaper, was significantly more damage to the fragile arctic environment than had been expected. Among the problems caused during construction were water pollution due to improperly run sewage-treatment facilities; massive oil spills at construction sites (one 65,000-gallon spill initially had been reported by Alyeska to be 100 gallons, the Journal said); uncontrolled erosion of the frozen tundra; blocked fish streams, and damage to fish spawning beds.

In May, two state-federal inspectors on the pipeline had written: "There is no environmental quality control in the area whatsoever. In our opinions, it is beyond belief that the quality-control program is so lax at this late stage of the game."

According to the Journal, Alyeska admitted violating some of the strict environmental regulations. "But overall," one official was quoted as saying, "I think we've done as good a job of protecting the environment as we could." Another company official said: "There are priorities, and we were involved in building a pipeline."

Government inspectors on the pipeline agreed that one of the biggest continuing problems was Alyeska's failure to control erosion, despite several warnings to do so, the Journal said. Erosion had led to stream and fish damage and large washouts of tundra, all violations of the right-of-way lease.

The Journal also reported that engineering experts were worried about the possibility of future environmental damage because of the "small probability of proper operation" of the pipeline's oil-leak detection system.

Alyeska had paid nearly $100,000 to settle some of the civil and criminal cases brought against it by the State of Alaska for violating environmental regulations. Other cases were pending.

Oil transport cost dispute—Seven of the Alyeska partners had asked the Interstate Commerce Commission (ICC) to set rates ranging from $6.04 to $6.44 a barrel for the transportation of oil through the Trans-Alaska pipeline, it was reported June 1, 1977. Federal officials had expected the tariff to be set at about $5.50 a barrel.

The ICC had jurisdiction over pipeline rates. The rates were supposed to allow the builders to recover the cost of the pipeline over a number of years and to make some profit.

The Justice Department filed a protest with the ICC June 15, contending that rates in the range of $4.40 to $4.60 a barrel would be fair. The department said the oil companies could net as much as $900 million in overcharges if the companies' proposed rates were approved.

The companies involved disputed the Justice Department's position June 16. In one of the strongest statements, Mobil Oil

Corp. called the department's brief "ludicrous" and asked who would risk capital to build essential facilities in the future if the government did not allow a return on investments.

In an unusual action, the ICC's Bureau of Investigation and Enforcement June 16 urged the commission to suspend the companies' proposed rates before they took effect and to allow a rate no higher than five dollars a barrel to be charged, pending an investigation. The bureau accused the oil companies of overstating costs, depreciation and the effect of taxes. It also said the Alyeska partners were asking for an "irrational" return on their investments.

The State of Alaska also had filed a brief with the ICC, it was reported June 17. Alaska said the commission should set interim rates ranging from $3.59 to $4.42 a barrel pending a "continuing investigation." The amount of the rate was of great importance to Alaska because the state was due a 12.5% royalty on North Slope oil production. The royalty was to be based on the price of the oil at the wellhead (before transport). Since the market price ceiling for domestic oil was fixed by the Federal Energy Administration, the higher the transportation tariff, the lower the price that could be charged at the wellhead and the less in royalties the state could collect. Alaskan officials had estimated that once oil was flowing at the rate of 1.5 million barrels a day, a 1¢ increase in the transportation tariff would mean the loss of $1 million in annual revenue to the state.

In a decision that was expected to be fought further, a three-judge federal appeals court in New Orleans upheld an ICC order to the oil companies to submit new tariffs that would not exceed $4.56 to $5.10 a barrel (reported July 29).

The delivery cost on the West Coast had already been fixed at $13.50 per 42-gallon barrel. The decision would not affect the consumer, but it would determine how much of the total cost would go to the oil firm and how much would go to the State of Alaska.

Oil starts through pipeline. Oil from the Prudhoe Bay field on Alaska's North Slope began flowing through the pipeline June 20, 1977. It reached the ice-free port of Valdez, the southern terminus of the pipeline, July 29 after a 38-day run during which the line was shut down six times.

Workers at the Valdez terminal next waited for 10 storage tanks at the terminal to fill with two million barrels of oil and then began pumping the crude into the *S. S. Arco Juneau*. When loaded, the *Arco Juneau* sailed for the Atlantic Richfield refinery at Cherry Point, Wash., where it arrived Aug. 5 with a load of 824,000 barrels.

Pipeline damaged by blast—Alyeska Pipeline Service Co. reported a series of deliberately set explosions had damaged insulation on part of the Alaska pipeline some 15 miles north of Fairbanks July 20. The flow of oil through the pipeline system continued unimpeded since the blast failed to penetrate the half-inch thick pipe.

The explosion damaged about 60 feet of four-inch-thick insulation and two supporting brackets on an above-ground section of the pipeline. An Alyeska spokesman said greater security measures would be taken in the future.

Three men were arrested July 27 in connection with the blasts. William J. Freeman 21, a cook and baker, and Donald E. Drum 19, a painter, were arrested and charged with malicious destruction of property for having placed dynamite near the pipeline. The third man, Larry D. Wertz 26, a fur trapper, was arrested on July 26 and charged similarly.

$1.5 billion waste charged. A seven-month $1-million investigation by the Alaska Pipeline Commission found that $1.5 billion had been wasted in the construction of the Alaska pipeline, it was reported Aug. 7, 1977. Poor management, unsound business practices and maneuvers to win congressional approval by the Alyeska Pipeline Service Co. had been the cause, according to the state commission.

Terry F. Lenzer, who prepared the report, said that of the $1.5 billion in question, $1.2 billion had been wasted in construction of the line and the supporting haul road. Another $300 million had been misspent in building pump stations and the Valdez terminal, according to Lenzer.

The report claimed losses had been incurred because the Alyeska Co. decided "not to hire an experienced construction management firm until a few months prior to construction in 1974." Pipeline owners did not want to appear committed to the project, the report said, and they hoped to obtain congressional approval of the pipeline by being able to threaten abandonment of the project if certain provisions proposed by environmental groups were included.

The report cited numerous examples of "goldbricking" by workers being paid for 140 to 160 hours work in a week. Alyeska's own documents showed it was management's poor supervision and utilization of the work force that were responsible for worker idleness, the report said.

Poor project supervision and owner management, unsatisfactory internal controls, and Alyeska's own ineptitude following its decision to manage the project on its own were found by the investigators to be responsible for financial waste. Inefficient use of funds made the pipeline the most expensive privately financed project in U.S. history, the report said.

Alyeska Pipeline Service Co. officials said the charges of the 690-page report "will prove to be unfounded." They chose to reserve comment on specific allegations in the report.

FEA opposes exporting surplus Alaskan oil. Frank G. Zarb, administrator of the Federal Energy Administration, said Nov. 30, 1976 that the FEA opposed proposals to export surplus oil from the Alaskan oil fields because the action would leave the U.S. vulnerable to a cutoff of vital Middle Eastern supplies.

As an alternative, Zarb said the FEA favored the shipping of oil from the West Coast through the Panama Canal to refineries on the Gulf Coast. It had been predicted that California refineries could handle no more than 50% of the expected 1.2 million barrels a day supply of Alaskan oil.

FPC splits on Alaska gas route. The Federal Power Commission May 2, 1977 recommended to President Carter that a trans-Canadian overland pipeline be built to carry natural gas from the Alaskan North Slope to the lower 48 states. The FPC commissioners split, 2–2, however, over which of two specific proposed pipeline routes they should recommend.

The two competing pipeline proposals had been submitted by Alaskan Arctic Gas Pipeline Co. (known as the Arctic Gas Study Group) and by Alcan Pipeline Co. The FPC said a third proposal, by El Paso Alaska Co., to transport natural gas in liquefied form by tanker from Alaska to California, should be adopted only if the Canadian government rejected both pipeline plans.

The commission found that all three proposals were technologically and economically feasible. The cost of each had been estimated at between $6.5 billion and $6.7 billion in 1975 dollars. (Inflation was expected to drive up the actual cost of any of them to $10 billion or more.)

The FPC said the two overland pipelines would deliver gas to the contiguous 48 states at about the same price: approximately 80¢ per thousand cubic feet over the next 20 years. Under the El Paso plan, delivered gas would cost about $1.10 per thousand cubic feet, the FPC said.

The Prudhoe Bay natural gas fields on the North Slope contained the largest single reserve yet discovered on the North American continent. Virtually every estimate exceeded 20 trillion cubic feet of proven producible reserves, enough to provide about 5% of U.S. natural gas consumption for the next 25 years. Production was expected to begin as early as 1982.

The commissioners who favored the Arctic Gas proposal said they would switch their support to the Alcan plan if Canada decided not to produce gas from the Mackenzie Delta fields, a rich gas reserve in Canada's Northwest Territories. The Arctic Gas pipeline would cut through the Mackenzie Delta to pick up about 1 billion cubic feet of Canadian gas a day, if permitted to do so. The line would stretch 4,000 miles from Prudhoe Bay across the northern Yukon and down through the Northwest Territories to a spot near Calgary, Alberta, where it would branch southeast and southwest to the U.S.

The Arctic Gas group comprised Panhandle Eastern Pipe Line Co., Texas Eastern Transmission Corp., Pacific Gas

& Electric Co., Columbia Gas Transmission Corp., Michigan Wisconsin Pipe Line Co., Natural Gas Pipeline Co. of America, Northern Natural Gas Co., Pacific Lighting Gas Development Co. and Canadian Arctic Gas Pipeline Ltd.

The Alcan project called for a 4,800-mile pipeline that would follow the Alyeska oil pipeline to Fairbanks and from there follow the Alaska Highway through the southern Yukon and British Columbia to James River Junction in Alberta. There it too would branch east and west to the U.S.

The Alcan Pipeline Co. was a group made up of Northwest Pipeline Corp. and the Canadian companies Foothill Pipe Lines Ltd., Westcoast Transmission Co. and Alberta Gas Trunk Line Ltd.

Chairman Richard L. Dunham and Vice Chairman James G. Watt voted to support the Alcan plan, while Don S. Smith and John H. Holloman 3rd favored the Arctic Gas route. A fifth seat on the FPC had been vacant for more than a year.

Among the points on which all four commissioners agreed was the conclusion that of the three plans, the Arctic Gas plan would involve the greatest damage to the environment because it would cross the Arctic National Wildlife Range in Alaska. The Alcan pipeline would bypass the range.

The FPC recommendation came after three years of hearings and consideration. It differed markedly from the recommendation of FPC Administrative Law Judge Nahum Litt, who in February had found the Arctic Gas proposal most favorable, El Paso second best and the Alcan route "neither efficient nor economic." Since Litt issued his report, the Alcan group had made significant adjustments in its design.

Oil Companies Under Scrutiny

Exxon paid $46–$49 million to Italian politicians. Exxon Corp., the world's largest oil firm, admitted July 12, 1975 that its political payments in Italy in 1963–72 totaled $46–49 million. Authorized payments made by Exxon's Italian subsidiary, Esso Italiana S.p.A., totaled $27 million. Another $19–$22 million in unauthorized payments, including $86,000 intended for the Italian Communist party, also were made through the Italian subsidiary, a company spokesman said.

Although all payments were ordered stopped in late 1971, the spokesman said, unauthorized donations continued to be made in 1972 and a company investigation of these contributions resulted in the resignation of Vincenzo Cazzaniga, then managing director of Esso Italiana.

Exxon's comptroller, Archie L. Monroe, testified on the Italian payments July 10 in a closed hearing the Senate Subcommittee on Multinational Corporations, which was investigating bribes and other payments made abroad by U.S. firms. Exxon's disclosure of the extent of its Italian payments was made in anticipation of public hearings before the subcommittee July 16.

In issuing the statement, a company spokesman conceded that Exxon had sought to keep Monroe's subcommittee testimony secret. "We asked for the confidentiality out of concern for the Italian political parties and citizens under investigation in Italy," the spokesman said.

According to the Exxon spokesman, annual authorized payments made through the subsidiary escalated from about $760,000 in 1963 to $5 million in 1968, when, he said, "our regional management [in London] got concerned and ordered the payments reduced."

In 1971, Exxon authorized the payment of $3 million in political contributions. Late that year, the spokesman said, "We discovered that the managing director [Cazzaniga] had made substantial unauthorized commercial arrangements, payments, and commitments over a period of years. Some of those, he indicated to us, were political contributions."

Exxon's spokesman said the authorized payments were made in checks and cash. Esso Italiana had indicated cash contributions were customary, he said, and added, "Some of the parties wanted it that way." "Most of the authorized contributions were entered on the books as payments to newspapers, publicity agencies, and the like," with the balance "entered under the broad category of administrative expenses. The unauthorized payments were concealed in various ways," he said.

When asked the purpose of Exxon's authorized payments, the spokesman quoted Monroe's response before the Senate subcommittee. The money was donated to "further the democratic process," he said.

Monroe amended that statement in subcommittee testimony July 16, conceding that the money was authorized to promote Exxon's "business objectives" in Italy, and that the money was earmarked on records toward efforts to reduce or defer taxes, obtain refinery licenses, win permission to import natural gas, and secure favorable locations for service stations. Monroe refused, however, to characterize the payments as bribes.

Questions at the subcommittee hearing July 16 centered on two internal Exxon audits that indicated the payments were intentionally camouflaged by Exxon officials. According to the first audit, dated August 1972, the authorized Italian payments were made out of a "special budget" Exxon "created to control payments, which by their very nature, couldn't be accurately recorded on the books of the Italian company" because Esso Italiana did not want them disclosed. "In addition to the $29 million disbursed through the special budget," the audit stated, "at least $30 million was expended in a similar manner over the last 10 years." Monroe stated that Cazzaniga, beside making unauthorized political contributions, also used the money to make "secret purchases of real estate" and "secret guarantees to banks in favor of other companies."

According to another report in October 1972 to the audit committee of Exxon's board, "the principal factor which permitted the irregularities to occur and remain undiscovered for such a long period of time was the fact that higher levels of management [regional headquarters in London and corporate headquarters in New York] condoned the falsification of records to obtain funds for confidential special payments. Bogus documentation and false accounting were accepted by many levels of management," the report stated.

(In a statement issued July 15, Exxon Chairman J. K. Jamieson admitted the firm had covered up political payments by its Italian subsidiary and conceded "this was a mistake." Management was "persuaded that it was necessary to make [political] contributions without disclosing the recipients, as it appears was the custom," Jamieson said. "This meant handling the payments so that they couldn't be identified as political contributions on the books of Esso Italiana.")

Although Monroe claimed Exxon had moved quickly to end the abuses in 1972, the August 1972 audit indicated that Exxon had continued to use phony accounting methods, including "dummy invoices," long after the problem had been identified several years earlier.

Monroe said Cazzaniga had assured him all the payments were legal. According to the Washington Post July 12, corporate contributions were legal in Italy. A 1974 disclosure law required the consent of the company's shareholders and the entry of payments on the company's books.

According to Exxon documents, the Christian Democratic party, which had dominated Italy's coalition governments in the post-war period, was allocated $11,-948,046 from 1963–1972; the Social Democrats $5,160,952; the Socialists $1,-245,028; the Liberals $591,531; the Republicans $267,521; the Italian Socialist Movement $236,106; and the Socialist Party of Proletarian Unity $71,111. Another $1,096,344 was disbursed to "others unknown."

The August 1972 audit identified the money earmarked in the special budget for obtaining favors from the government, including $2,350,000 for the "deferred payment" of taxes from 1969–1971; $500,-000 for "tax legislation" in 1971; and $1,-150,000 to secure refinery licenses from 1969–1971.

According to the Wall Street Journal July 14, Esso Italiana was a major factor in Italy's oil industry, with revenues in 1974 of $2 billion, and whole or part ownership of four refineries there. Its gasoline was marketed through more than 3,500 service stations.

Mobil's Italian payments totaled $2.1 million. A Mobil Oil Corp. official told the Senate Subcommittee on Multinational Corporations July 17, 1975 that $2.1 million was donated to Italy's three major center-left political parties in 1970–73 by the U.S. company's affiliate Mobil Oil Italiana.

Everett S. Checket, executive vice president of Mobil's international division, conceded the disbursements were disguised on company books, but insisted the payments were made to "support the democratic process."

"All such contributions were made from Mobil's normal business accounts, by company check or bank remittance," Checket said. "These contributions were not paid from any secret fund and did not involve out-of-country financial transactions [laundering of funds]," he said. However, under questioning, Checket admitted that the payments were recorded as advertising, research or other expenses on Mobil Oil Italiana's books, at the request of the parties.

Subcommittee Chairman Frank Church (D, Ida.) charged that Mobil had "falsified" its accounts to conceal the payments, but another Mobil official objected to use of the term, contending that nobody was "defrauded or deceived" by the accounting procedure.

Checket acknowledged that the contributions were actually assessments made by Unione Petrolifera, Italy's trade association of petroleum companies. The association took a bank loan to make contributions and then set assessments on various oil companies to repay the loan based on the amount of oil each had sold Italy's state-owned electricity company, according to the testimony.

Payments to Canadian parties. Imperial Oil Ltd. of Canada, which was 70%-owned by Exxon, disclosed July 15, 1975 that its political contributions to Canadian parties at the provincial and federal level had totaled $234,000 annually over the past five years.

The money was limited to three parties: the Progressive Conservative, Liberal and Social Credit parties. No individuals received contributions from Imperial, a spokesman said. The company refused to say how much money was received by each party.

SEC lists corporate payments. In a report filed May 12, 1976 with the Senate Banking Committee, the Securities & Exchange Commission identified 79 companies that had admitted making, or had been formally charged with making, foreign payments of questionable legality.

The SEC named another 24 companies that had admitted making domestic political contributions.

The first part of the three-part SEC report was a list of 89 companies that had filed with the agency public documents relating to foreign and domestic payments.

The second part summarized reports prepared by six special-review commissions that had been established under court order by American Ship Building Co., Ashland Oil Inc., Gulf Oil Corp., 3M Co., Phillips Petroleum Co. and Northrop Corp. The SEC had filed charges against the six companies, charging that they had violated federal securities laws by concealing their domestic and/or foreign payments. All six consented to court injunctions barring future violations of securities laws, but the firms neither admitted nor denied the SEC charges.

In the third part of the report, the SEC summarized allegations made in enforcement actions against eight other companies whose special review panels had not yet completed their reports.

According to the SEC's tabulations of the 95 companies that had made disclosures of questionable or illegal corporate payments, 66 were manufacturing companies. The two largest groups were drug manufacturers and oil companies, each represented by 12 firms.

According to the SEC, 19 of the 95 companies specifically said that the cessation of their payment practices "would have no material effect on their total revenues or overall business." For this reason, the SEC said, an end to the practice of making foreign payments "will not seriously affect the ability of American business to compete in world markets."

Continental Oil fires 2 officials. Continental Oil Co. Dec. 15, 1976 fired two top operating officials because, the company said, they had known of, or participated in the making of, questionable domestic and foreign political payments since 1971. The company also accused the officials—Wayne E. Glenn and Willard H.

Domestic and Foreign Corporate Payments

(From reports filed with the SEC by April 21, 1976)

Corporation	Domestic Political Contributions or Other Domestic Payments	Questionable Foreign Payments
Ashland Oil Inc.[1]	Political contributions totaled $850,000 from 1967-72. A total of $25,700 expended from 1972-74 constituted legal contributions. The following sums were reported but not identified as legal: 1967—$66,500, 1968—$239,600, 1969—$46,300, 1970—$71,700, 1971—$54,500, 1972—$256,815. A subsidiary paid $15,000 in 1970 in response to an extortionate demand by a local government. Between 1967-72, another $71,700 was "presumed to have been used" for political contributions.	679,500 (1967-74)
Cities Service Co.	Not indicated	645,000 (1973-75)
Coastal States Gas Corp.	None yet discovered	8,000,000
Exxon Corp.	Not indicated	56,771,000 (1963-75)
Gulf Oil Corp.[1]	Political contributions totaled $1.4 million from 1960-72. Another $5.4 million returned to U.S. from foreign countries in off-book transactions to be used for political contributions. Disposition of $4 million of this total not determined.	6,900,000 (1960-73)
Phillips Petroleum Co.[1]	Contributions totaled $585,000 from 1964-72.	1,258,000
Public Service of New Mexico	Grand jury investigation regarding possible violation of federal law in connection with $9,656 paid to private company that may have been passed on to candidate.	Not indicated
Standard Oil Co. (Indiana)	Probably illegal state contribution of $10,000 in 1970; $10,000 payment in 1970 to trade association for political contribution. Aggregate of $289,000 in promotional allowances from supplies not recorded as assets.	1,359,400 (1970-75)
Tenneco Inc.	Subsidiaries made lawful contributions of $180,000 in California. Some $3,000 contributed illegally by subsidiary in Louisiana. Payments of $2,000 a month to Louisiana sheriff presently under investigation. Contribution to U.S. Senator may have been paid to obtain influence in General Service Administration decision.	865,480
UOP Inc. (A subsidiary of Occidental Petroleum Corp.)	None	290,000 (1971-75)

Source: Securities and Exchange Commission and Congressional Quarterly, Aug. 28, 1976.

[1]Information based on court-ordered report prepared by special review commission.

Quotes indicate excerpt from company report.

Burnap, both vice chairmen—of withholding information about the payments from a directors' committee that was investigating them.

Burnap Dec. 16 denied the company's charges.

In a filing with the Securities and Exchange Commission Dec. 16, Continental Oil said C. Howard Hardesty Jr., also a vice chairman, had not taken "adequate steps to investigate the possible misuse" of domestic political contributions channeled through a slush fund operated by a subsidiary between 1970 and 1972. Other senior management personnel "were aware or and/or authorized" improper transfers of more than $80,000 from 1970–73 "outside normal accounting channels" to pay unnamed foreign tax officials, Continental Oil said.

The company said that its internal investigation, which was "substantially completed," had uncovered a total of $85,000 in corporate contributions to U.S. politicians and $148,000 in payments to foreign political parties or officials.

Sales unaffected by bribery crackdown. The government's crackdown on bribes and other illegal or questionable payments made abroad by U.S. companies apparently had not resulted in the loss of U.S. sales overseas to free-spending foreign competitors, the Wall Street Journal reported Feb. 28, 1977.

The report was based on the Journal's survey of 25 large U.S. corporations that had admitted making large questionable payments abroad. None of the 25 firms said it had lost a significant portion of its foreign sales after these payments were stopped.

"There may, indeed, be a new and higher corporate morality," the Journal said, adding that its "survey suggests that, if there is, it hasn't hurt."

A management consultant told the Journal that "a lot of the payments were unnecessary and reflected a lack of confidence in our ability to compete. In many cases we were far superior to foreign competitors and we didn't need to pay off." Other analysts said the payments had been used not against foreign companies but against other U.S. competitors, particularly in the weapons industry.

Gulf Oil Corp. chairman Jerry McAfee, who replaced the ousted Bob R. Dorsey in the aftermath of Gulf's payments scandal, cited another reason why sales seemed to be unaffected by the halt in the payments. At Gulf, McAfee said, the payments had been aimed not at spurring sales but at protecting existing assets.

California trust suit. California filed antitrust charges June 26, 1975 against 11 major oil companies, alleging they conspired and eliminated competition in the production of crude oil and the sale of refined petroleum products, fixed prices for crude oil and refined products, and allocated customers among themselves in bidding on petroleum contracts.

The state asked that the defendants be ordered to divest themselves of their crude oil production interests in California.

The antitrust suit was filed on behalf of the state, which purchased $10 million in petroleum products annually, and 450 "political subdivisions and special districts." The oil companies' alleged restraint of trade forced California to pay "substantially higher prices for refined petroleum products" than if no violations of antitrust laws had occurred, the suit contended. The plantiffs sought treble damages for the unspecified amount of their financial injuries.

The defendants' share of crude oil production in California in 1973, the year in which charges were based, was about 72%, but crude oil produced by other oil firms also was purchased or otherwise acquired by the defendants, state officials said.

In 1973, according to the complaint, the defendants had 86% of the total crude oil refinery capacity in California, 93% of the total gasoline refinery capacity, 85% of the retail gasoline market, and 85% of the retail market for other refined petroleum products.

Washington state case—The state of Washington, in a similar complaint, sued 10 major oil companies Aug. 16, 1977 on charges of conspiring to monopolize the production and marketing of petroleum products in Washington, Oregon, California, Alaska, Hawaii and Arizona.

Named in the suit were Standard Oil Co. of California, Texaco Inc., Union Oil Co. of California, Atlantic Richfield Co.,

Exxon Corp., Getty Oil Co., Gulf Oil Corp., Mobil Corp., Phillips Petroleum Co. and Shell Oil Co.

The Washington suit was the second major test of the 1976 federal antitrust law, which permitted state attorneys general to recover damages on behalf of individual citizens in price-fixing cases. The provision allowing recovery of treble damages for consumers was the so-called *parens patriae* clause.

Six other states had brought antitrust charges against major oil companies since the 1973 oil embargo. The suits filed by California, Oregon, Arizona, Florida, Connecticut and Kansas had been consolidated in U.S. District Court in Los Angeles. They had added the *parens patriae* claims to their original complaints, according to Business Week Sept. 12.

Like the other six states, Washington was seeking treble damages for the state, its political subdivisions and all citizens who had purchased refined oil products from the defendants.

Oil overcharges may total $165 million. Twelve U.S. oil companies may have overcharged the public up to $165 million during the 1973–1974 Arab oil embargo by inflating their foreign crude oil costs, the Federal Energy Administration declared July 10, 1975. The firms were not charged with any "wrongdoing," the FEA said, but would be required to justify the "transfer prices" paid their overseas subsidiaries in purchases of foreign oil for resale in the U.S.

The 12 companies and the amount of possible overcharges: Gulf Oil Corp. $58 million; Standard Oil Co. (Indiana) $27.7 million; Atlantic Richfield Co. $24.1 million; Sun Oil Co. $17.6 million; Mobil Oil Corp. $14.2 million; Phillips Petroleum Co. $13.1 million; Union Oil Co. $4.4 million; Continental Oil Co. $2.7 million; Standard Oil Co. (California) $2.1 million; Ashland Oil Inc. $796,000; Getty Oil Co. $497,000; and Texaco Inc. $336,000.

In May 1974, the FEA had accused Gulf of inflating transfer prices paid its African affiliates.

Gulf Oil indicted. Gulf Oil Corp., its U.S. subsidiary and Z. D. Bonner, president of the U.S. firm and director of the parent company, were indicted by a federal grand jury in Pittsburgh Aug. 5, 1975 on charges that they willfully violated the Federal Energy Administration's crude oil entitlements program.

The defendants were accused of failing to comply with the program during the first five days of February when refineries with access to more than the average amount of price-controlled "old" oil were ordered to make "entitlement" payments to firms with less than the average amount of the cheaper fuel.

The program was designed to equalize the cost of crude oil so that small independent refineries, heavily dependent on higher-priced imported oil, would not be at a competitive disadvantage with the major oil companies that had large sources of cheaper domestic oil available to them. The controversial program took effect Jan. 31.

According to the five-count criminal indictment, Gulf failed to buy entitlements valued at $3.9 million. Gulf currently was in compliance with the rule, FEA spokesmen said. According to the oil company's 2nd quarter report, the government's entitlements and allocation programs cost Gulf $100 million.

A Gulf suit challenging the FEA rule was pending in U.S. District Court in Pittsburgh. In a statement Jan. 29, just before the payments deadline, Gulf said it had "elected voluntarily not to comply" with the FEA rule, which, the statement charged, required Gulf to "pay its competitors for the right to refine its crude oil in its own refineries." Two other legal challenges to the FEA order, filed by Exxon Corp. and Marathon Oil Co., were rejected Feb. 12.

Trust case dismissed. A New York State judge March 10, 1976 dismissed a criminal indictment against three of the nation's largest oil companies. The indictment had charged that Exxon Corp., Mobil Oil Corp. and Gulf Oil Corp. violated state antitrust laws during the 1973 energy crisis.

According to the indictment, filed in July 1974 by the state, the defendants had engaged in "an arrangement" to restrain competition in bidding on gasoline contracts with both New York City and New York state to drive up prices.

The state charged that between May and September 1973, the defendants had bid only on city and state contracts similar to those that the firms had won the year before. The 1973 contracts were obtained at prices that averaged 37% higher than in the previous year's contracts. The oil companies did not deny these charges, but they did deny that their behavior constituted collusion.

State Supreme Court Justice Burton B. Roberts dismissed the criminal charges on a defense motion after the state had presented its case. Roberts ruled that the state had failed to prove the existence of "an arrangement" to rig prices.

Roberts criticized the oil companies' actions, but said that their behavior could be explained as compliance with the federal government's voluntary allocation guidelines. The program, instituted in May 1973 in response to the nation's growing fuel shortage, permitted oil companies to deal only with existing customers and not accept new ones.

The New York case was the first major criminal action against oil companies growing out of the energy crisis. Still pending was a separate price-fixing case brought by the state against seven major oil companies. Three defendants, Amoco, Shell and Exxon, had entered a consent decree in the case, neither admitting nor denying the charges but promising not to violate price-fixing laws in the future, the New York Times reported March 11. Mobil, Sun Oil, Gulf and Texaco faced trial on the charges.

5 firms guilty. A federal jury in Baltimore Aug. 30, 1977 convicted five independent oil companies, an oil industry trade association and its director of violating federal antitrust laws. They were found guilty of conspiring to fix retail gasoline prices in six Mid-Atlantic states and Washington, D.C. They had been indicted June 1, 1976.

The 17 billion gallons of gas they had sold in New York, New Jersey, Pennsylvania, Delaware, Maryland, Virginia and Washington from 1967 to 1974 were worth $4 billion. The defendants were accused of agreeing to keep their gas prices no more than 2¢ below those charged by major oil companies.

Convicted were Amerada Hess Corp., Ashland Oil Inc., The Meadville Corp., a unit of Commonwealth Oil Refining Co., a subsidiary of Continental Oil Co., the Society of Independent Gasoline Marketers of America and Robert R. Cavin.

Two oil companies were acquitted of the antitrust charges: Continental and Crown Central Petroleum Corp. Three oil company executives had been acquitted earlier during the three-month trial.

Several class action suits filed by consumer groups were pending against a number of the defendants.

The executives acquitted earlier were oil company Vice Presidents Norman Goldberg of Amerada Hess, Charles J. Luellen of Ashland and W. H. Burnap of Continental.

The government charged that the trade association had served as a clearinghouse for information to "coordinate price increases and to eliminate discounts and settle pricing disputes" among the oil companies.

'Rebrand' settlement. A trust case against six oil companies was concluded Aug. 31, 1977 when a final settlement was entered in U.S. District Court in Los Angeles.

The settlement, proposed by the Justice Department, barred the six defendants from fixing prices on "rebrand" gasoline sold in California, Oregon, Washington, Nevada and Arizona. Rebrand gasoline was sold under a brand name other than the refiner's.

Consenting to the settlement were Phillips Petroleum Co., a subsidiary of Continental Oil Co., Powerine Oil Co., Fletcher Oil and Refining Co., MacMillan Ring-Free Oil Co. and a unit of Ultramar Co. Ltd. of London.

They previously had pleaded no contest to the charges and had been fined $50,000 each, the maximum amount.

Other Oil Developments

Colorado oil-shale leases suspended. Interior Secretary Thomas S. Kleppe Aug. 23, 1976 suspended for a year the development of two oil-shale tracts in Colorado.

The four companies that held leases on the two tracts—Standard Oil Co. (Indiana), Gulf Oil Corp., Shell Oil Co. and Ashland Oil Co.—had asked Kleppe for a two-year suspension so they would not have to make installment payments on the leases, obtained in 1973.

Standard Oil (Indiana) and Gulf, which had jointly bid $210.3 million for one tract, already had paid $126 million. Shell and Ashland, the two remaining partners in a four-company consortium that had bid $117.8 million for the other tract, had paid about $70 million each in installments.

The companies said that the oil-shale project had been delayed by "environment, legal and technical problems." Kleppe said that he granted the one-year delay because studies showed that air pollutants at the tracts sometimes exceeded federal clean-air standards. He said that the companies also had encountered problems of rock strength that limited the quantity of shale that could be extracted by underground-mining methods.

(Kleppe also was considering the request by four holders of two leases in Utah for a two-year suspension of their oil-shale project.)

An Interior Department official said that the "suspension sets back our oil-shale program, but it doesn't kill it." Kleppe added, "We haven't relinquished our goal of demonstrating the feasibility of shale-oil production on a commercial scale."

Suspension of the Colorado leases was the latest in a series of setbacks to the oil-shale program:

In December 1975, Atlantic Richfield Co. and Oil Shale Corp. had ended their partnership with Shell and Ashland to develop the consortium's Colorado tract.

The companies had cited congressional inaction, environmental roadblocks and a shortage of capital as reasons for their withdrawal from the consortium. A chief factor in their decision was the House of Representative's rejection earlier that month of a $6-billion loan-guarantee program for industries participating in a synthetic-fuels demonstration project.

Until the congressional rebuff and industry pullback in late 1975, the Ford Administration had made rapid development of the Western oil-shale leases a top energy priority. It was believed that 600 billion barrels of recoverable oil were located in the Green River shale formations beneath Colorado, Utah and Wyoming.

Before Kleppe suspended development of the two Colorado leases, the government had hoped to obtain 100,000 barrels of oil a day from the two sites by 1985.

'Synfuels' plan defeated. In 1975 and again in 1976 Congress rejected proposals to provide loan guarantees for the commercial development of "synthetic fuels" (synfuels). Such a proposal had been inserted by the Senate in mid-1975 in the fiscal 1976 authorization measure for the Energy Research & Development Administration (ERDA), but the House rejected the plan. A House Rules Committee rule to allow consideration of a version of the plan was then rejected by the House Sept. 23, 1976.

In December 1975 the Congressional Budget Office prepared an analysis of Senate and Administration synfuel-aid proposals. It said in its summary:

> The President has proposed a program to bring synthetic fuels—oil and gas produced from coal, oil shale, and urban waste—into commercial production in the near future. The program would consist of price supports, loan guarantees, and construction grants designed to achieve an interim synthetic fuel production target of the equivalent of 350,000 barrels of oil per day, with an option of expanding the program to 1,000,000 barrels per day by 1985 if the initial phase were successful—350,000 barrels per day of oil is about 1.3% of 1974 total U.S. consumption of oil and gas. Projects eligible for assistance would include:
>
> Conversion of coal to oil or gas
> Extraction of oil from shale
> Production of oil or gas from urban wastes
>
> The Senate version (S. 598) of the FY 1976 ERDA Authorization bill (H.R. 3474), currently in conference, provides for guarantees of up to $6 billion in loans, but does not contain the other incentive provisions. Loan guarantees in this amount would be sufficient for approximately 350,000 barrels per day.
>
> *Decisions facing Congress*
>
> Congress may decide to:
>
> Accept or reject a synthetic fuels commercialization proposal as a whole, independent of other actions related to energy production or conservation.
>
> Accept the goal of speedy commercialization but choose other target production levels, alternative process mixes, or different levels or types of production incentives (e.g., limit incentives to the loan guarantees of S. 598).

Delay decision so as to consider the program in the context of the proposed Energy Independence Authority or other broad energy policies.

Postpone commercialization and pursue instead further research, development, and demonstration projects relevant to synfuels.

Criteria for decision

A decision to proceed with a program of incentives to stimulate synthetic fuel (synfuel) production by private industry requires answers to four questions: (1) will synthetic fuels production be justified before 1985? (2) does private industry require government incentives to produce them in that time frame?, (3) what is the best package of incentives, considering program objectives and cost?, and (4) are the costs of that package acceptable in light of the factors which justify synfuel production?

Answers to each of the four fundamental questions depend on a number of considerations.

1. The justification of synthetic fuel production between 1975 and 1985 depends on economic and non-economic factors. Quantifiable economic benefits include the value of the fuel themselves, a degree of embargo protection, and reduced cost of future synfuel production. Unless world oil prices rise substantially above their current levels, the economic costs of synfuel production will probably exceed these economic benefits. However, non-quantifiable and non-economic considerations could tip the balance either way. Synfuel production capability could provide insurance against large increases in world oil prices and might influence OPEC nations to restrain price increases. On the other hand, large-scale synfuel production would create significant environmental risks and social costs.

2. It is highly unlikely that private industry will produce significant quantities of synthetic fuels before 1985 without government support. Factors cited include lack of profitability, technological and economic risk, difficulty in raising capital, and constraints imposed by the government.

It is clear that some synfuel production would be unprofitable at current oil and gas prices. If profitability were achieved through use of government incentives, the remaining factors might still discourage investment.

3. An incentive program should address all significant constraints. The incentives proposed by the President are targeted at making synfuel production profitable, shifting risks from investors and producers to the government, and alleviating shortages of capital. If these are not the real constraints to synfuels development, a different level and mix of incentives may be appropriate. (For example, if risk alone is the problem, loan guarantees might be relied on exclusively.) An alternative program could include regulatory reform, tax incentives, government ownership, or measures to increase the cost of imported fuels. Although comprehensive evaluation of the costs and effectiveness of alternative incentive packages is beyond the scope of this report, examination of these candidates suggests that a program substantially better than that proposed by the Administration may be difficult to design.

4. A final decision on synfuels commercialization can only be made after an incentive program has been designed and evaluated. Then the costs attributable to the incentives themselves must be added to the previously estimated costs of synfuel production to determine if acquiring synfuel production capacity through government action is justified. Although potentially effective in achieving production targets, the Administration's proposed incentives may reduce competition, increase costs of producing synthetic fuels, and adversely affect private capital markets. No alternative is entirely free of comparable disadvantages.

Other issues

If a decision to proceed with a commercialization program at some level is made, three other issues—the targeted production level, the mix of processes to be encouraged, and protection of the environment—must be considered.

1. A program with a target of 350,000 barrels per day production by 1985 has substantially lower costs than one aiming at 1,000,000 barrels per day, yet the smaller program would provide much information on technology, process economics, and environmental consequences.

The benefits of synfuel commercialization come in two broad classes—production of energy and acquisition of information. If information is the primary goal, a program with a target as low as 125,000 barrels per day might be chosen.

2. Alterations in the emphasis given different processes can also alter program costs and benefits. S. 598 provides that not less than $2.5 billion of the $6 billion in loan guarantees would be used for high-BTU gasification processes, and that funds may be used for solar, geothermal, and other unconventional processes. The Administration proposal contains no such provisions.

3. Major uncertainties concerning environmental impact cloud the synfuels decision process; their resolution could be an important objective of a small synfuels program. The principal known and potential impacts include large-scale land disruption from oil shale and coal mining, disposal of wastes from oil shale processing, consumptive use of water in water-short regions, air pollution from processing, potential carcinogen formation in the processes dealing with liquids, and the socioeconomic impacts of the influx of workers, their families, and associated developments on sparsely populated regions. Some such impacts may be mitigated, at the lower levels, by a strict Environmental Protection Strategy (the Administration's proposal contains the outlines of such a

strategy) and by grants or loan guarantees to impacted communities. Nevertheless, aggregate impacts of a larger, 1,000,000 barrel per day, level could be severe.

Potential budget impact

For FY 1976, the Administration requests borrowing authority of $1.5 billion for loan guarantees, plus $1.0 billion for price guarantees, and appropriation of $0.6 billion for construction grants.

Even if passed immediately, however, a synthetic fuels commercialization program would be unlikely to lead to Federal outlays in Fiscal Year 1976. Rather, there would be a time stream of outlays whose extent and timing would depend on the program level, the mix of incentives, and the riskiness of synfuels technologies pursued. Tax revenues could also be decreased from levels predicted in the absence of the synfuel program.

The cost to the government of price supports, construction grants and foregone taxes implicit in the Administration's 350,000-barrel per day program is expected to total from $1.6 billion to $13.3 billion by the year 2005, when obligations to support prices will have expired. During the 1980's annual costs might reach $600 million per year. Additionally, a maximum of just over $2 billion in guaranteed loans would be outstanding in 1985. Costs would become higher—$3 to $29 billion over the life of the program—if a decision were made to proceed to 1,000,000 barrel per day capacity by 1985.

Timing of decision

If there were a decision to proceed on January 1, 1976, it would be at least the end of 1980 before there could be a year of operating experience with a synfuel plant. But ERDA has several second-generation synfuel processes ready for demonstration which, if successful, could make obsolete a synfuel plant based on current technology. Delay in synfuel commercialization could actually improve the economics of the program.

Congress will determine whether an immediate decision is required. It has the option simply to defer decision—possibly postponing production targets beyond 1985—or to rely upon research, development, and demonstration to lay the groundwork for an expanded synthetic fuel production capability after 1985. Whatever the decision, it should be made in the context of a larger perspective on the proper role of the Federal government with respect to the continuum of energy activities from research through development and demonstration to commercialization.

In the body of the analysis, the Budget Office provided the following data on the proposals and issued involved:

Technologies for producing synthetic fuels have long been known, and there is some synthetic fuel production in foreign countries. However, much of that production is subsidized, and the scale of production is considerably smaller than that envisioned for the proposed programs.

The Energy Research and Development Administration (ERDA) conducts an extensive research program aimed at developing "second-generation" processes for producing synfuels which offer promise of being more economically attractive, efficient in use of resources, and environmentally acceptable. Nevertheless, such technologies are not yet available, and economics of existing processes have not, in the past, been sufficient to induce industry to produce synthetic fuels commercially in the United States.

To alter this situation, the Administration proposes three goals for 1985: (1) development of technical, environmental, and economic information on synfuel production processes; (2) accumulation of experience with synfuel production in American industry; and (3) production of significant quantities (the equivalent of at least 350,000 barrels per day, and possibly as much as 1 million barrels per day) of synthetic oil and gas. The goals would be achieved by measures which would shift risks of synfuels development from private industry to the government and would involve net government subsidies to synthetic fuels production at least through 1985.

The proposal itself is reported in detail in a four-volume interagency study recently made available to the Congress. (Draft report by the Synfuels Interagency Task Force to the President's Energy Resources Council, "Recommendations for A Synthetic Fuels Commercialization Program", June, 1975.)

The Administration's report examined five production levels:

No program, with zero synthetic fuels production in 1985. (The Administration's report does not consider those alternative ways of achieving domestic energy balance objectives which do not involve synfuels.)

An "information" program, achieving production of the equivalent of 350,000 barrels per day of crude oil by 1985.

A "two-stage nominal" program, initially targeted to reach 350,000 barrels per day before 1985, with a decision in the late 1970's whether to proceed to 1,000,000 barrels per day by 1985. (This is the Administration's proposal).

A "one-stage nominal" program, achieving 1,000,000 barrels per day by 1985.

A "maximum" program, achieving 1,700,000 barrels per day by 1985 and several types of synthetic fuels:

Fuels produced from coal:

Gas with low to medium heat content compared to natural gas ("low to medium-BTU gas"). (The BTU, or British Thermal Unit, is a common measure of heat.)

Gas with approximately the same heat content as natural gas (high-BTU gas").

Synthetic crude oil. ("syncrude")

Crude oil extracted from oil shale.

Gas and oil produced from urban waste or other biological materials.

The incentives recommended by the Administration are:

For oil shale and syncrude: Non-recourse loan guarantee and price support. (A "non-recourse" loan would have as security only the assets of the proposed venture itself; in event of default, the government would have no recourse to the assets of larger corporations sponsoring the venture. Those corporations would retain any patents granted in the course of the project.)

For high-BTU gas: Non-recourse loan guarantee.

For low and medium-BTU gas:

a. For regulated industries, construction grants.

b. For unregulated industries, a non-recourse loan guarantee and price support.

For fuel from urban waste: Non-recourse loan guarantee.

A General Accounting Office report, released Aug. 24, 1976, dealt a new blow to the Administration's plan to develop synthetic fuels as an alternative energy supply.

The GAO, the auditing arm of Congress, concluded that synthetically produced oil and gas would be too costly to be competitive with foreign energy supplies. For that reason, the GAO said, the government should not commit billions of dollars for development projects.

The GAO estimated the cost of oil produced from coal or shale at $18 a barrel. Noting that imported crude oil currently cost about $12 a barrel, the GAO warned that if the U.S. committed itself to "uneconomical, high-cost supply technologies," crude-oil prices set by the Organization of Petroleum-Exporting Countries could become a floor, rather than a ceiling.

(The GAO estimated the cost of synthetically produced gas at $3 per 1,000 cubic feet, which was more than twice the rate of $1.42 set by the Federal Power Commission for natural gas produced at new wells.)

Conservation, rather than synthetic-fuel development, was the most "cost-effective" way to reduce U.S. dependence on foreign energy supplies, the GAO said.

As alternatives to the synthetic-fuel project, the GAO urged the government to spend more money on research into municipal waste combustion, geothermal energy and solar heating. Although the supplies of energy available from these sources might be smaller than the potential from shale or coal conversion, the GAO said, the costs were more predictable and therefore government support was more justified.

Rep. Olin Teague (D, Tex.), who had sponsored the 1975 loan-guarantee program for synthetic fuels, had introduced a similar, but scaled down version in 1976.

Robert C. Seamans Jr., head of the Energy Research and Development Administration, which sponsored synthetic-fuel development, criticized the GAO report. Conservation and renewable resources, such as solar and geothermal energy, "cannot obviate the indicated need in the 1990s for substantial quantities of synthetic fuel," Seamans said.

Oil industry pollution-control costs analyzed. The U.S. oil industry was expected to spend $3–$4 billion by 1983 meeting federal pollution-control requirements, the Council on Economic Priorities, a nonprofit research group, reported July 21, 1975.

Most of these added costs, which represented about 1% of the industry's profits, were expected to be passed on to consumers because, according to the study, the industry could not absorb pollution-control costs without its profit margin suffering, making it difficult for oil companies to attract investment capital. The total costs of pollution curbs were expected to add less than one-third of a cent to the price of gasoline.

The report stated that the oil industry was in near-compliance with federal standards for controlling refinery emission of sulphur oxides, hydrocarbon gases, and other air pollutants, but was far from meeting government requirements for the treatment of refinery waste that was harmful to marine life.

Most of the projected outlays by 1983 would be spent on water-pollution control, according to the report. The "remainder will be needed to control the planned expansion and capacity [of refineries], assuming continued growth at an annual rate of 3.5%."

The council's study was based on pollution control records of eight major oil companies from 1972–1974. Atlantic Richfield Co. had the best record of the group in curbing pollution, the report declared, while Gulf Oil Corp. and Texaco Inc. "displayed the poorest overall control performances."

Auto Fuel-Economy Ranking—1977 Model Year*

Fuel Economy (MPG) Comb.	City	Hwy.	Manufacturer	Car Line	Engine Size (CID/Cylinders)	Trans.
			Two-Seaters			
21	18	27	Datsun	280Z	168/6	M
21	17	31	Porsche	924	121/4	M
20	18	22	Datsun	280Z	168/6	A
18	15	24	Porsche	911S	164/6	M
18	16	21	Porsche	911S	164/6	S
17	15	20	Chevrolet	Corvette	350/8	A
17	14	24	Porsche	Turbo Carrera	183/6	M
15	14	18	Chevrolet	Corvette	350/8	M
			Sub-Compact Cars			
44	39	52	Volkswagen	Rabbit Diesel	90/4	M
41	36	49	Toyota	Corolla	71/4	M
40	35	47	Volkswagen	Dasher Diesel	90/4	M
38	35	42	Mazda	808	78/4	M
36	31	43	Chevrolet	Chevette	98/4	M
35	29	45	Dodge	Colt	98/4	M
34	29	41	Datsun	B-210	85/4	M
34	29	41	Datsun	F-10	85/4	M
34	29	43	Volkswagen	Rabbit	97/4	M
34	29	43	Volkswagen	Scirocco	97/4	M
33	28	42	Chevrolet	Chevette	85/4	M
33	28	42	Plymouth	Cricket	98/4	M
32	28	42	Subaru	Subaru	97/4	M
32	28	39	Toyota	Corolla	97/4	M
31	26	39	Dodge	Celeste	98/4	M
31	26	39	Plymouth	Arrow	98/4	M
30	26	36	Chevrolet	Chevette	98/4	A
30	26	35	Dodge	Celeste	98/4	A
30	26	35	Dodge	Colt	98/4	A
30	26	37	Ford	Pinto	140 (2.3L)/4	M
30	26	37	Lincoln-Mercury	Bobcat	140 (2.3L)/4	M
30	26	35	Plymouth	Cricket	98/4	A
30	26	35	Plymouth	Arrow	98/4	A
30	26	37	Pontiac	Astre	151/4	M
30	26	37	Pontiac	Sunbird	151/4	M
29	25	35	Chevrolet	Chevette	85/4	A
29	26	33	Datsun	B-210	85/4	A
28	24	36	Audi	Fox	97/4	M
28	24	33	Audi	Fox	97/4	A
28	24	33	Chevrolet	Vega	140/4	M
28	24	33	Chevrolet	Monza	140/4	M
28	24	33	Oldsmobile	Starfire	140/4	M
28	24	33	Pontiac	Astre	140/4	M
28	24	37	Volkswagen	Rabbit	97/4	M
28	24	33	Volkswagen	Dasher	97/4	A
28	24	36	Volkswagen	Dasher	97/4	M
28	24	37	Volkswagen	Scirocco	97/4	M
27	23	36	Buick	Opel by Isuzu	111/4	M
27	23	33	Mazda	808	97/4	M
27	24	32	Pontiac	Astre	151/4	A
27	24	32	Pontiac	Sunbird	151/4	A
27	25	31	Toyota	Corolla	97/4	A
27	24	33	Volkswagen	Rabbit	97/4	A
27	24	33	Volkswagen	Scirocco	97/4	A
26	24	30	Buick	Opel by Isuzu	111/4	A
26	23	32	Ford	Pinto	140 (2.3L)/4	A
26	23	33	Ford	Mustang II	140 (2.3L)/4	M
26	23	32	Lincoln-Mercury	Bobcat	140 (2.3L)/4	A
26	23	30	Mazda	808	97/4	A
26	24	31	Subaru	Subaru	97/4	A
26	21	35	Toyota	Celica	134/4	M
26	23	33	Volkswagen	Beetle	97/4	M
25	20	32	Mazda	Cosmo	80/2	M
25	20	32	Mazda	RX-4	80/2	M
25	21	35	Toyota	Corona	134/4	M
25	22	29	Toyota	Celica	134/4	A
24	21	28	Chevrolet	Vega	140/4	A
24	21	28	Chevrolet	Monza	140/4	A

Fuel Economy (MPG) Comb.	City	Hwy.	Manufacturer	Car Line	Engine Size (CID/Cylinders)	Trans.
			Sub-Compact Cars			
24	21	28	Dodge	Celeste	122/4	A
24	20	33	Dodge	Celeste	122/4	M
24	21	28	Dodge	Colt	122/4	A
24	20	33	Dodge	Colt	122/4	M
24	21	29	Ford	Mustang II	140 (2.3L)/4	A
24	21	28	Ford	Maverick	250/6	M
24	21	28	Ford	Maverick	200/6	M
24	21	28	Lincoln-Mercury	Comet	200/6	M
24	21	28	Lincoln-Mercury	Comet	250/6	M
24	21	28	Oldsmobile	Starfire	140/4	A
24	21	28	Plymouth	Cricket	122/4	A
24	20	33	Plymouth	Cricket	122/4	M
24	21	28	Plymouth	Arrow	122/4	A
24	20	33	Plymouth	Arrow	122/4	M
24	21	28	Pontiac	Astre	140/4	A
24	22	29	Toyota	Corona	134/4	A
23	20	27	American Motors	Gremlin	232/6	M
23	20	27	Ford	Mustang II	171 (2.8L)/6	M
21	19	26	Buick	Skyhawk	231/6	A
21	18	29	Buick	Skyhawk	231/6	M
21	18	26	Mazda	Cosmo	80/2	A
21	18	26	Mazda	RX-4	80/2	A
21	19	26	Oldsmobile	Starfire	231/6	A
21	18	29	Oldsmobile	Starfire	231/6	M
21	19	26	Pontiac	Sunbird	231/6	A
21	18	29	Pontiac	Sunbird	231/6	M
20	18	24	American Motors	Gremlin	232/6	A
20	17	26	American Motors	Gremlin	258/6	M
20	17	25	Chevrolet	Monza	305/8	A
20	18	25	Chevrolet	Camaro	205/6	M
20	18	23	Ford	Pinto	171 (2.8L)/6	A
20	18	24	Ford	Maverick	200/6	A
20	18	23	Lincoln-Mercury	Bobcat	171 (2.8L)/6	A
20	18	24	Lincoln-Mercury	Comet	200/6	A
20	17	25	Pontiac	Firebird	231/6	A
19	17	23	American Motors	Gremlin	258/6	A
19	16	22	Chevrolet	Camaro	305/8	M
19	17	22	Chevrolet	Camaro	250/6	A
19	17	22	Ford	Mustang II	302/8	A
19	17	23	Ford	Mustang II	171 (2.8L)/6	A
19	17	22	Ford	Maverick	250/6	A
19	17	22	Ford	Maverick	302/8	A
19	17	22	Lincoln-Mercury	Comet	302/8	A
19	17	22	Lincoln-Mercury	Comet	250/6	A
19	17	23	Pontiac	Firebird	301/8	A
19	16	26	Pontiac	Firebird	231/6	M
18	16	22	Chevrolet	Monza	305/8	M
18	16	21	Chevrolet	Camaro	305/8	A
18	16	21	Ford	Mustang II	302/8	M
18	15	23	Pontiac	Firebird	301/8	M
18	16	22	Pontiac	Firebird	350/8	A
17	15	20	Chevrolet	Camaro	350/8	A
17	15	20	Pontiac	Firebird	400/8	A
15	14	18	Chevrolet	Camaro	350/8	M
15	12	19	Pontiac	Firebird	400/8	M
			Compact Cars			
26	22	34	Pontiac	Ventura	151/4	M
24	21	28	Ford	Granada	250/6	M
24	21	28	Ford	Granada	200/6	M
24	21	28	Lincoln-Mercury	Monarch	250/6	M
24	21	28	Lincoln-Mercury	Monarch	200/6	M
24	21	29	Pontiac	Ventura	151/4	A
23	20	29	Dodge	Aspen	225/6	M
23	20	29	Plymouth	Volare	225/6	M
22	19	27	Chevrolet	Nova	250/6	M
21	18	27	Audi	100LS	114/4	M
21	19	26	Oldsmobile	Omega	231/6	A

Fuel Economy (MPG) Comb.	City	Hwy.	Manufacturer	Car Line	Engine Size (CID/Cylinders)	Trans.
21	18	26	Pontiac	Ventura	231/6	A
20	18	23	American Motors	Pacer	232/6	A
20	18	23	American Motors	Pacer	232/6	M
20	18	23	American Motors	Hornet	232/6	A
20	18	23	American Motors	Hornet	232/6	M
20	18	25	Buick	Skylark	231/6	A
20	18	23	Chevrolet	Nova	250/6	A
20	17	24	Dodge	Aspen	225/6	M
20	18	24	Dodge	Aspen	225/6	A
20	18	23	Ford	Granada	250/6	A
20	18	23	Lincoln-Mercury	Monarch	250/6	A
20	16	27	Oldsmobile	Omega	231/6	M
20	17	24	Plymouth	Volare	225/6	M
20	18	24	Plymouth	Volare	225/6	A
20	17	27	Pontiac	Ventura	231/6	M
19	17	24	American Motors	Pacer	258/6	M
19	17	23	American Motors	Pacer	258/6	A
19	17	24	American Motors	Hornet	258/6	M
19	17	23	American Motors	Hornet	258/6	A
19	17	23	Audi	100LS	114/4	A
19	16	26	Buick	Skylark	231/6	M
19	17	23	Buick	Skylark	301/8	A
19	16	22	Chevrolet	Nova	305/8	M
19	15	25	Dodge	Aspen	318/8	M
19	17	23	Oldsmobile	Omega	260/8	A
19	16	22	Oldsmobile	Omega	305/8	M
19	15	25	Plymouth	Volare	318/8	M
19	17	23	Pontiac	Ventura	301/8	A
19	16	23	Pontiac	Grand Prix	301/8	A
18	16	21	Buick	Skylark	305/8	A
18	16	21	Chevrolet	Nova	305/8	A
18	16	21	Dodge	Aspen	225/6	A
18	16	22	Ford	Granada	302/8	A
18	16	24	Ford	Granada	302/8	M
18	16	22	Lincoln-Mercury	Monarch	302/8	A
18	16	24	Lincoln-Mercury	Monarch	302/8	M
18	16	21	Oldsmobile	Omega	305/8	A
18	16	21	Plymouth	Volare	225/6	A
18	15	23	Pontiac	Ventura	301/8	M
18	16	22	Pontiac	Ventura	305/8	A
17	15	20	Chevrolet	Nova	350/8	A
17	16	20	Chevrolet	Monte Carlo	305/8	A
17	15	20	Dodge	Aspen	318/8	A
17	15	19	Ford	Thunderbird	302/8	A
17	15	20	Plymouth	Volare	318/8	A
17	14	21	Pontiac	Grand Prix	400/8	A
			Large Station Wagons			
19	16	23	Pontiac	Pontiac Safari Wagon	301/8	A
18	15	21	Buick	Estate Wagon	403/8	A
18	16	21	Buick	Estate Wagon	350/8	A
18	15	21	Oldsmobile	Custom Cruiser Wagon	403/8	A
18	16	21	Oldsmobile	Custom Cruiser Wagon	350/8	A
18	15	21	Pontiac	Pontiac Safari Wagon	403/8	A
17	16	20	Chevrolet	Chevrolet Wagon	305/8	A
16	14	19	Chevrolet	Chevrolet Wagon	350/8	A
15	13	17	American Motors	Matador Wagon	304/8	A
15	13	18	Ford	Ford Wagon	400/8	A
15	13	18	Lincoln-Mercury	Mercury Wagon	400/8	A
14	13	16	American Motors	Matador Wagon	360/8	A
13	11	16	Ford	Ford Wagon	460/8	A
13	11	16	Lincoln-Mercury	Mercury Wagon	460/8	A
12	10	16	Chrysler	Chrysler Wagon	400/8	A
12	10	16	Chrysler	Chrysler Wagon	440/8	A
12	10	16	Dodge	Royal Monaco Wagon	400/8	A
12	10	16	Dodge	Royal Monaco Wagon	440/8	A
12	10	16	Plymouth	Gran Fury Wagon	400/8	A
12	10	16	Plymouth	Gran Fury Wagon	440/8	A

*Ratings are for 1977 models exclusive of those sold in California, where stricter exhaust-pollution standards were required, resulting in generally lower miles-per-gallon figures.

EPA curb on lead in gas upheld—The U.S. Court of Appeals in the District of Columbia upheld by a 5–4 decision March 19, 1976 the Environmental Protection Agency's authority to regulate against the amount of lead in gasoline.

The ruling reversed a December 1974 opinion by a three-member panel of the same court against the proposed EPA regulations requiring a substantial phase-out of the lead content in gasoline.

About 90% of the gasoline produced in the U.S. contained lead, chiefly as an additive to reduce knocking in the engine. The court said there was no dispute that lead-particulate emissions from cars accounted for 90% of the lead in the air, that lead could be absorbed into the body and that lead in high concentrations in the body was toxic.

Lead manufacturers and oil companies challenged the EPA proposed regulations on the ground that the health danger from lead could not be specifically linked to car emissions.

The court majority agreed there was no "certainty" that lead in gasoline was a danger to the public health but it said that "awaiting certainty will often allow for only reactive, not preventive regulation." Written by Judge J. Skelly Wright, the opinion said that where a statute was "precautionary in nature," the regulation was designed to protect the public health and the decision was "that of an expert administrator, we will not demand vigorous step-by-step proof of cause and effect."

The court itself, he said, was not an expert on the subject but a reviewing panel "exercising our narrowly defined duty of holding agencies to certain minimal standards of rationality."

Therefore, it found that the EPA could use the Clean Air Act to regulate gasoline additives whose emission products involved a risk to public health, and it found that the EPA administrator was authorized "to assess risks of harm and where the risk is found to be significant, to act to prevent the harm from happening."

Only six months later, however, the EPA announced Sept. 24 a relaxation of its timetable for reducing the lead content of gasoline.

The agency acted in view of possible gasoline shortages during the next two years. EPA Administrator Russell E. Train said that the new regulations "will meet the same ultimate goal as the original standards but will not do so at the expense of another gas shortage."

More crude oil was required to produce unleaded gasoline that had the same octane rating as leaded gas.

The new regulations required a reduction of the lead content to no more than 0.5 gram per gallon, effective Oct. 1, 1979. The original standards had called for reductions by phases, to 1.4 grams a gallon after Oct. 1, 1 gram a gallon in 1977, 0.8 gram a gallon in 1978 and the 0.5 gram level on Jan. 1, 1979.

Auto fuel economy rises. The Environmental Protection Agency reported Sept. 22, 1975 that fuel economy of 1976-model cars averaged 12.8% higher than that of 1975 models. The average for foreign and domestic cars was 17.6 miles a gallon compared with 15.6 miles a gallon for 1975 models.

The data on 1976 models showed a 26.2% gain over the 1974 figure of 13.9 miles a gallon. "This means that auto makers already have gone more than halfway towards achieving the 40% fuel economy improvement President Ford asked for between 1974 and 1980," EPA Administrator Russell Train commented in releasing the data.

Exactly one year later the EPA reported that 1977-model cars were averaging 18.6 miles a gallon, which represented another 6% gain in fuel economy.

The first-place rating, 44 miles a gallon in EPA tests simulating combined city and highway driving, went to a Volkswagen Rabbit with a diesel engine having a 90-cubic-inch displacement. In second place was a Toyota Corolla, 71-cubic-inch engine, 41 miles a gallon. The Volkswagen Dasher diesel, 90-cubic-inch engine, was rated third at 40 miles a gallon. In fourth was a Mazda 808, 78-cubic-inch engine, which averaged 38 miles a gallon.

Fifth place was taken by the top American car, Chevrolet's Chevette, 98-cubic-inch engine, 36 miles a gallon.

Natural Gas

Ford, governors discuss gas shortage. President Ford informed 16 governors at the White House Aug. 28, 1975 that natural gas would be in short supply during the coming winter by 1.3 trillion cubic feet. The shortage would be 30% greater than that experienced during the previous winter's shortfall, Ford warned.

As a short-run answer to the problem, governors from the nation's three leading producing states—Texas, Louisiana, and Oklahoma—said they would encourage producers to divert excess supplies to interstate needs, but they warned that deregulation and higher prices promised the only long-range solution to the worsening supply situation.

As a step toward the eventual deregulation of interstate gas prices, most of the governors attending the meeting endorsed a plan drawn up by Oklahoma's Gov. David Boren (D) calling for a five-year suspension of controls on newly-developed gas reserves. Boren said the governors also supported legislation lifting federal price controls for an emergency 180-day period.

The permanent deregulation of natural gas prices was a focal point of the Administration's energy plan. The White House argued that higher prices would provide incentives to oil and gas producers to intensify their exploration and recovery efforts. The Democratic majority in Congress generally opposed relying on higher prices to expand the nation's energy supply.

Pennsylvania Gov. Milton J. Shapp (D) was the only governor who refused to support the temporary decontrol proposal. Shapp called on the White House to investigate whether the current gas shortage was "real or contrived." "The public has been ripped off or thinks it has been ripped off by the energy companies," Shapp said.

The governors of four states—Massachusetts, California, New York, and Illinois—where there was strong consumer opposition to energy price decontrol, did not attend the White House meeting.

Louisiana Gov. Edwin D. Edwards (D) criticized those states and others that also were opposed to offshore exploration of oil and gas reserves. Edwards warned that the coastal industrial states that would "suffer most" from the expected shortage "had better realize there is untapped gas and oil off their Atlantic Coast and it should have been tapped five years ago. We [in Louisiana] have gas because we pay $1.50 and not 52¢ [per 1,000 cubic feet]," the federally regulated price.

The governors also called for an excess profits tax on natural gas producers, with rebates if the profits were used for new exploration. In other resolutions July 23, they opposed federal taxation of energy supplies as a way to reduce consumption and favored federal regulations for strip mining on federal lands.

Midwest governors oppose gas controls—Midwestern governors urged deregulation of newly developed natural gas July 23. They expressed their view by resolution adopted at their annual conference, in Cincinnati, with nine of the region's 15 governors in attendance.

Price controls eased on gas. The Federal Power Commission ruled Aug. 28, 1975 that certain industries could bypass pipeline distributors of their natural gas supplies and buy gas directly from the "field" producers at unregulated prices if industrial gas supplies were curtailed during an expected winter shortage. The ruling only affected industries unable to switch to alternative fuels if their gas supplies were interrupted.

"While this policy will not solve the gas shortage," the FPC said in its 2-1 decision, "direct sales may result in increased producer revenues, which would promote increased exploration for and development of gas supplies."

FPC member William L. Springer dissented from the majority opinion, contending that deregulation was a policy that could be enacted only by Congress.

Under current allocation regulations, households would have top priority for gas, which was expected to be in especially short supply for a 10-state area centered in the mid-Atlantic region from the East Coast to the Mississippi River. Industrial users would be hardest hit by a gas shortage.

The FPC-controlled price of gas sold interstate currently was 52¢ per 1,000 cubic feet. (The rate was raised from 50¢ in January when the commission made its biennial price adjustments.) Gas sold intrastate was not subject to federal price controls; it currently sold for $1 to $2 per 1,000 cubic feet.

An earlier FPC ruling decontrolling the interstate price of natural gas during an emergency period had been overturned in an appeals court.

In another 2-1 decision affecting the price of natural gas, the FPC ruled Aug. 28 that small producers, whose annual output was less than 10 billion cubic feet a year, could charge 30% more than large producers for gas sold across state lines. Small producers accounted for 12% of the nation's annual production.

The aim of the new policy, the FPC said, was to strengthen the competitive position of small producers who faced higher risks than their larger rivals and needed to raise more capital for exploration.

House rejects deregulation. Liberal Democrats Feb. 5, 1976 mustered just enough votes—205 to 201—to win approval of an amendment, offered by Rep. Neal Smith (D, Iowa), that continued regulation of natural gas prices of the major producers, while exempting small companies from Federal Power Commission (FPC) control. A last attempt by supporters of deregulation was beaten back by a 204–198 vote. The natural gas bill, as amended, was then approved 205–194, and sent to a House-Senate conference committee. The Senate Oct. 22, 1975 had approved, on a 58–32 vote, a deregulation bill like that just defeated in the House. The legislation, however, died without final action.

In the House version, the bill would have ended FPC regulation of prices of natural gas sold by producers of less than 100 billion cubic feet a year. Regulation would have continued for 25 to 30 companies that sold more than that amount and would have been extended to cover their intrastate sales.

The major companies which, under the bill, would be regulated, accounted for about 75% of the gas sold nationally. Although deregulating prices for several thousand independent producers, the bill would, according to the claims of a supporter of full deregulation, Rep. Clarence J. Brown (R, Ohio), actually increase the percentage of gas under regulation. The percentage at present, Brown said, was about 65%.

In a concession to the interests of the major gas companies, the bill provided that the FPC would use new procedures, more generous to producers, in determining prices.

Rep. Robert C. Krueger (D, Tex.) had been the leader of the fight for deregulation. The bill he proposed would have immediately deregulated prices on new sales of natural gas from on-shore wells and deregulated prices on sales from off-shore wells after Dec. 31, 1980. Long-term contracts made under FPC regulations involving the purchase of gas from on-shore wells would have been kept in force until their expiration date.

Supporters of deregulation argued that without it gas producers would not have the incentive to explore and develop new supplies. They acknowledged that higher prices for consumers would result from deregulation, but maintained that the increases would be slight, since transportation costs accounted for 75% to 80% of the retail price of natural gas. They said that regulation was responsible for gas shortages in states which must import gas, since producers preferred to sell to intrastate markets where they could obtain two to four times as much as the FPC determined rate of 52¢ per thousand cubic feet for interstate sales.

Also, they argued that if gas prices were not deregulated, consumers would end up paying more than if deregulation were approved, because eventually they would have to turn to more expensive fuels when natural gas became short in supply.

Deregulation was supported by the Administration, which had argued that it was necessary in order to avert winter shortages of natural gas.

Opponents of deregulation said that the shortages predicted by the Administration had not occurred, or at least had not been acute. They argued that the only effect of deregulation would be to transfer money from consumers to producers, and that such shortages as had occurred had been artificially created by gas companies, holding back supplies in the expectation of higher prices after deregulation.

The Feb. 5 vote in favor of continuing regulation of major gas companies came as a surprise to Congressional observers. An earlier vote, on Feb. 3, had been interpreted as indicating fairly strong support for Rep. Krueger's deregulation measure. The House had voted then, 230-184, to allow Krueger's version of the bill to be presented immediately on the floor, instead of being kept in committee.

FPC hikes 'new' natural gas rate; court orders refund tie. The Federal Power Commission July 27, 1976 revamped the national rate structure for gas sold in the interstate market and ordered the largest rate increase in the agency's history for "new" gas.

The FPC ruled, 3-1, that producers of gas from wellheads discovered or committed to interstate pipelines after Jan. 1, 1975 could charge $1.42 per 1,000 cubic feet, nearly triple the existing 52-cent rate that the FPC had authorized Dec. 31, 1975. The ruling, which also raised other gas rates, was effective immediately.

A coalition of consumer-interest groups July 28 challenged the ruling in the U.S. Court of Appeals in Washington and won an immediate temporary stay of the rate hike. The court Aug. 9 issued a unanimous ruling that significantly modified the FPC action.

The court ruled that producers of new gas could raise their interstate prices immediately only if they also agreed to refund to pipeline customers any portions of the FPC's new rates that later might be found illegal. Producers refusing to include the refund provision in tariffs filed with the FPC were barred from raising their prices.

The consumer coalition, which included the Consumer Federation of America, the U.S. Conference of Mayors, a number of labor unions, the state of Minnesota and several state public utility commissions, hailed the court decision.

They had petitioned the court to prevent the new rates from becoming effective until the FPC made its order final. That process, which included a rehearing during which critics could ask the agency to reconsider its ruling, could take about 60 days.

The FPC had opposed the temporary stay, saying that the added cost to consumers resulting from the higher rates would not come due until the 60-day appeals process had ended. The court rejected this argument, noting that the obligation of consumers to pay the higher rates would begin immediately even though their bills would not reflect this at once.

The consumer group contended that the FPC rate increases represented "de facto deregulation of natural gas without authorization from Congress." According to the coalition, the new rates would cost consumers $2.2 billion annually, rather than $1.5 billion annually that the FPC had claimed. The agency had said that the new rates would raise the average consumer's gas bill $15.60 a year, or 5%-6%.

The only commissioner on the four-member FPC board to dissent from the majority ruling was Don S. Smith. He called the new rates "too high" and said they were not justified by the cost of producing gas. "The impact is excessive," Smith said.

The commissioners who supported higher rates said that the increases would provide producers with incentives to increase exploration for and development of new natural-gas deposits. Without added supplies, the FPC warned, the U.S. faced a gas shortage during the 1976-77 winter. Disincentives to consumption, which would thereby conserve short supplies, also were built into the rate structure through higher prices, the agency said.

The FPC had spent nearly a year considering the rate increases. In March, the FPC staff had submitted various rate-increase proposals for "new" gas that ranged from 56.45¢ to $1.69 per 1,000 cubic feet. Industry groups had suggested a rate as high as $2.85.

Instead of setting a single higher price for "new" gas, the commission revamped its two-tier rate structure that had existed since 1974 so that there were now four tiers.

The two-tier system had set different rates for "old" gas—that which was committed to the interstate market before Jan. 1, 1973 and "new" gas—that which was discovered or produced after Dec. 31, 1972.

In 1974, the FPC had set a national ceiling price of 42¢ per 1,000 cubic feet for "new" gas. The rate subsequently was raised to 50¢ and then to 52¢ at the end of 1975.

Prior to Jan. 1, 1976, rates for "old" gas had varied according to producing regions. In a Dec. 31, 1975 ruling, the commission had set a single national rate of 23.5¢ per 1,000 cubic feet for "old" gas, and had authorized a subsequent increase to 29.5¢ on July 1. The July rate increase for "old" gas was needed to compensate producers for increased taxes stemming from repeal of the depletion allowance in 1975, the FPC said.

Under the four-tier arrangement, "new" gas was sold at two separate prices rather than at one national rate and "old" gas was sold at two price levels. The concept was known as "vintaging."

For "new" gas discovered or committed to the interstate system after Jan. 1, 1973 but before Jan. 1, 1975, the rate was set at $1.01 per 1,000 cubic feet. The rate for "new" gas discovered or committed to interstate pipelines after Dec. 31, 1974 was raised to $1.42 per 1,000 cubic feet. The FPC also authorized an additional rate increase, beginning Oct. 1, of one cent per quarter for the life of the contract between producer and pipeline customer.

Gas in production before Jan. 1, 1973 would continue to sell at 29.5¢ per 1,000 cubic feet. Under what constituted the fourth tier, expiring contracts on "old" gas would be renewed at 52¢ per 1,000 cubic feet. (Old gas accounted for 75% of all gas sold in the interstate market.)

Natural gas sold within the state where it was produced was not subject to federal regulation. The unregulated price for intrastate gas averaged $1.54 per 1,000 cubic feet. Market factors helped set this price.

Natural gas provided about 30% or the nation's primary energy, according to the American Gas Association. Gas provided 41% of the energy used by industry and 42% of the energy used for residential and commercial purposes.

FPC reaffirms rate raise—The FPC voted to reaffirm its July decision on natural gas prices, the Wall Street Journal reported Nov. 3, 1976.

The commission also voted to modify another portion of the July rate-ruling by lowering the national rate on gas discovered or contracted for between Jan. 1, 1973 and Dec. 31, 1974 to 93¢ per 1,000 cubic feet. In its initial decision on the 1973–74 gas, the FPC had set the national rate at $1.01 per 1,000 cubic feet.

When the FPC's new votes on the higher gas rates were made public Nov. 5, the commission revealed that it had received 24 petitions to intervene in the case and 27 applications for rehearing, reconsideration or clarification from "gas producers, interstate pipelines, gas distributors, consumer-protection groups, 15 members of the House of Representatives, four senators and several trade associations."

After a legal challenge to the July rate ruling was filed by several groups representing consumer interests, the FPC Sept. 2 agreed to reconsider its initial decision.

During hearings Sept. 16–17, consumer spokesmen reiterated their charge that the FPC had underestimated the cost impact of the new higher rates. They also contended that the FPC had based its decision on unreliable information about gas supplies collected by the gas industry. Opponents of the higher rates also said that the commission was permitting the industry to reap windfall profits by allowing excessive costs in the rate base for such items as federal income tax. (For example, the FPC had included an allowance of 43¢ per 1,000 cubic feet for payment of federal income taxes at the maximum rate of 48%. If actual, rather than assumed costs had been calculated, the rate increase would have been 84¢ per 1,000 cubic feet, rather than $1.42, some critics contended.)

Dissatisfied spokesmen for the gas industry urged the FPC to consider setting even higher rates so that producers would have additional incentives to undertake exploration and development of new gas supplies.

After evaluating the testimony and new information supplied by producers and pipeline companies, the FPC Oct. 20 conceded that its cost estimate had been too low. According to the new data, the commission said, consumers would pay an additional $2.05 billion annually as a result of the rate increases authorized in July.

To lessen that impact, the commission voted to tighten its definition of newly discovered gas that would qualify for the $1.42 rate. Consumers had argued that under the existing definition, producers could convert "old" gas, selling at a lower rate, into "new" gas by drilling shallow wells in known fields.

The definition was tightened by reclassifying "recompleted" gas—that which was taken from different levels of an existing well—as "old" gas, regardless of when production began.

At its Oct. 20 meeting, the commission also agreed to consider a proposal to reduce the rate on gas discovered or newly sold between Jan. 1, 1973 and Dec. 31, 1974 to 93¢ per 1,000 cubic feet from the $1.01 rate set in July. (Formal approval of the proposal was announced Nov. 5.)

When the effects of the restricted definition and rate reduction were incorporated in the upwardly revised cost calculation, the FPC said the net impact of the overall rate action on consumers would be about the same as the original estimate. Consumers were expected to pay an additional $1.59 billion in gas bills, instead of $1.5 billion as originally forecast.

Consumer groups were not mollified by the FPC action Oct. 20, saying that some utilities had estimated that the actual impact of the higher rates would be as much as $4 billion annually.

Rep. John E. Moss (D, Calif.), chairman of the House Commerce Committee's subcommittee on oversight and investigations, charged Oct. 30, that the FPC was not justified in tripling the rate for new gas flowing after Jan. 1, 1975. Moss based his charge on a subcommittee staff report that accused the FPC of overestimating exploration costs in determining rate increases.

The report said that if the commission had used reports filed by producers rather than data supplied by the American Gas Association (AGA), the FPC would have raised the rate to about $1, rather than to $1.42 per 1,000 cubic feet.

The Natural Gas Act, passed by Congress in 1938, required the commission to base prices on actual costs. The single most important cost in determining rate increases was the projected productivity rate on successful drillings. If productivity estimates were low, then costs would be high and rates would be increased. The FPC, in its July decision, had contended that a declining productivity rate was the "cornerstone" of its decision to triple the national rate on new gas.

The subcommittee report contended that the AGA information on exploration costs was less reliable than producers' reports because producers were required to submit sworn statements to the FPC. (The FPC had adopted the stiffer standards in 1975 because it distrusted AGA data.)

Moss had threatened Sept. 1 to impeach FPC chairman Richard L. Dunham and the two other commissioners, John H. Holloman 3rd and James G. Watt, who had voted the record rate increase. Moss said that the majority had disregarded the Natural Gas Act's intent, allowing the FPC to set rates based on a "just and reasonable" return to producers. "No record anywhere," Moss said, showed that the majority had made "any determination of cost" in setting the higher rates.

Reserves fell in '75. The FPC Jan. 28, 1977 released a staff report showing that domestic natural gas reserves committed for sale to interstate gas pipeline companies had declined in 1975 for the eighth consecutive year.According to the report, reserves had declined by 13.7 trillion cubic feet, or 11.4%, to 106.8 trillion cubic feet from 120.5 trillion cubic feet.

The FPC in December 1976 had said that the average price paid by interstate pipeline companies for all natural gas in 1975 was 42.3¢ per thousand cubic feet, 13.5¢ higher than the 1974 price of 28.8¢. The average price for domestically produced gas was 36.17¢ per thousand cubic feet, up 9.28¢ from 1974. Imports from Canada were sold at an average price of $1.20 per thousand cubic feet, up 65.52¢ from 1974. There were no imports from Mexico in 1975.

The pipeline companies bought gas from producers in 24 states and Canada in 1975, the FPC said. The five states that produced 88% of all gas sold on the interstate market in 1975 were, in order of volume of interstate sales, Louisiana, Texas, Oklahoma, New Mexico and Kansas.

FPC acts to ease fuel shortages. The Federal Power Commission Jan. 14, 1977 unanimously approved emergency sales of natural gas by the Houston Pipe Line Co. to two pipelines serving states in the East and on the Gulf Coast. With the decision, the commission in effect authorized extension of emergency sales beyond the 60-day limit that had long been the standard for such sales.

(Emergency sales were permitted at prices of up to $2.50 per 1,000 cubic feet of gas. In normal practice, there was a price ceiling of $1.42 per 1,000 cubic feet on sales of new gas by FPC-regulated companies.)

Under a previous ruling, Houston Pipe Line had supplied the same two companies with emergency supplies in November and December 1976, selling up to 150 million cubic feet of gas daily to Transcontinental Pipe Line Corp., a division of Transco Cos., and 85 million cubic feet a day to United Gas Pipe Line Co. at $2.15 per 1,000 cubic feet. Because of continued shortages, the companies had asked the FPC to approve another round of emergency sales, but with the amounts of gas switched. The FPC Jan. 5 had authorized the sales as extensions of the November-December sales. But Houston Pipe Line had refused to make deliveries under that interpretation because the company was not currently regulated by the FPC and feared that extended interstate sales would force it under FPC jurisdiction.

The Jan. 14 decision met the company's request that the new sales be considered brand-new emergency sales. The FPC, as commission members openly acknowledged, thus opened the door to allowing an unregulated pipeline company—one that sold gas only in the state in which it was produced—to make a series of 60-day emergency interstate sales by switching among various interstate pipelines.

The precedent-setting action came as already low heating fuel supplies around the nation were being taxed by overconsumption as the result of the coldest winter weather in five years. Widespread concern about the fuel situation was demonstrated by the fact that more than 200 industry executives and government officials attended an unusual all-day hearing on the Houston Pipe Line request. Moreover, the full commission was present at the hearing, which normally would have been conducted by administrative law judges.

Natural gas supplies had already been depleted during an unusually cold fall in 1976. Industry sources estimated that residential gas usage alone had been 37.3% higher in November 1976 than in the previous November.

(American Petroleum Institute figures, reported Jan. 14, showed that nationwide inventories of heating oil and other distillates had fallen by three million barrels during the first week in January despite record output by U.S. refineries and a 62.7% week-to-week spurt in imports.)

The FPC said that as of Jan. 15, pipelines serving the eastern half of the country had 317 billion cubic feet of natural gas stored, compared with 541 billion cubic feet one year earlier. Reserves had plummeted by 127 billion cubic feet in the first two weeks of January alone. Carter Administration officials said the pipelines had already exhausted gas reserves slated for use in February and March.

Natural gas normally constituted roughly half the fuel for U.S. industrial production and about 30% of the nation's total energy supply.

In an attempt to keep up with record demand (22% higher for heating oil than the previous winter, for instance), U.S. refineries had turned out a record 1.9 million barrels of residual fuel a day during the week that ended Jan. 14, compared with slightly under 1.9 million barrels the week before and less than 1.4 million barrels a day a year before. Residual fuel reserves were reported holding steady at 72.5 million barrels during the same week, equal to levels of the preceding week and just under the 75.2-million barrel reserve of January 1975.

(Residual fuel was used mainly for industrial purposes, for generating electricity and for heating office and apartment buildings.) Also during the week that ended Jan. 14, oil companies imported 460,000 barrels of heating oil per day, an increase of 25% over the previous week and nearly double the level of the comparable period in 1975.

Sustained cold weather in the South, where temperatures had been more than 40% below averages of previous years, had caught many gas suppliers unprepared for increased heating demands. (Temperatures in Shreveport, La., for instance, were reported to be averaging 84% below normal.) The same was true in the Northeast, where December 1976 temperatures had averaged 10–11 degrees below normal, and the Great Plains and Midwestern states, where January temperatures were averaging 12–19 degrees below normal. Ice had closed the Mississippi River to traffic below St. Louis for the first time in 30 years.

The short supply had forced large gas and power suppliers to curtail industrial deliveries throughout those areas, forcing numerous factories, schools and other nonresidential facilities to close for short periods of time. More than 50,000 employes were laid off by auto manufacturers Jan. 17 due to plant closings. Some 20,000 had not yet been recalled two days later.

Thousands of other workers had been laid off in the textile industry in the South and in the steel industry in Pennsylvania and Alabama. Other effects of fuel curtailments included electric power outages, brownouts and voltage reductions in many areas.

Meanwhile, heating fuel prices were rising. Figures for November 1976 showed fuel prices were as much a four-cents-a-gallon higher than peak levels during the winter of 1975–76. Natural gas prices were up 18.7% and electricity costs were up 7.1%. (Over 60% of U.S. homes were heated with gas, about 24% with oil and 12% with electricity.)

As a result of increased demand, gas and electric companies in the South, Midwest and East were reporting new record business and earnings each week. On Wall Street, the Dow Jones average of utility stock prices rose to a three-year high Jan. 24.

The FPC Jan. 19 approved emergency purchases of natural gas by a unit of Columbia Gas System Inc. from a Canadian company for delivery to areas in West Virginia, Kentucky, Maryland, New Jersey, Virginia, New York, Pennsylvania and Ohio. The Canadian National Energy Board had authorized emergency shipment of up to 250 million cubic feet of natural gas a day for two months. The two-month total of 15 billion cubic feet was to be over and above the one trillion cubic feet of Canadian gas usually exported to the U.S. each year under long-term export arrangements. (The Canadian energy board also had authorized the emergency export of 50,000 barrels of heavy fuel oil to the Detroit Edison Co. of Detroit, where several days of near-zero temperatures had forced the utility to pulverize manually coal that was too frozen for its generating plant's supply conveyors.)

Emergency conservation, fuel-sharing and delivery-curtailment programs allowed restoration of normal electric service in the Midwest, South and East by Jan. 19. Supplies of natural gas in those areas also had been increased.

Moreover, several plants that had closed down Jan. 17 had been able to resume production. But weather-induced layoffs and school and transportation disruptions were still widespread as the federal government prepared further emergency measures to cope with record demands for natural gas during a prolonged cold wave that affected states as far south as Florida.

Carter Administration acts. The new President, Jimmy Carter, called on "all Americans Jan. 21, 1977 to set their thermostats at 65° during the day and lower at night. He did this after his energy adviser, James R. Schlesinger, had met for two hours with more than two dozen executives of natural gas pipelines to urge them to share natural gas voluntarily.

Five days later, on Jan. 26, Carter sent Congress a proposal for emergency legislation to give the federal government temporary authority to reallocate gas from one interstate pipeline to another to relieve shortages. The measure would not allow gas to be taken from intrastate sellers. It would, however, ease the current Federal Power Commission (FPC) restrictions on higher-priced emergency purchases of natural gas by interstate pipelines in intrastate markets. It also would provide so-called interim authority for intrastate pipelines to serve as conduits between interstate pipelines without bringing the intrastate carriers under federal jurisdiction, a status that they feared.

Carter the same day repeated his call for lowered thermostats, saying "half the shortages" could be made up through conservation in homes and offices.

In other federal government responses to what Carter referred to as an energy "crisis":

■ The FPC Jan. 20 issued new emergency authorizations to several companies to reallocate natural gas to hard-hit areas, to use and make purchases of other types of fuel on a temporary basis and to allow intrastate gas into the interstate market. Among the orders was a 60-day authorization for East Tennessee Natural Gas Co., which served Tennessee and Virginia, to purchase 1.5 billion cubic feet of gas a day from producers in Louisiana and Texas at prices ranging from $2.25 to $2.37 per thousand cubic feet, well above the $1.42 ceiling on regulated gas. (But the FPC Jan. 21 refused to approve an emergency purchase by Tennessee Natural Gas Lines Inc. of Nashville from the intrastate market on the ground that the prices involved, $2.50 and $2.75 per thousand cubic feet, were unjustified.) Another order authorized the Transcontinental Pipe Line Corp. of Houston to provide additional supplies to customers by drawing "base gas" from one of its natural gas storage fields. (A certain amount of gas, "base gas," was normally required to be kept in storage fields to ensure sufficient pressure to get working gas out of the fields.)

■ The FPC Jan. 20 also required 29 major pipeline companies to report daily any changes in supplies, curtailment of deliveries or operating problems in order to help the agency act quickly to deal with emergency situations.

■ The Interstate Commerce Commission, according to press reports Jan. 24, issued orders easing trucking regulations and giving priority on railroads to movement of essential commodities and energy-related items.

■ The Interior Department Jan. 24 authorized natural gas producers on federal lands, including offshore oil fields, to exceed usual production limits imposed for conservation reasons.

■ The Federal Energy Administration (FEA) Jan. 24 ordered oil refineries in four north-central states to produce more home heating oil for customers in Michigan, Wisconsin, Minnesota and North Dakota. The FEA also revised its allocation rules to assure residential and nonindustrial users of propane, mainly in the South, of adequate supplies of that fuel.

Meanwhile, varying states of emergency had been declared in Ohio, Pennsylvania, New York, New Jersey and Florida. Ohio Gov. James A. Rhodes Jan. 23 called for voluntary conservation of fuel and ordered the legislature into special session Jan. 26. Earlier in the week, on Jan. 20, Rhodes had ordered all schools to close and all businesses to observe a 40-hour week in a 24-county area around Dayton, where sustained sub-zero temperatures—21 degrees below zero at one point—had caused severe natural gas shortages. Rhodes rescinded the order the next day, saying he lacked the authority to enforce it.

In Pennsylvania, Gov. Milton Shapp Jan. 18 declared a state of emergency and Jan. 26 ordered all the state's public and private schools closed until at least Jan. 31. Pennsylvania officials estimated that 51,000 workers in the state had already been laid off due to curtailments of fuel deliveries to industrial plants. Shortages in Ohio and Southwestern Pennsylvania were compounded by shipping problems on frozen rivers.

Schools also remained closed in much of Kentucky and Tennessee and fuel cutbacks were keeping thousands of workers idle in Tennessee, Georgia, Mississippi, New York and Ohio. The FPC estimated Jan. 22 that from 200,000 to 300,000 workers had been laid off nationwide as a result of gas shortages.

Blizzard strikes gas-short states. A fierce blizzard with 60 mile-per-hour winds swept across the already frozen Midwest into the East Jan. 28–29, 1977.

Full states of emergency were declared in New York and New Jersey Jan. 27. In Ohio, one of the other hardest-hit states, Gov. James A. Rhodes Jan. 29 led a 15-minute noontime prayer service in the state capitol, asking God to end the most severe fuel crisis in Ohio history.

President Carter flew into Pittsburgh Jan. 30 to see the effects of the blizzard and warned the nation of the possibility of a "permanent, very serious energy shortage." (Emergency legislation requested by the President to deal with the current regional shortages was passed by both houses of Congress and signed Feb. 2.)

Despite new emergency infusions of natural gas from Canada, the Midwest was still suffering severe fuel shortages because ice on all the major river systems in the central U.S. had stranded hundreds of barges loaded with fuel oil and other badly needed items.

Conditions in the East were only slightly better: a major Texas supplier, Transcontinental Gas Pipe Line Corp., Jan. 28 asked its 69 customer utilities along the East Coast to end immediately deliveries to "non-essential" customers to insure that enough gas remained available for home heating.

The gas shortage actually eased slightly—but only temporarily—Jan. 30, when pipelines suffering the most critical shortages managed to stem the drain on their reserves for the first time in more than two weeks. One reason cited by pipeline officials for the easing of the crisis was the orders in 11 states that schools and factories be closed. Another was that the blizzard of the days before had not struck farther south than New York and Pennsylvania. Therefore, demand in the South had not outstripped supplies.

■ The California Public Utilities Commission Feb. 1 imposed tough restrictions on the use of natural gas in the state in order to free emergency supplies for shipment east on a loan basis.

FPC approves further gas imports—The Federal Power Commission (FPC) Jan. 27 authorized Columbia Gas of Pennsylvania Inc. to import approximately 250 million cubic feet of liquefied natural gas from Canada at a final cost to customers of about $5 per thousand cubic feet. The Canadian National Energy Board approved the transaction the next day.

Law waived to ease gas transport—Treasury Secretary Michael Blumenthal Feb. 2 granted a rare waiver of the Jones Act to allow Columbia Gas System Inc. to make two shipments of liquefied natural gas to Massachusetts from Alaska on a foreign-registered vessel. The Jones Act required that U.S. ships be used for transport of goods between U.S. ports. It could be waived only for reasons of national defense.

Blumenthal Feb. 3 again waived the act to authorize five shipments of liquefied propane gas by Tropigas International Corp. to Virginia from Texas on a foreign tanker.

Company buys own gas field—Libbey-Owens-Ford Co., one of the many companies that had been forced to lay off workers because of natural-gas curtailments, acquired its own gas field Jan. 27 from Damson Oil Corp. for $4 million. The Ohio-based company had been actively searching for additional energy supplies and had 60 gas wells under operation in Ohio alone.

Emergency gas bill signed. President Carter Feb. 2 signed the Emergency Natural Gas Act of 1977, which was aimed at keeping natural gas supplies flowing to homes and essential services by temporarily authorizing reallocation of interstate gas from surplus areas to shortage areas. The bill also allowed the President to lift, until July 31, price ceilings on interstate gas.

Carter had submitted the bill Jan. 26. Congress gave it expedited consideration, with final passage coming Feb. 2 when the Senate approved the bill by voice vote and the House, 336 to 82. Neither Carter nor congressional leaders claimed the bill would provide an answer to long-term energy problems or even do much towards helping factories closed by weather-induced gas shortages to reopen. However, the bill attempted to ensure that heating fuel would be available to residential users and such places as hospitals and small stores in the Northeast and midwestern states particularly hard-hit by the winter.

As finally passed, the bill followed Carter's proposal. But when it had first emerged from the House, it contained a price ceiling on interstate gas sales. The ceiling had been backed by representatives from Texas who feared that out-of-state purchasers not bound by a ceiling would bid up prices. What would result, they said, would be shortages or "potentially desperately high [price] levels" on the intrastate natural gas market. Texas was a major producer of natural gas.

Although the House provision would have allowed the President to raise the price ceiling if he determined it was necessary, the White House objected that it would "reduce [the Administration's] flexibility." The provision was dropped in House-Senate conference committee. However, language was put in the bill to prevent a rise in interstate prices from triggering renegotiation clauses in intrastate natural gas contracts.

The authority in the bill enabling the President to reallocate supplies of interstate natural gas to areas in need would expire April 30.

Dunham handles emergency law—President Carter Feb. 2 designated FPC Chairman Richard Dunham to administer the new Emergency Natural Gas Act.

Dunham Feb. 3 issued his first order under the terms of the act, requiring that 150 million cubic feet of gas be moved daily from western states to the Transcontinental Gas Pipe Line Corp. system, which extended from Texas to the New York City area. In addition to ordering several other shifts in supplies over the next few days, Dunham Feb. 5 took action to free supplies of gas currently being used to generate electricity. He announced that the FPC would permit electric utilities in Texas temporarily to link their systems with any other electric companies in or outside the state without coming under FPC jurisdiction.

In a related action, the FPC Feb. 4 authorized imports of natural gas from Mexico and Algeria. The price of the Algerian gas would be about $3.34 per thousand cubic feet delivered. The Mexican imports, up to 40 million cubic feet a day for 60 days, according to the Mexican government, would cost approximately $2.25 per thousand cubic feet. (Mexico, which at one time had sold substantial volumes of natural gas to the U.S., had all but halted exports in recent years because of growing demand within Mexico.)

Gas continued to flow in from Canada. The FPC Feb. 5 approved an arrangement whereby 50 million cubic feet of gas would be shipped from Canada to the U.S. in exchange for 600 megawatts of electric power per hour.

The emergency actions were expected to prevent cutoffs of gas to residences during the winter, but Dunham Feb. 3 said the measures alone would probably not end the shortages that had caused so many factories to close. These shortages were expected to continue well into the summer in some areas, as pipeline companies continued to curtail deliveries in order to replenish their reserves. Moreover, industry officials Feb. 3 pointed out that the new gas act did not compel producers to sell to the interstate market. While gas supplies could be shifted from one pipeline to another, interstate supplies were limited everywhere, they added.

Gas consumption surged—FPC statistics reported Feb. 9 showed that natural gas customers had drawn about 100 billion cubic feet of gas a day from interstate pipelines in December 1976 and January 1977. Normal usage, the FPC said, was about 57 billion cubic feet a day. The commission had said Feb. 2 that the volume of natural gas in storage for interstate sales had dropped 19% between Jan. 15, 1976 and Jan. 15, 1977.

Inventories of heating fuel and other petroleum distillates dropped to 141.1 million barrels nationwide during the week that ended Feb. 4, according to the American Petroleum Institute Feb. 9. That was a drop of about 10.2 million barrels from the previous week and down 23.7 million barrels from the comparable week in 1976.

Institute figures, reported Feb. 8, showed that U.S. oil imports had averaged more than 8.7 million barrels a day during January, an increase of more than 30% over January 1976 and the highest monthly import average in the nation's history. January imports had just about matched U.S. domestic oil production, which usually totaled about 8.8 million barrels a day.

The Federal Energy Administration announced Feb. 4 that a record average of 5.2 million barrels of heating oil had been burned daily during January, 21% more than during January 1976.

Industrial output down 1% in January—Production in the nation's factories, mines and utilities declined a seasonally adjusted 1% in January, the worst drop since February 1975, according to the Federal Reserve Board Feb. 15. The shortage of natural gas was blamed.

Gas-shortage crisis abating. Several large natural gas suppliers in the East and Midwest during the week of Feb. 13–19 resumed industrial deliveries that had been curtailed because of the weather-induced high demand. Billions of cubic feet of natural gas continued to flow into the U.S. from Canada during the week to help replenish stocks depleted by what the National Weather Service Feb. 16 said had been the coldest winter "since the founding of the Republic."

An estimated one million workers idled earlier in the winter because of natural-gas shortages had returned to their jobs, the Federal Energy Administration (FEA) reported Feb. 18.

Gas withholding charges denied. Sen. James B. Pearson (R, Kan.) Jan. 28, 1977 denied the "numerous allegations [made over the past several years] that large volumes of natural gas, otherwise available for sale and delivery in the interstate marketplace, are being withheld." In his statement, as it appeared in the Congressional record, Pearson said:

"... The individuals who have made these accusations claim that the natural gas industry is consciously withholding these available supplies of natural gas in order to accentuate shortages being experienced in consuming States. The goal of this so-called conspiracy is the deregulation of Federal price controls over wellhead sales of natural gas in interstate commerce.

"Withholding, according to those who have identified this practice, is accomplished in various ways. The natural gas industry is accused of underreporting proved reserves of natural gas in order to create momentum for incentives that will improve the domestic reserve situation. Producers are accused of intentionally refusing to deliver natural gas to their pipeline customers. This producer action, it is alleged, is in direct violation of contractual agreements and FPC certificates of public convenience and necessity which authorizes producer sales. Government agencies charged with responsibility for policing the withholding situation are accused of incompetence and, in certain instances, conspiring with the natural gas industry by promoting withholding.

"Mr. President, I refute these accusations and allegations. These charges are void of any substantial proof. They are made by those who claim to be protectors of the public interest. They are made by individuals who refuse to accept the economic logic of and national need for deregulation of natural gas sold in interstate commerce. They are made by those who must now explain why a million employees are out of work, thousands of schools are closed, and millions of residences are threatened with loss of heat.

"The withholding theory is nothing more than bad fiction. The General Accounting Office and the National Research Council have concluded that industry reserve estimates are very close in comparison to Government estitmates made by the U.S. Geological Survey, Federal Energy Administration, and Federal Power Commission. The dispute with respect to reserve estimates is caused by difference in terminology and variance in frames of reference rather than willful underreporting.

"Investigations have found only isolated incidents of failure by producers to meet certificated delivery requirements. These exceptions to the rule are being corrected in the agencies and in the courts.

"The FPC has exercised its information gathering authority to the maximum extent practicable under existing law with respect to natural gas reserves. That agency has consistently asked Congress for expansion of its information gathering authority for the past 20 years.

"Mr. President, the natural gas shortage is real. It cannot be hidden by the rhetoric of those who claim it is a conspiracy to achieve price deregulation. The enormous economic disruption and human suffering now being experienced are not due to withholding or to the record cold weather gripping much of the Nation. Natural gas shortages which have been with us for 5 years have precipitated the immediate crisis. These shortages will not go away until we face up to our responsibility to the American public and provide a meaningful, long-term solution."

Data lacking—Sen. John A. Durkin (D, N.H.) told the Senate Jan. 28, 1977 that "the only sad, true knowledge we do have [in the controversy over allegations of withholding gas production] is that we do not know how much gas is out there—in our reservoirs, offshore and in privately or federally owned and managed onshore U.S. gas fields." Durkin said:

"Over the past 6 years, numerous studies of offshore gas fields, shut-in wells, and other Federal gas reserves have been made. In more than a couple of instances, there have been glaring discrepancies to the tune of trillions of cubic feet of natural gas, between the figures supplied to the American Gas Association by the gas producers, and the results of the Government studies conducted by the FPC staff or the U.S. Geological Service. . . ."

Industry accused. Rep. William H. Harsha (R, Ohio) told the House Feb. 4, 1977 of "shocking revelations made during hearings being conducted by the Interstate & Foreign Commerce Subcommittees on Oversight & Investigations and Energy & Power":

"First of all, there is considerable reason to doubt the validity of natural gas reserve figures reported by the American Gas Association to the Federal Power Commission. AGA field-by-field estimates for offshore south Louisiana were 37 percent less than U.S. Geological Survey data for the same fields.

"The USGS studies were compared with individual gas field estimates which had to be obtained under subpena from the AGA.

"Earlier last year, a staff report from the Federal Trade Commission recommended to the Interstate and Foreign Commerce Committee that the AGA and 11 major oil companies be sued on charges of consistently underreporting the Nation's natural gas reserves. The committee subpena the records of seven uncooperative natural gas producers. Federal Trade Commission investigators learned from 'proved reserve ledgers' that the AGA figures on reserves were 24 percent less than the companies' estimates of their proved reserves.

"This had tremendous significance for American consumers, since the FPC regulates the price of natural gas and bases its decisions on rate increase requests partly on estimates of reserves.

"There is more: More than half of the Federal leases in the Gulf of Mexico were found, during a random sample, to be producing on an average 21 percent lower than producers' own estimates of 'reasonable' and 'efficient' production.

"Two producers were discovered to have failed to undertake work projects necessary to maintain deliverability from a large gas field in south Louisiana. Furthermore, it failed to establish production from untapped reservoirs.

"In November of last year, the Commerce Subcommittee on Energy and Power uncovered evidence which showed that natural gas producers had for years been failing to deliver gas to their pipeline customers in accordance with their contracts.

"The committee's hearings have disclosed further evidence that producers are withholding natural gas in anticipation of higher prices. An internal document from Getty Oil Co. projected a 281 percent rise in the price of natural gas should deregulation of old and new gas be approved. . . . "

Utilities blamed. Sen. Hubert H. Humphrey (D, Minn.) Jan. 31, 1977 blamed the gas utility companies for the speedy depletion of the U.S.' natural gas resources. He told the Senate:

"Our gas utilities have treated non-reweable gas though it were electricity. They oversold it to consumers, to businesses and to industry with quantity discounts, rebates and subsidized equipment and installation costs. But unlike electricity—which can be produced by a variety of fuels, even including water—natural gas is not inexhaustible.

"And that is the other side of the coin. Out most attractive, easily reached gas deposits have already been discovered and exploited. Major finds, like Alaska or Gulf Coast geopressure zones, will be made, but infrequently.

"Utility overselling and dwindling new gas deposits—a deadly combination—are at the root of our present situation. . . ."

Gas resources not used. A study of four Gulf of Mexico gas fields leased by 10 companies appeared to reinforce suspicions that natural gas supplies might have been withheld deliberately from customers during the 1976–77 winter gas shortage.

The gas-production study, prepared by representatives of the Interior Department, the Federal Power Commission (FPC), the Senate Judiciary Committee's antitrust and monopoly subcommittee and a private consulting firm in Washington, found that at the end of 1976, 225 gas reservoirs with proven reserves totaling 981.5 billion cubic feet were not being tapped. Those reserves amounted to nearly one half of the total needed to offset the winter shortage, previously estimated at between two trillion and 2.5 trillion cubic feet.

The 10 companies that held the leases were Union Oil Co. of California; Shell Oil Co.; Gulf Oil Corp.; Amoco Production Co., a subsidiary of Standard Oil Co. of Indiana; Continental Oil Co.; Exxon

Total Expenditures by State By All Gas Utility Customers[1]

[Dollar amounts in millions]

State	1974 expenditures[2]	1980 deregulation expenditures[3 4]	Percent increase	Annual natural gas deregulation rate of inflation[5]
Total	$19, 088. 2	$43, 444. 1	127. 6	25. 5
New England:				
Connecticut	396. 7	600. 9	51. 5	10. 3
Maine	267. 8	379. 0	41. 5	8. 3
Massachusetts	364. 3	551. 8	51. 5	10. 3
New Hampshire	366. 3	607. 9	65. 9	13. 2
Rhode Island	337. 4	519. 6	54. 0	10. 8
Vermont	327. 7	403. 1	23. 0	4. 6
Middle Atlantic:				
New Jersey	275. 3	463. 0	68. 2	13. 6
New York	247. 8	427. 8	72. 6	14. 5
Pennsylvania	404. 7	802. 9	98. 4	19. 7
East North Central:				
Illinois	380. 3	822. 2	116. 2	23. 2
Indiana	392. 8	908. 2	131. 2	26. 2
Michigan	471. 9	985. 8	108. 9	21. 8
Ohio	411. 6	894. 5	117. 3	23. 5
Wisconsin	430. 0	939. 4	118. 5	23. 7
West North Central:				
Iowa	383. 3	946. 7	147. 0	29. 4
Kansas	375. 7	1, 240. 6	230. 2	46. 0
Minnesota	440. 2	988. 2	124. 5	24. 9
Missouri	286. 1	673. 9	135. 6	27. 1
Nebraska	367. 3	964. 7	162. 6	32. 5
North Dakota	337. 6	771. 2	128. 4	25. 7
South Dakota	304. 5	721. 8	137. 0	27. 4
South Atlantic:				
Delaware	369. 4	698. 3	89. 0	17. 8
District of Columbia	276. 4	477. 1	72. 6	14. 5
Florida	349. 0	806. 9	131. 2	26. 2
Georgia	318. 6	760. 4	138. 7	27. 7
Maryland	300. 3	551. 3	83. 6	16. 7
North Carolina	450. 4	959. 1	112. 9	22. 6
South Carolina	502. 2	1, 133. 8	125. 8	25. 2
Virginia	336. 7	652. 7	93. 8	18. 8
West Virginia	368. 4	873. 3	137. 1	27. 4
East South Central:				
Alabama	327. 4	814. 6	148. 8	29. 8
Kentucky	277. 3	653. 2	135. 6	27. 1
Mississippi	322. 8	869. 7	169. 3	33. 9
Tennessee	384. 1	984. 8	156. 4	31. 3
West South Central:				
Arkansas	331. 7	995. 2	200. 0	40. 0
Louisiana	298. 3	972. 2	225. 9	45. 2
Oklahoma	309. 7	961. 2	210. 4	42. 1
Texas	459. 5	1, 296. 2	182. 1	36. 4
Mountain:				
Arizona	261. 5	636. 7	143. 5	28. 7
Colorado	284. 8	781. 1	174. 2	34. 9
Idaho	504. 0	1, 100. 9	118. 4	23. 7
Montana	353. 3	920. 3	160. 5	32. 1
Nevada	552. 7	1, 310. 3	137. 1	27. 4
New Mexico	343. 2	977. 6	184. 8	37. 0
Utah	258. 4	766. 7	196. 7	39. 3
Wyoming	379. 4	1, 323. 9	248. 9	49. 8
Pacific:				
Alaska	778. 0	2, 153. 7	176. 8	35. 4
California	243. 4	569. 8	134. 1	26. 8
Hawaii	543. 0	661. 1	21. 8	4. 4
Oregon	479. 0	957. 9	100. 0	20. 0
Washington	556. 0	1, 208. 2	117. 3	23. 5

[1] Excludes customers purchasing for resale and sales for resale.
[2] Last full year for which data available; AGA, "Gas Facts," 1975, pp. 64 and 112.
[3] Assumes $1.70 per Mcf is the average deregulated wellhead price; assumes 1974 consumption volumes.
[4] Based on definitions as provided for by the Krueger amendment.
[5] Assumes 1974 expenditures equal 1975 expenditures; assumes 5-yr deregulation period.

Corp.; Texaco Inc.; Tenneco Oil Co., a unit of Tenneco Corp.; Pennzoil Co., and Phillips Petroleum Co.

The study found that production had fallen sharply over the previous two years in three of the four fields. Within individual fields, the study said, several wells had been shut down for no apparent reason. In all four fields, the production rate at the end of 1976 was 64% below the maximum capability. The study showed further that only 19 of 188 non-producing reservoirs identified by the FPC in 1974 had since been brought into production.

Interior Secretary Cecil Andrus said Feb. 17, 1977 that the study, in the wake of accusations that the petroleum industry was withholding gas from production, had raised enough questions and turned up enough discrepancies in company figures to necessitate a thorough probe of all Gulf of Mexico gas production. However, Andrus stopped short of accusing the producing companies of deliberately curtailing production to await a higher price for gas, saying, "Today isn't the day to point a finger of blame."

"But, in my opinion," he continued, "had [the Ford Administration] taken the necessary steps six months or a year ago to urge accelerated production, we might have helped out on this winter's gas problems." (One of the study group members said much of the information in the report came from documents previously in the possession of the U.S. Geological Survey, a branch of the Interior Department, according to a report Feb. 22.)

An earlier FPC study, reported Feb. 11, had concluded that only as much as 300 billion cubic feet of gas from Gulf of Mexico fields had been withheld by producers, enough for about a three-day supply for the nation at current consumption rates. The FPC report said that most of the gas had been withheld for what were described as sound economic reasons.

But a private study commissioned by leading gas utilities alleged that producers had failed to bring to market proven reserves in the Gulf of Mexico that would have been sufficient to offset almost all of the increased winter demand, according to a report Feb. 26. A confidential "executive summary" of the report, inadvertently sent to a New York State Assembly committee, said that of 105 offshore Louisiana leases that had been held by producers for at least five years, only 35 had been brought into production to date.

In a related development, staff investigators for the oversight and investigations subcommittee of the House Commerce Committee Feb. 22 accused Texaco Oil Co. of "currently holding in abeyance" the development of close to 530 billion cubic feet of proven reserves in the Gulf of Mexico because of its "desire to maximize its profits."

A Texaco spokesman Feb. 23 denied the gas was being withheld solely for economic reasons. Richard Palmer, senior vice president in charge of Texaco's western hemisphere operations, told the committee that the U.S. only had a 10-year supply of natural gas remaining and that premature development of the Gulf of Mexico reserves would amount to a "drain America now policy."

A subcommittee report released Feb. 23 accused Gulf Oil Corp. of reneging on a contract to deliver up to 625 million cubic feet of natural gas a day to the Texas Eastern Transmission Corp., which served 16 southern and northeastern states. The report alleged that Gulf and Texas Eastern, beginning in 1971, "may have engaged in a conspiracy in violation of federal law to withhold gas from interstate markets until the price increased."

Gulf chairman Jerry McAfee immediately labeled the charge "absurd."

Behind the allegations, press commentators noted, lay the suspicion that producers were holding back in the hope that the federal government would deregulate the price of natural gas on the interstate market, currently set by the FPC at $1.44 per thousand cubic feet. (The price was frequently quoted at $1.42, but the FPC order setting that price in July 1976 had provided for a one-cent increase per quarter.)

Since the onset of the winter gas crisis, there had been many press reports about onshore producers who were waiting for deregulation before they would tap their reserves. Science News Feb. 26 reported that domestic production of natural gas had dropped nearly 12% since 1973. Exploratory drilling had slowed to the point where the U.S. was burning gas at twice the rate at which new sources were being discovered, the magazine said.

U.S. Natural Gas Production & Producing Wells, By States—1975

Area, State	Production (million cubic feet)	Producing wells (12–31)	Production per year (million cubic feet)
Appalachian			
Kentucky	60, 511	7, 386	8. 2
Maryland	93	15	6. 2
New York	7, 628	900	8. 5
Ohio	84, 960	10, 382	8. 2
Pensylvania	84, 676	17, 500	4. 8
Tennessee	27	5	5. 4
Virginia	6, 723	186	36. 1
West Virginia	154, 484	21, 700	7. 1
Subtotal	399, 102	58, 074	6. 9
Mid-Continent			
Arkansas	116, 237	1, 128	103. 0
Kansas	843, 625	8, 865	95. 2
Missouri	30	19	1. 6
Nebraska	2, 565	3	855. 0
Oklahoma	1, 605, 410	9, 769	164. 3
Subtotal	2, 567, 867	19, 784	129. 8
Southeast			
Alabama	37, 814	9	4, 201. 5
Florida	44, 383	NA	NA
Mississippi	74, 345	248	299. 8
Subtotal	156, 542	257	NA
Midwest			
Illinois	1, 440	41	35. 1
Indiana	346	478	. 7
Michigan	102, 113	209	488. 6
Subtotal	103, 899	728	142. 7
Rockies			
Arizona	208	1	208. 0
Colorado	171, 629	1, 662	103. 3
Montana	40, 734	1, 235	33. 0
North Dakota	24, 786	18	1, 377. 0
Utah	55, 354	271	204. 2
Wyoming	316, 123	950	332. 8
Subtotal	608, 834	4, 137	147. 2
Other States			
Alaska	160, 270	61	2, 627. 4
California	318, 308	1, 585	4, 617. 2
Louisiana	7, 090, 645	9, 182	772. 2
New Mexico	1, 217, 430	10, 352	117. 6
Texas	7, 485, 764	26, 184	285. 9
Subtotal	16, 881, 251	47, 364	356. 4
U.S. total	20, 108, 661	130, 364	154. 2

Source: U.S. Bureau of Mines.

According to a spokesman for the Natural Gas Supply Committee, a Washington-based lobbying group for the nation's gas producers, there were currently about 7,000 producers sitting on undrilled reserves "waiting for a signal that will provide them with the incentive" to drill for the gas, it was reported Feb. 3. The "signal," according to the Wall Street Journal that day, would be federal deregulation.

Tenneco reports diversion. Tenneco, Inc. said March 7, 1977 that it might have improperly diverted to intrastate customers natural gas earmarked for interstate sales. The company, which ran one of the nation's largest natural-gas operations, said in a filing with the Securities and Exchange Commission that it had uncovered the possible improprieties while preparing its defense against lawsuits brought by two corporate customers for failure to deliver gas.

Involved was about 350 billion cubic feet of gas sold by Tenneco since 1965 to its Channel Industries Gas Co. subsidiary in Texas. Tenneco said only "some" of that amount had been released from contracts for interstate sale through procedures "which may have failed to meet all [federal] regulatory requirements." (Even had the entire amount been diverted over 10 years, it would have represented well under 1% of total U.S. consumption, currently about 20 trillion cubic feet a year.)

FPC 'reprisals' charged. Federal Power Commission lawyers who had been vocal in their criticism of natural gas producers had been given transfers or made to feel unwelcome by superiors, several former FPC natural-gas attorneys said March 10.

The lawyers made the claims before the House Commerce Committee's oversight and investigations subcommittee, which was studying allegations that the FPC had taken retaliatory action against its own staff lawyers who had spoken out publicly against the gas industry. One lawyer, Russell B. Mamone, said he had been abruptly transferred to another section of the FPC following his congressional testimony accusing some major producers of withholding gas on their Gulf of Mexico drilling leases.

FPC gas-rate rulings upheld. A three-judge panel of the U.S. Court of Appeals June 16, 1977 upheld FPC decisions that had nearly tripled the price of some natural gas. The FPC actions, taken July 27, 1976, had been challenged in Washington by a coalition of consumer groups led by the American Public Gas Association, a group of 200 municipal gas systems.

Upheld were FPC actions that:

- Set the ceiling price for gas discovered or sold on the interstate market after Jan. 1, 1975 at $1.42 per thousand cubic feet. The old ceiling had been 52¢ per 1,000 cubic feet.
- Raised to $1.01 per thousand cubic feet the ceiling price for gas discovered or committed to the interstate market between Jan. 1, 1973 and Jan. 1, 1975. (The ceiling was later lowered to 93¢.)
- Raised to 52¢ per thousand cubic feet from 29¢ the rate for gas in production before Jan. 1, 1973, effective as existing contracts expired.

The appellate panel said that especially at a time of national supply emergency, the FPC had wide latitude in setting rates in order to ensure that interstate pipelines could acquire natural gas. The court noted the pipelines were having difficulty getting gas because gas sold within the producing state was unregulated and currently commanded a higher price than so-called interstate gas.

LNG import approved. The FPC April 29, 1977 authorized, 3-1, importation of about 168.4 billion cubic feet of liquefied natural gas (LNG) a year for 20 years from Algeria by Trunkline Gas Co., beginning in 1980. The price of the regasified LNG sold in interstate commerce in the U.S. would be $3.37 per thousand cubic feet. (The current fixed price of domestically produced gas on the interstate market was $1.45 per thousand cubic feet.)

As part of the import scheme, the FPC approved construction of a $163.4-million terminal and regasification facility near Lake Charles in Louisiana, 25 miles north of the Gulf of Mexico. The gas would be transported to Trunkline's main pipeline system through a proposed $28.8-million, 45.8-mile pipeline extension.

The FPC required that Trunkline, a subsidiary of Panhandle Eastern Pipe Line Co., charge its customers in such a way that only those who received the Algerian LNG paid the higher price for it. The commission said this would avoid "the inefficient use of the gas because the LNG will be subject to the market test of whether its users value the LNG enough to pay the true cost of supplying" it.

Solar Energy

Plans & prospects. The sun was expected to meet 7% of the nation's energy needs by 2000 and up to 25% by 2020, the Energy Research & Development Administration (ERDA) said in a report to Congress Aug. 13, 1975. U.S. use of solar energy could save 6.5 million barrels of oil daily by 2000 and 22.5 million barrels daily by 2020, when U.S. output of oil and natural gas was expected to be negligible, according to the report. (The current oil consumption rate was 17.8 million barrels daily.)

ERDA's budget for the development of solar energy technology during the current fiscal year was $89 million, compared with about $1 million in fiscal 1972. Private industry was expected to spend $50 million on solar energy projects in 1975, officials added.

ERDA's budget was directed toward developing three different applications of solar energy:

- Direct thermal applications such as in the heating and cooling of buildings and in the use of the sun's heat for farming and industrial purposes.
- Systems to convert the sun's energy to electricity, which could provide an inexhaustible supply of electric power.
- Solar power to convert farm wastes into useful fuel gases, such as methane, methanol, and hydrogen.

An American Institute of Aeronautics & Astronautics assessment entitled "Solar Energy for Earth" reached the following conclusions (as published in the November 1975 issue of Astronautics and Aeronautics):

(1) Technical feasibility has been demonstrated for a number of solar-powered energy systems designed to provide terrestrial heat, electric power, or both.

(2) Solar energy can begin to make significant contributions to the nation's energy supply sometime in the period 1985–2000. Its present economic disadvantage as compared with alternative energy sources can be reduced or eliminated altogether if (a) a vigorous program of federal research and development support is provided, (b) the implementation of pilot-plant, demonstration, and prototype solar-powered plants is actively promoted, and (c) the prices of fossil fuels remain high or fluctuate unpredictably and nuclear power costs continue to rise.

(3) Identifiable environmental and sociological impacts of solar energy systems are far less severe than those associated with fossil-fueled and nuclear fission-powered systems. In particular, solar-powered systems do not deplete natural energy sources.

(4) In contrast to fossil-fueled and nuclear-fission sources, most solar energy systems depend critically on the availability of either energy storage facilities or supplementary power sources. This mismatch between available and demand power is not a problem for ocean thermal energy conversion powerplants or biomass energy and is of only minor concern for satellite solar power stations.

(5) Premature implementation of solar-powered energy systems without adequate research, technology, and development support, or demonstration efforts involving economically or operationally unsuitable components or systems, could lead to an undesirable "backlash" effect. Adequate R&D pilot-plant, and demonstration projects should be accelerated but not bypassed.

Housing programs block solar energy. Rep. Max S. Baucus (D, Mont.) told the House Aug. 3, 1976 that "certain provisions of federal housing programs ... discourage the use of solar thermal equipment by homeowners." He said in a statement in the Congressional Record:

"Until recently, the Federal Housing Administration, FHA, categorically denied Government loans or loan guarantees to applicants who desired to use solar rather than conventional systems to heat or cool the property on which they sought to secure a loan. Due to a lack of sufficient research, FHA felt that it could not approve any solar systems for its minimum property standards—standards which must be met in order to qualify for Federal housing loan programs.

"With the information provided by the solar heating and cooling demonstration program—set up by Public Law 93-409—FHA recently formulated intermediate standards for solar equipment, and final standards will be available by the end of the decade. Solar systems which meet these standards will now be approved by FHA and can be included in applications for its loan programs.

"There is still a feature of these programs, however, which discourages potential mortgagors from "going solar." Almost every Federal mortgage program contains a limit or limits on the size of loans which are covered. Because these limits do not, in my opinion, realistically reflect the higher initial cost of solar as opposed to conventional thermal equipment, they tend to turn potential program participants away from the former. . . ."

Solar energy research lab set up. The U.S. Energy Research and Development Administration said March 24, 1977 that it was setting up a solar energy research center in Golden, Colo. The center, to be known as the Solar Energy Research Institute, would operate on a budget of $4 million–$6 million in its first year.

Other Developments

Fuel costs raise electricity rates. Electricity rates increased an "unprecedented" 30% during 1974, the Environmental Protection Agency reported July 25, 1975. An 82.2% rise in the cost of fossil fuels over the year accounted for three-fifths of the annual increase, the agency said.

Anti-pollution measures taken by utilities accounted for no more than 5% of the 1974 increase in rates. Conversion of fuels, from high-sulfur to low-sulfur coal and from coal to oil or coal to gas, represented only half of that percentage gain.

Nonfuel electricity costs were up 16.8% over the year, chiefly because of a rise in interest rates which made borrowing more expensive. Revenues in the electric industry rose $9 billion to total $40 billion in 1974, according to the report.

"Consumers along both the Atlantic and Pacific coasts were the most seriously affected by the higher rates," the EPA said. "New England and the Middle Atlantic regions, which historically have had the highest electric rates, also had the highest rates of increase" in 1974.

New River dams barred. President Ford Sept. 11, 1976 signed a bill incorporating a 26.5-mile stretch of the New River in North Carolina in the National Wild and Scenic River System. According to a White House fact sheet, the bill would have the effect of nullifying a Federal Power Commission license for two hydroelectric dams on the river.

The dams would have been located in Virginia close to the North Carolina border. Environmentalists opposing the project claimed that the dams would upset the ecology of the river-valley area. The North Carolina congressional delegation had solidly opposed the power project.

Final congressional clearance of the bill had come Aug. 30 when the Senate passed it 69–16. The bill had passed the House, 311–73, Aug. 10.

The New River was thought by geologists to be about 100-million-years old, second only to the Nile in age.

The District of Columbia U.S. Court of Appeals March 24 had upheld the FPC license to build the two dams. The license had been granted to the Appalachian Power Co. of Roanoke, Va. in 1974. North Carolina had challenged the license, petitioning the interior secretary to include the North Carolina section of the New River in the wild and scenic river system. Interior Secretary Thomas S. Kleppe had announced his agreement to the petition March 12.

Kleppe's decision, which had been anticipated, was protested by Appalachian Power's parent company, American Electric Power Co. of New York, the largest privately-owned electric utility in the country. AEP accused Kleppe March 11 of making the decision "on a crash basis in order to . . . influence the vote" in the North Carolina presidential primary March 23.

3-inch fish stops $116-million dam. The U.S. 6th Circuit Court of Appeals in Cincinnati Jan. 31, 1977 ordered construction halted on the $116-million Tellico Dam project of the Tennessee Valley Authority. The court said the dam on the Little Tennessee River near Knoxville posed a peril to a fish on the endangered species list of the Interior Department.

The dam, begun in 1966, was 90% completed and ready for an opening of the

Technologies Currently Available For Pursuing Major Energy Technology Goals

[The last column of this table presents data from ERDA–48. It represents the maximum impact of the technology in any scenario measured in terms of additional oil which would have to be marketed if the technology were not implemented. Basis for the calculation is explained in app. B of ERDA–48. These data are being reexamined, and changes will be made when analysis is completed. In a number of cases, revised projections of the impacts will be lower.]

Technology	Term of impact [1]	Direct substitution for oil and gas [2]	R. D. & D. status	Impact in year 2000 in quads
Goal I: Expand the domestic supply of economically recoverable energy producing raw materials:				
Oil and gas—enhanced recovery	Mid	Yes	Pilot	13.6
Oil shale	Near	Yes	Study/pilot	7.3
Geothermal	Mid	No	Lab/pilot	3.1–5.6
Goal II: Increase the use of essentially inexhaustible domestic energy resources:				
Solar electric	Long	No	Lab	2.1–4.2
Breeder reactors	Long	No	Pilot/demo	3.1
Fusion	Long	No	Lab	
Goal III: Efficiently transform fuel resources into more desirable forms:				
Coal—Direct utilization utility/industry.	Near	Yes	Pilot/demo	24.5
Waste materials to energy	Near	Yes	Comm	4.9
Gaseous and liquid fuels from coal.	Mid	Yes	Pilot/demo	14.0
Fuels from biomass	Long	Yes	Lab	1.4
Goal IV: Increase the efficiency and reliability of the processes used in the energy conversion and delivery systems:				
Nuclear converter reactors	Near	No	Demo/comm	28.0
Electric conversion efficiency	Mid	No	Lab	2.6
Energy storage	Mid	No	Lab	
Electric power transmission and distribution.	Long	No	Lab	1.4
Goal V: Transform consumption patterns to improve energy utilization:				
Solar heat and cooling	Mid	Yes	Pilot/demo	5.9
Waste heat utilization	Mid	Yes	Study/demo	4.9
Electric transport	Long	Yes	Study/lab	1.3
Hydrogen in energy systems	Long	Yes	Study	
Goal VI: Increase end-use efficiency:				
Transportation efficiency	Near	Yes	Study/lab	9.0
Industrial energy efficiency	Near	Yes	Study/comm	8.0
Conservation in buildings and consumer products.	Near	Yes	Study/comm	7.1

[1] Near—now through 1985. Mid—1985 through 2000. Long—Post-2000.

[2] Assumes no change in end-use device.

Source: Energy Research & Development Administration (ERDA) April 19, 1976

flood gates that would fill the reservoir. It was expected to generate about 200-billion kilowatt hours of electricity a year, which was less than 1% of the total electricity-generating capacity of the TVA system.

The endangered fish was the snail darter, a three-inch member of the perch family. It fed on snails at the bottom of the Little Tennessee River. A 17-mile stretch of the river scheduled to become part of the reservoir was the only known habitat of the snail darter. The fish had been discovered in August 1973 by zoologist David A. Etnier of the University of Tennessee.

The court, in a unanimous ruling, found that "our responsibility under the Endangered Species Act is merely to preserve the status quo where endangered species are threatened, thereby guaranteeing the legislative or executive branches sufficient opportunity to grapple with the alternatives."

Hydroelectric power cuts. The federal Bonneville Power Administration (BPA), the country's largest marketer of hydroelectric power, Feb. 11, 1977 asked its customers to voluntarily reduce electricity use by 10% because of the severe drought. (BPA was one of 20 major utilities that made up the Northwest Power Pool.)

The energy-consuming aluminum industry, which relied heavily on relatively inexpensive electricity generated by water power, was the hardest hit by the requested reduction. The aluminum division of Anaconda Co. Feb. 16 announced a 10% curtailment of production at its giant aluminum smelter in Columbia Falls, Montana, which normally produced three million pounds of aluminum per month. An executive of Kaiser Aluminum & Chemical Corp., which Feb. 20 closed down two primary aluminum-producing potlines in Mead, Washington, said resultant aluminum shortages would probably occur in 1978.

Small-dam use urged. Rep. Richard L. Ottinger (D, N.Y.) called May 24, 1977 for "tapping the immense hydroelectric potential of the tens of thousands of existing small dams across the nation." He said, according to the Congressional Record:

"By utilizing the capacity for hydroelectric production from the more than 49,000 existing small dams, the Nation could replace up to 1 million barrels of oil per day and save American consumers about $5.5 billion annually. . . .

"The city of Vanceburg, Kentucky's municipal utility, is installing 140,000 kilowatts of capacity in existing dams on the Ohio River at a cost of about $60 million. If Vanceburg were to install 1 million kilowatts of hydroelectric capacity—the equivalent of a nuclear power station—the cost would be $480 million, or less than half the going rate for nuclear plants at $1 billion.

"In fact, the Federal Power Commission estimates that by tapping existing dams in the Ohio River Basin alone, 2 million kilowatts of new capacity could be developed, producing 8,800,000,000 kilowatt hours of electricity per year, enough to serve the total electricity needs—commercial and industrial, as well as residential—of 800,000 people—cities the size of Cleveland, Dallas, the District of Columbia, Indianapolis, and Sacramento.

"The potential hydroelectric capacity in New England and New York is such that if only 10 percent of the presently untapped dams were developed for hydroelectric power production at an average capacity of 5,000 kilowatts, we could have 1,745,000 kilowatts of new capacity, all of which would be independent of any fuel requirement. And the savings in fuel would be equivalent to 12,570,000 barrels of oil per year, with a cost saving of $188,500,000 annually. The output would equal 7,640,000,000 kilowatt hours of electricity per year, or enough to supply the total needs—commercial and industrial, as well as residential—of 764,000 people.

"Following is a list of the six New England States and their present dams, both tapped and untapped:"

[Energy output in millions of kilowatt-hours per year]

State	Number of dams	Dams with power	Installed capacity (kilowatts)	Energy output
Connecticut	530	18	128, 920	390
Maine	439	72	536, 497	2, 740
Massachusetts	1, 040	51	223, 245	840
New Hampshire	472	30	417, 992	1, 250
Rhode Island	139	1	1, 500	4
Vermont	241	50	198, 308	83

Magnetohydrodynamics. Rep. Max S. Baucus (D, Mont.) inserted in the Congressional Record Feb. 18, 1976 a statement on magnetohydrodynamics (MHD), which he described as "a comparatively new technology for generating electricity."

In the MHD technique, coal is heated to about 4,000°F.; the resultant gas is "seeded" with a salt to effect ionization; the mixture is propelled through a magnetic field, and electric power is produced "with dramatically higher thermal efficiencies than those possible through conventional technologies." Baucus said that "an added advantage is greatly reduced thermal and gaseous pollution."

Baucus included a Feb. 17 letter to Chairman Ken Hechler of the House Science & Technology Subcommittee on Energy Research, Development & Demonstration. In the letter Baucus said:

> Actual progress in MHD to date is impressive. Improved working knowledge of the operation of slagging electrodes in the arc mode phase of MHD power generation has been gained within the last two years. Usable data on the solubility of "seed" (a salt employed for ionizing the hot coal gas generated by combustion) in the residual slag has also been obtained. An American research team operated high temperature electrodes continuously for one hundred hours in the U.S.S.R.'s U-02 MHD facility. These and other notable accomplishments resulted from efforts by firms and organizations such as Avco Everett Research Laboratory, the University of Tennessee Space Institute, Westinghouse Research Corporation, and others.
>
> The first week of February, 1976, saw a major advance in MHD power generation. On February 9, Dr. Philip White, Assistant Administrator—Fossil Fuels, ERDA, announced the achievement of four days' continuous operation of Avco's Mark VI MHD generator. The Mark VI, America's largest operational MHD facility, logged one hundred hours of power generation, ninety-five of which were continuous at power levels of 200 kilowatts or more. Prior to this time, no continuous test lasting more than twenty hours had occurred at significant power levels. This test was a major step toward commercial use of MHD because the Mark VI was operated at voltage gradients and current densities equal to those that would occur during full-scale utility plant generation. Coal ash was introduced into the combustor to simulate conditions encountered during coal-fired operations. A major objective of the test was to evaluate the resistance of the channel and electrodes to coal slag.

Regional electric power systems urged. An independent study published by the Federal Energy Administration (FEA) and reported March 23, 1977 proposed that legislation be enacted to "federally charter regional bulk power corporations to operate on a multistate basis and market electricity at wholesale" to local distribution companies.

The study, prepared under contract for the FEA by Gordian Associates of New York, further recommended establishment of regional regulatory agencies to ensure effective regulation and "a minimum of multi-jurisdictional conflicts."

An FEA spokesman said that while the dangers of consolidation in the power industry had been widely discussed, the existing problems stemming from the other extreme—fragmentation—generally had been overlooked. These problems, according to the study, included the inability adequately to diversify the financial risks of new projects and inability to take full advantage of economies of scale associated with large-system operations. The study also noted the current difficulties of many utilities, especially in urban areas, in siting new facilities within their service areas.

Without reforms, the industry would not be able to cope with environmental costs, might have difficulty satisfying demand and might have to charge higher rates than would otherwise be the case, the study maintained, suggesting that all these factors could lead to "an increasing tendency toward government ownership."

International Developments

Oil Production & Trade

World oil output down in '75. World oil production dropped 11.5% in 1975, compared to 1974, the Middle East Economic Survey reported April 25, 1976. Mideast output fell 10.5% in 1975, with Kuwait registering the largest decrease, 18.2%. Next was Saudi Arabia, down 16.5% and Iran, off 11.1%. The exception was Iraq, where production rose 17.1%, the survey reported.

The magazine said that Venezuela's production cutback of 26.2% was the largest, followed by Nigeria down 20.7%.

Mideast output up in '76—The nations of the Middle East produced an average of 21.63 million barrels of crude oil a day in 1976, 12.3% more than in 1975, according to the Middle East Economic Survey March 21, 1977. Saudi Arabia, the world's largest exporter of crude oil, registered a 22.2% rise. Large increases were also posted by Libya (29.3%) and Nigeria (16%). The only country with declining production was Iraq, which registered a drop of 1.4%.

U.S.S.R. becomes top oil producer—The U.S. Bureau of Mines had reported Sept. 3, 1975 that the Soviet Union had replaced the U.S. during 1974 as the world's leading oil producer. The Soviet figure, for both crude oil and field condensate, was 3.4 billion barrels compared with 3.2 billion for the U.S.

U.S. reserves continue to drop—The U.S. Geological Survey was reported by the New York Times June 20, 1975 to have further lowered its estimates of recoverable oil reserves available to the country. The new total for offshore reserves stood at 10–49 billion barrels; the amount of onshore oil was estimated at 37–81 billion barrels.

As reported in the Wall Street Journal Nov. 5, the Federal Energy Administration put total recoverable U.S. reserves of crude oil as of Dec. 31, 1974 at 38 billion barrels and natural gas reserves at 240.2 trillion cubic feet. The Journal noted that the figures for oil were 11% higher than those published in April by the American Petroleum Institute and that those for gas were 3% above estimates made by the American Gas Association.

The International Energy Agency declared Nov. 27 that the U.S. rated at the bottom of the 17-member organization in terms of policies for conserving energy. The report said the U.S. had "no standards, no incentives, almost no taxes" that would cut down on energy consumption.

OECD oil imports cut in '75. Imports of petroleum by the 24 nations of the Organization for Economic Coopera-

Capacity & Production For Various Producing Countries

	(1)	(2)	(3)
Saudi Arabia	11.50	7.08	62
Kuwait	3.50	2.08	59
Libya	2.50	1.52	61
Iraq	3.00	2.25	75
Abu Dhabi	2.00	1.40	70
Algeria	1.00	0.92	92
Qatar	0.70	0.44	63
Egypt	0.35	0.25	71
Syria	0.20	0.16	80
Total OAPEC [4]	24.75	16.17	65
Iran	6.80	6.07	89
Venezuela	3.00	2.35	78
Nigeria	2.50	1.78	71
Indonesia	1.70	1.31	77
Dubai	0.30	0.25	83
Gabon	0.25	0.21	84
Ecuador	0.25	0.17	68
Sharjah	0.04	0.04	100
Total OPEC [5]	39.04	27.16	70

[1] Mid-1975 capacity (million barrels per day).
[2] 1975 production (million barrels per day).
[3] Production as a percent of capacity.
[4] Excludes Bahrain.
[5] Includes all OAPEC countries except Bahrain, Egypt and Syria.

Source: Data supplied by the U.S. Treasury Department.

tion & Development fell by 8% in 1975 to about 1.2 billion tons, according to an OECD report published Dec. 6, 1976. The decline in imports reflected a 3.9% drop in oil consumption by the OECD countries in 1975 to about 1.6 billion tons.

The impact of reduced oil consumption varied over the three main areas where OECD countries were concentrated. Consumption fell 5% in Europe, where GDP declined 1.6%. North American consumption fell 2% while its GDP fell 1.6%. In Japan, oil consumption was down 8.9% and GDP rose 2.1%.

The reduced consumption had different effects on the three areas' demand for imports. In North America, where domestic production of oil fell, there was a 2.3% rise in imports of crude oil and oil products in 1975, despite the overall OECD drop in oil imports. In Japan, oil imports were down 4.6%, while in Europe oil imports fell 13.4%, reflecting the effect of North Sea production and massive stockdrawing from oil reserves.

Banks see OPEC surplus shrinking. Two major New York banks, First National City and Morgan Guaranty Trust Co., predicted that the current petrodollar surplus held by members of the Organization of Petroleum Exporting Countries (OPEC) would shrink rapidly by the 1980s and possibly become a deficit.

According to Citibank's analysis, published June 11, 1975, OPEC's cumulative surplus most likely would peak at about $196 billion by 1979 and then decline rapidly to less than $40 billion by 1985. (Following the quadrupling of oil prices in 1973-1974, experts had predicted that oil-producing nations' surplus revenues would total more than $500 billion, but Citibank's economists saw the cumulative surplus reaching $295 billion at most.)

If other members of OPEC tried to wrest control of the market from Saudi Arabia and Kuwait, which currently dominated the producing states, retalia-

Production as a Percentage of Capacity For Various Producing Countries, First Half 1975

	(1)	(2)
Saudi Arabia	6.82	59
Kuwait	2.13	61
Libya	1.14	46
Iraq	2.18	73
Abu Dhabi	1.20	60
Algeria	.93	93
Qatar	.45	64
Egypt	.24	69
Syria	.15	75
Total OAPEC [3]	15.80	62
Iran	5.43	80
Venezuela	2.53	84
Nigeria	1.71	68
Indonesia	1.24	72
Dubai	.25	83
Gabon	.21	84
Ecuador	.14	56
Sharjah	.04	100
Total OPEC [4]	26.40	68

[1] Production (million barrels per day).
[2] As a percentage of capacity.
[3] Excludes Bahrain.
[4] Includes all OAPEC countries except Bahrain, Egypt and Syria.

Source: Data supplied by the U.S. Treasury Department.

tory price-cutting could ensue, Citibank said. Free market oil prices would drop to about $5–$6 a barrel (in constant 1975 dollars) and the cartel would break up. Other factors, such as rapid conversion to nuclear fuel, also could turn OPEC's surplus into a deficit, the bank added.

According to Citibank, the problem of recycling petrodollars from the newly-rich OPEC states to the financially-pressed oil-importing nations had proved less severe than anticipated because OPEC's demand for imported goods produced by industrial states had been heavier than expected and the worldwide recession had reduced the overall demand for oil.

Morgan Guaranty noted these same factors in scaling down its estimate of OPEC's current and anticipated surplus. According to the Wall Street Journal May 23, the bank reported that OPEC's oil production averaged 26 million barrels daily during the first quarter, down 14% from the average for all 1974 and down 19% from the production peak set in the third quarter of 1974.

Imports by OPEC nations climbed 75% during 1974 compared with 1973, and were still accelerating, the bank stated. Morgan Guaranty predicted that OPEC's revenue surplus for 1975 would total less than $40 billion, compared with a $60–$70 billion surplus estimated for 1974. By 1980, the bank forecast a cumulative OPEC surplus of $179 billion.

Despite an increase in the revenue surplus of OPEC nations in 1976, the surplus was likely to decline in 1977 and continue to drop for years, according to a Morgan Guaranty study reported Jan. 28, 1977.

The OPEC nations' surplus rose to an estimated $38 billion in 1976, about $5 billion higher than had been predicted. The study attributed the increase to year-end stockpiling by customers in anticipation of 1977 price increases.

The study said world demand for oil from OPEC nations would likely grow at an annual rate of only 1.4% between 1976 and 1980. The rate was so small, it said, because oil from the Alaskan North Slope, the North Sea and Mexico was expected to satisfy nearly two-thirds of the growth in demand through 1980.

At the same time, the study said, OPEC nations' imports would probably grow at about a 10% annual rate after inflation. Accordingly, the study concluded, the OPEC surplus could decline to less than $20 billion by 1980, from an estimated $32 billion in 1977.

Levy sees OPEC surplus rising—Another analysis of OPEC's monetary reserves, conducted by oil economist Walter J. Levy of New York, contradicted the banks' studies and predicted a continued rise in petrodollar surpluses.

In a report published June 16, 1975, he forecast an annual OPEC surplus of $47 billion (in current dollars) for 1975, $73 billion in 1977, and $47 billion in 1980. Levy said OPEC's cumulative surplus would rise from $122 billion in 1975 to $264 billion by 1977 and $449 billion in 1980, nearly twice the current world monetary reserves, according to IMF statistics.

According to Levy, an important and increasing portion of OPEC's surplus would derive from investment income. Levy estimated that investment income would rise from about $7 billion in 1975 to about $30 billion in 1980, when it would account for 65% of the projected annual surplus, and thereafter account for most of the total OPEC surplus.

In making his forecasts, Levy assumed that the growth rate for OPEC's volume of imports would slow to about 15% during the next few years. He predicted that OPEC oil exports would rise from about 26.5 million barrels daily in 1975 to 32.5 million in 1978 and 1979, and fall to 31.5 million in 1980. The forecasts also were based on an assumption that oil prices would rise about as much as the rate of inflation for goods imported by OPEC nations. The inflation rate currently was estimated at 12% annually and 7% in succeeding years.

Recycling of petrodollars would remain a "persistent problem," Levy warned. "The oil-importing countries not only face collective total payments deficits, year after year, into the 1980s, but the burden of oil-import costs will fall unevenly upon industrialized nations since a favored few are likely to benefit more than others from both OPEC imports of goods and services and OPEC investments," Levy said.

OPEC exports & investments abroad. OPEC exports were estimated at a total of $128 billion in 1976. This figure was given by Assistant State Secretary (for economic and business affairs) Julius L. Katz May 14, 1977 in a letter answering questions of Chairman Charles A. Vanik (D, Ohio) of the House Ways & Means Subcommittee on Trade. Among Vanik's questions and Katz' answers on OPEC operations:

Q. What was the 1976 export figure for the combined OPEC countries? What is it projected to be in 1980? 1985?

A. We estimate that total exports from OPEC countries were approximately $128 billion in 1976. The size of their exports in future years will be largely determined by their oil prices and the rate of oil imports in the OECD countries. Assumptions can be made about these two factors, giving a forecast of the size of future OPEC exports. Assuming (a) constant real oil prices and 6 percent inflation per year and (b) OECD and LDC average growth rates from 1977 to 1980 of 4.2 percent and 4.5 percent respectively, we estimate that exports from OPEC countries would be $184 billion in 1980. Other assumptions would, of course, produce different forecasts. We have made no forecast of OPEC exports in 1985. Too much uncertainty would be involved in such a long range projection to make it meaningful.

Q. What was the OPEC current account surplus in 1976? What is it estimated to be in 1980? In 1985?

A. We estimate that the aggregate current account surplus of OPEC countries was approximately $40 billion in 1976. The size of their combined surpluses in future years will be largely determined by their oil prices, the rate of oil imports in the OECD countries, and the rate of growth of OPEC imports. Assumptions can be made about these factors, giving a forecast of the size of future OPEC current account surpluses. Of these factors, OPEC's oil pricing policy will be the most important. Assuming (a) constant real oil prices and 6 percent inflation per year and (b) OECD and LDC average growth rates from 1977 to 1980 of 4.2 percent and 4.5 percent respectively, we estimate that the aggregate current account surplus for OPEC countries would be roughly $26 billion in 1980. Other assumptions would, of course, produce different forecasts. We have made no forecast of the OPEC surplus in 1985. Too much uncertainty would be involved in such a long range projection to make it meaningful.

Q. A critical factor in the financing of higher priced energy imports has been OPEC investment. What have been the changes in OPEC investment in the U.S. in the past five years? How much OPEC investment is there today in the U.S.? What is the level of OPEC investment that the Administration assumes for 1980? 1985? Of total OPEC investment, how much is placed in OECD countries? How much in non-oil LDCs?

A. The State Department does not itself collect data on OPEC investment. The following information has been provided to the State Department by the U.S. Treasury Department.

Oil exporting countries have added roughly $31.5 billion to the approximately $1 billion of investments they held in the U.S. at the end of 1972. As of March 31, 1977 placements in the U.S. by oil exporting countries were as follows:

	Amount $, billion	*Share in percent*
U.S. Treasury securities	13.8	42.5
Federal agency and corporate bonds	3.7	11.5
U.S. corporate stocks	4.1	12.5
Bank deposits	8.8	27.0
Liabilities of U.S. nonbanking concerns	2.1	6.5
Total	32.5	100.0

In recent months an increasing proportion of oil exporting country investments in the U.S. have been long-term in nature with an emphasis on debt securities. Investments in U.S. corporate stocks have been relatively small.

Rough U.S. Treasury estimates of the worldwide disposition of OPEC surpluses in 1973–1976 are as follows:

ESTIMATED DISPOSITION OF OPEC INVESTIBLE SURPLUS 1974–76

[In billions of dollars]

	1974	1975	1976 [1]
United States	12	12	11½
Short-term banks deposits and Treasury bills	(9.3)	(.3)	(.3)
Long-term banks deposits	([2])	(.8)	(.3)
U.S. Treasury bonds and notes	(.2)	(2.0)	(4.2)
Other domestic bonds and notes	(.9)	(1.6)	(1.2)
Equities	(.4)	(1.6)	(1.8)
Subtotal (banking and portfolio placements)	(10.8)	(6.3)	(7.8)
Other (including real estate and other direct investment, prepayments on U.S. exports, debt amortization, etc.)	(1.2)	(3.7)	(3.7)
Eurobanking market	22½	8	10½
United Kingdom	7½	¼	−1
Other developed countries	6	7¾	8
Less developed countries	4	6	6
Nonmarket countries	½	2	1¼
International financial institutions (including IMF oil facility)	3¾	4¼	1¾
Total allocated	56¼	38¼	38
Estimated cash surplus plus borrowings	59	40	42
Error of estimates of surplus and unidentified investments	2¾	1¾	0

[1] Preliminary.
[2] Less than $500,000.

Q. What were the OPEC financial assets in 1976? What are they projected to be in 1980? In 1985?

A. We estimate that the accumulated financial surpluses of OPEC countries were between $135 billion and $145 billion at the end of 1976. The size of this aggregate surplus in future years will be determined by the current account surpuses of OPEC countries. We expect these surpluses to remain large for the foreseeable future. The range of forecasts which have been made of the surpluses is very broad. As none of these forecasts can be used with much confidence, particularly the longer-term forecasts, we doubt that projections of OPEC's accumulated financial surpluses in 1980 or 1985 would be useful.

Oil Price Controversy

Ford opposes rise. President Ford warned at a news conference June 25, 1975 that a new rise in Middle East oil prices would be "very disruptive and totally unacceptable." If the threatened increase of $2-$4 a barrel were imposed, he said, the U.S. and other consuming nations would have to "find some answers other than OPEC [Organization of Petroleum Exporting Countries] oil."

A $2-$4 per-barrel increase would have a negative impact on the U.S. economy, Ford said. But he added that it would have "a much more significant impact" on the economies of Western Europe and Japan, probably an "even more adverse impact" on the economies of developing nations and "an adverse impact worldwide." He said "it would be very unwise for the OPEC countries to raise their prices under these circumstances because an unhealthy economy in the United States and worldwide is not in their best interests."

OPEC raises oil price by 10%. The Organization of Petroleum Exporting Countries agreed Sept. 27, 1975, at the end of a 4-day Vienna meeting, to a 10% increase in the price of oil, effective Oct. 1, and to maintain that cost level until June 30, 1976.

The OPEC action raised the market price of a barrel of oil from $10.46 to $11.51 and was expected to increase the world oil bill by about $10 billion annually. On the basis of 1974 imports, the U.S. was expected to pay an added cost of more than $2 billion for imported oil, Western Europe more than $5 billion, and Japan between $1.7 billion and $2 billion.

The OPEC decision on the 10% price increase represented a compromise that followed several days of sharp wrangling between member nations favoring a considerable higher boost and others advocating a lesser rise. Saudi Arabia sought to continue prices at their current level or to hold any rise to 5%. Iran led a group that at first backed a 28% increase and later scaled down its demands to 15%. The Saudis argued that further increases would slow world economic recovery. The compromise had been worked out by Venezuela, supported by Kuwait, the United Arab Emirates and Algeria. There also was a dispute over the extent of any price freeze once a new level was set, with Saudi Arabia pressing for a freeze that would continue through 1976.

At the Vienna meeting, the oil ministers agreed to lower crude oil prices for developing countries. The proposal had been submitted by Kuwait, Venezuela, Iraq and Iran. Regardless of OPEC's decision on this matter, Teheran intended to set aside 10 cents on every barrel of oil it exported to distribute to the developing countries, which would come to a total of $200 million a year, Iranian delegate Jamshid Amouzegar announced.

Saudis back oil-price freeze—Saudi Arabian Petroleum Minister Sheik Ahmed Zaki Yamani said Oct. 14 that his country favored extending the current nine-month freeze on higher oil prices until the end of 1976, six months longer than approved by the Organization of Petroleum Exporting Countries at its meeting in Vienna in September.

Speaking at a news conference in Bonn, where he attended an energy meeting sponsored by the West German Social Democratic Party, Yamani said his country would want to review the world economic situation at the start of 1977 before making further decisions on prices. Yamani said he expected opposition to the Saudi stand when the issue would be reopened at a scheduled OPEC meeting in the summer of 1976. Saudi Arabia, he said, had tried unsuccessfully to get its position accepted at the Vienna talks but finally endorsed the mid-1976 date as a compromise. His delegation, however, had rejected a proposal that tied a potential price increase in June 1976 to a reduction of oil-producer purchasing power resulting from inflation, Yamani said.

Oil price shifts—Saudi Arabia increased some of its petroleum prices less than the 10% agreed to by OPEC, it was reported Nov. 3, 1975.

Although the cost of Arab light oil was raised by the full 10%, the price of three other grades (Arab Berri, Arab Medium and Arab heavy) was increased between 6.8% and 9.6%.

Kuwait announced Nov. 12 that it would reduce the price of its crude from $11.40 a barrel to $11.30 a barrel. All firms taking crude in Kuwait, except British Petroleum Co. and Gulf Oil Corp., were notified of the reduction.

Saudi oil production in October had totaled 5.6 barrels a day, a 2.5 million-barrel-a-day drop compared with the September rate, the lowest monthly output since July 1972, the Arabian American Oil Co. (Aramco) reported Nov. 7. The decrease was attributed by industry sources from the rush to amass stocks in the consuming countries during August–September to beat the deadline on the Oct. 1 price increase.

Iran Feb. 16, 1976 announced a cut in the price of its heavy crude oil by 9.5 cents to $11.40 a barrel. The decision, following a week of talks between the National Iranian Oil Co. and a consortium of major American and British oil companies, was aimed at spurring Iran's sagging output. However, an official of one of the Western companies claimed the price reduction would not be "enough to stimulate much in new sales for Iranian oil," which, he said, "is still at least 30 cents a barrel out of line."

Guerrillas raid OPEC conference. When the OPEC oil ministers met again in Vienna, six pro-Palestinian guerrillas burst in on them Dec. 21, 1975. The intruders shot three persons to death and wounded seven others. The raiders seized 81 hostages, including 11 OPEC ministers, flew to Algeria Dec. 22 on a comandeered Austrian DC-9 jetliner after freeing all 41 Austrian hostages, released some of the other prisoners on arriving in Algiers, turned over several more captives after a stopover in Tripoli, Libya and returned Dec. 23 to Algiers, where the remaining hostages were set free. The guerrillas, five men and a woman, surrendered to Algerian authorities.

The raiders had interrupted the second day of a two-day meeting on membership applications and price reductions at OPEC headquarters. The three persons they killed in an initial battle with four Austrian policemen guarding the facilities were one of the policemen, a Libyan official and an Iraqi employe of OPEC. One of the attackers was wounded, given emergency treatment in a Vienna hospital and transferred to a hospital in Algiers when the plane landed there the following day.

Credit for the OPEC attack was claimed in a French-written communique broadcast in Geneva Dec. 22 by a group calling itself the Arm of the Arab Revolution. The statement termed the raid "an act of political contestation" aimed at "the alliance between American imperialism and the capitulating reactionary forces in the Arab homeland." The communique assailed Israel, asserted that the Palestine Liberation Organization was too conservative, denounced Iran as "an active imperialist tool," referred to Egyptian President Anwar Sadat as "one of the leading traitors," called for abrogation of the Sinai accord, demanded a role "for the Arab people and other people of the third world" in dealing with oil resources and urged formation of a resistance front of Syria, Iraq and the Palestinians.

Austrian Chancellor Bruno Kreisky, Algerian Foreign Minister Abdelaziz Bouteflika and an Iraqi diplomat had negotiated the accord with the guerrillas, permitting them to depart Vienna on a plane in exchange for their release of the Austrian hostages, with the other captives signing a statement that they were voluntarily leaving Austria.

Among the OPEC delegates seized by the guerrillas and later released in Algiers Dec. 22 were Saudi Arabian Oil Minister Sheik Ahmed Zaki Yamani and Iranian Interior Minister Jamshid Amouzegar. The other released representatives were from Gabon, Nigeria, Venezuela, Libya,

Iraq, Kuwait, Qatar, the United Arab Emirates and Ecuador.

The guerrillas surrendered to Algerian officials, but they were reported Dec. 29 to have been granted political asylum and flown Dec. 30 to another Arab country (identified later as Libya).

Algerian official sources said Dec. 29 that the six gunmen had been given asylum because Algeria supported "Palestinian revolutionaries or people acting for the same cause." Algerian investigators had identified the nationalities of the six as two Palestinians, a Lebanese, two West Germans and a Venezuelan. Wiesbaden police Dec. 26 identified one of the West Germans as Hans-Joachim Klein of Frankfurt.

An official of the Cairo newspaper Akhbar al-Yom, board chairman Ali Amin, charged in an article Dec. 27 that Libyan leader Muammar el-Qaddafi had financed the OPEC assault, paying the attackers $1 million in advance and promising more money later. Qaddafi and a group calling itself the Popular Front had arranged the mission along with "Carlos," the international terrorist leader identified as Ilich Ramirez Sanchez of Venezuela, according to Amin. (Venezuelan Mines Minister Valentin Hernandez Acosta, one of the OPEC delegates released in Algiers Dec. 23, had identified the guerrilla leader as Ramirez.)

OPEC keeps price freeze. OPEC oil ministers met in Bali, Indonesia May 27–28, 1976 and agreed to retain the current nine-month oil price freeze with adjustments justified by market conditions.

Dr. Mohammed Sadli, the Indonesian minister of mines, referred several times in presenting the final communique to the "lack of unanimity" among delegates on the issue of a price increase. The Saudi Arabian oil minister, Sheik Ahmed Zaki Yamani, had argued against a rise in prices, insisting that industrialized countries had not yet fully recovered from the recent world inflation and claiming that they would greet such a rise by going to their stockpiles of crude oil.

(The marker crude on which OPEC's current price was based was Saudi Arabian light oil. That country produced nearly twice as much oil as Iran, the next largest exporter in the organization.)

Price adjustments—Adjustments were made in June under the May decisions.

Kuwait June 7 announced a reduction in the price of its crude oil by seven cents a barrel retroactive to June 1. This brought the price of a barrel down to $11.23.

Arabian American Oil Co. (Aramco) said June 9 that Saudi Arabia had reduced the cost of its medium and heavy crude oil by five and 10 cents a barrel, respectively. As a result, medium crude would sell for $11.28 a barrel and heavy crude for $11.04.

Iran announced June 10 a drop of between five and seven cents in the price of its heavy crude oil, which previously sold for $11.40 a barrel.

Libya had said June 8 that it would increase the price of its light crude by 10 cents or more a barrel. In announcing the decision in Kuwait, Libyan Petroleum Minister Izzedin Alu Mabrouk told a news conference that his country favored Saudi Arabia's ouster from OPEC if it continued to oppose price boosts by OPEC member states.

Differing price increases. OPEC delegates, meeting in Qatar Dec. 15–17, 1976, agreed to let members set differing price increases for their oil exports in 1977, Venezuelan Mines Minister Valentin Hernandez said Dec. 17.

Saudi Arabia and the United Arab Emirates would raise prices by 5% for the first six months of 1977, while OPEC's 11 other members would increase prices by 10%, Hernandez said. The 11 would also raise prices another 5% after July 1, 1977, according to OPEC sources cited by the New York Times Dec. 17.

The unprecedented split increases represented a compromise between Saudi Arabia, which wanted to extend OPEC's current price freeze for another six months, and other members which sought 15%–25% increases. The compromise was only a "temporary resolution" of the disagreement, Hernandez declared.

Sheik Ahmed Zaki Yamani, Saudi Arabia's oil minister, had argued against a price increase since he arrived in Qatar Dec. 14, asserting that the economies of Western nations had not recovered sufficiently to absorb higher oil costs. The United Arab Emirates gave Yamani qualified support. They said they would vote to extend the price freeze if other OPEC members demanded more than a 10% increase. Iraq and Libya reportedly asked for a 25% increase, while Qatar, Iran and Venezuela asked 15%.

Yamani stalked out of the OPEC meeting Dec. 16 after failing to persuade other delegates to extend the price freeze. He flew home to confer with Saudi Arabia's King Khalid and other leaders, and returned to Qatar eight hours later. The compromise on prices was worked out late that night.

Yamani's dramatic walkout was "a big game," according to Iraqi delegate Tayeh Abdel Karim. Yamani had walked out of an OPEC meeting in Vienna in 1975 after arguing for a price freeze. He flew to London, ostensibly to telephone Saudi Arabia from there, then returned to Vienna and agreed to a 10% price increase.

In other action, the OPEC delegates elected Ali Jaideh, Qatar's oil minister, as OPEC secretary general for 1977–78, it was reported Dec. 17. It had been Ecuador's turn for the post, but Qatar took precedence because it had skipped its turn four years previously because it lacked a suitable candidate.

Ford scores increases—U.S. President Gerald Ford Dec. 17 praised Saudi Arabia and the United Arab Emirates for their restraint on price increases. He denounced the 11 other OPEC nations as "irresponsible."

Ford had made an appeal for moderation by the OPEC delegates in an interview Dec. 1 with Ignacio Iribarren Borges, Venezuela's ambassador in Washington. Venezuelan Mines Minister Valentin Hernandez said Dec. 9 that the delegates would take Ford's request into account, but he added the next day that OPEC's greatest responsibility was to the Third World.

The U.S. State Department also had argued against a price increase, asserting Dec. 9 that any increase "could have damaging consequences for the world economy." A State Department study, cabled Dec. 8 to U.S. embassies around the world, said a 5% increase would cost the "big seven" industrial nations a total of nearly $4 billion per year and add more than $1 billion to the import costs of smaller developed countries and a similar sum to the import costs of developing countries that did not produce oil.

A 5% increase would cost American consumers $1.9 billion in 1977, including gasoline and heating-oil increases of about 0.8¢ per gallon, the State Department study added.

Another appeal for moderation was issued Dec. 16 by U.S. President-elect Jimmy Carter, who asked OPEC to "hold down the price of oil."

In other developments related to the OPEC price hike, the International Energy Agency in Paris said Dec. 20 that the higher cost of the producers' oil would force a sharp reduction in demand and this might stem any growth of the cartel's revenues. An agency official said that the demand would be restrained by the anticipatory purchase by oil companies in advance of the OPEC decision. Western oil companies had stocked up an additional three million to four million barrels daily in the quarter immediately preceding the price rises, IEA officials said.

Iraq & Iran score Saudis—Iraq and Iran assailed Saudi Arabia for not agreeing with other OPEC members to raise oil prices by 15% in 1977.

On returning to Baghdad Dec. 18, Iraqi Oil Minister Tayeh Abdel-Kerim said his delegation had "unmasked Saudi Arabia as a defeatist and compromising reactionary cell working inside and outside OPEC against the interests of its people and against the interests of the oil-producing and other developing states." Abdel-Kerim expressed confidence that "the pressure of the liberated world and Arab public opinion" would force the Saudis to bring their price increase of 5% in line with the other OPEC members' price.

Iran's government-controlled news media Dec. 20 took Saudi Arabia to task for not opting for the higher price. The newspaper of the Rastakhiz Party, the only legal party in the country, said, "The Third World and all progressive nations everywhere are angry and detest Saudi Arabian Oil Minister Sheikh [Ahmed Zaki] Yamani for having sold the real interests of his country and of OPEC to imperialism."

A government broadcast noted that the Iranian press had denounced Yamani as a traitor who had stabbed OPEC in the back.

Split price hurts Iran. Abdol Majid Majidi, Iranian minister of state for planning and budget, said Jan. 11, 1977 that Iran would attempt to conduct its direct petroleum sales on a barter basis, swapping crude oil for commodities as it currently did with Soviet bloc countries, because of a sharp decline in crude oil exports in early January. Majidi said Iran would also reduce its foreign aid in the face of expected petroleum revenue losses of about $23.5 million per day. Iran had extended more than $2 billion in foreign aid since the quadrupling of crude oil prices in 1973–74.

Following the December OPEC meeting, Iran had increased the price of light crude oil $1.19 per barrel, to $12.81, a full 10% hike. By agreement with Kuwait, the price rise of heavy crude had been somewhat smaller, however.

Majidi said his country's total petroleum exports had dropped 38%, to 4.2 million barrels a day, since the split price increase had taken effect. He said direct sales by the National Iranian Oil Co. (NIOC), the state company, had dropped to 651,000 barrels a day in early January from the December 1976 average of 1.3 million barrels, a 51% reduction. Sales through a consortium of Western oil companies that handled about 80% of Iranian crude oil marketing had fallen 35%, to under 3.6 million barrels a day, he said.

NIOC Jan. 5 warned that oil companies that did not fulfill their contractual purchase commitments would be put on an OPEC blacklist and that the Iranian government would take any action deemed necessary against them. NIOC said 25 new customers, under contract for 1.2 million barrels of crude oil a day in 1977, had protested when informed of the 10% price hike.

Kuwait Jan. 3 announced it too would reduce crude-oil production because customers were cutting their purchases. Heavy crude, of which Iran and Kuwait were major suppliers, had been considered overpriced and in oversupply in world markets recently. Demand for the oil had also fallen off as oil companies' pre-price-increase stockpiling ended Jan. 1. (An estimated 4 million barrels a day had been stockpiled around the world.) In addition, a proposed increase in Saudi Arabian production of heavy crude for sale at a lower price was expected to cut into Iran's and Kuwait's sales.

The oil minister of Venezuela, another OPEC member, said Jan. 5 that his country did not intend to cut production since all buyers of Venezuelan oil had accepted the 10% price hike and were not buying lower-priced Saudi Arabian oil.

Other OPEC price increases—The Arabian American Oil Co. (Aramco), a consortium of U.S. oil companies that produced and sold Saudi Arabian oil, said Jan. 5 the price of Arabian light crude had been raised 5% to $12.09 per barrel from $11.51. Arabian light crude was the benchmark oil on which other OPEC oil prices were based. The price of Aramco's high-quality Berri crude had also increased 5%, to $12.48 per barrel.

The United Arab Emirates, led by Abu Dhabi, Jan. 9 formally raised their prices an average of 4.9%.

Meanwhile, the Middle East Economic Survey said the Saudi government had authorized an increase in oil production, although it had not specified the amount of the increase, according to a report Jan. 2. Saudi Petroleum Minister Sheik Ahmed Zaki Yamani had previously said production would be increased 40% to the current maximum potential of 11.9 million barrels a day from 8.5 million. He had been quoted as saying the increase was intended to neutralize the higher price increases implemented by other OPEC members.

Abu Dhabi officials said Jan. 10 the emirate had no production increase planned, other than a previously announced 12% boost to 1.75 million barrels a day. That increase had been planned before the December 1976 meeting and was intended to help local industry. The Abu Dhabi combination of holding production levels constant while raising prices only moderately was reportedly the result of a political decision to avoid deepening the split among OPEC members.

Indonesia Jan. 1 raised the prices of six varieties of crude oil by amounts ranging from 6.05% to 10.77%. The price of the low-sulphur Minas benchmark crude was hiked by 5.86%, to $13.55 a barrel from $12.80. Indonesian officials said the Minas had been overpriced and would now be back in line with demand. They said the relatively small price boosts on some other grades did not represent a departure from the OPEC majority pricing decision, which they said Indonesia backed "in principle."

Buyers of Nigerian crude oil had said they were told Nigeria had raised the prices of its better-quality, lighter-grade crude oil $1.06 a barrel, to $14.24, and its medium-grade crude $1.19, to $13.73, according to a report Jan. 5. (Most other OPEC producers had raised lighter-grade oil prices by larger amounts than prices of their heavier, lower-quality oils.)

Overall, world crude-oil prices were reportedly an average of 8% higher in the wake of the split price increase. Prices had risen an average of about $1 a barrel.

It was reported later (Jan. 31) that Oman had decided to increase its oil prices by 9%, retroactive to Jan. 1. The move raised the price of Omani light oil to $12.75 a barrel, up from $11.70.

Oman, which was not a member of the Organization of Petroleum Exporting Countries, had produced an average of 365,000 barrels of oil daily in 1976, about 2% of the Middle East total.

Saudis ask consumer price break—In a statement published in the Middle East Economic Newsletter Jan. 8, the Saudi Arabian government said it would require U.S. oil companies to submit reports to prove they were passing on to consumers the benefits of the lower Saudi oil price increases. The newsletter said the stringent Saudi instructions on marketing and reporting were unprecedented in the oil industry.

Saudis bar price compromise. Saudi Arabia Feb. 6, 1977 rejected a compromise plan proposed by Qatar's oil minister to end the two-tier OPEC pricing system.

Sheik Abdul Aziz Bin Khalifa al-Thani of Qatar had been touring the OPEC nations in recent days in an attempt to win support for his plan. Under al-Thani's proposal, Saudi Arabia and the United Arab Emirates would raise their oil prices to a level 10% above 1976 prices and the rest of the OPEC members, most of whom had already raised prices 10%, would forego an additional 5% increase scheduled to take effect in July.

Saudi Arabian oil minister Sheik Ahmed Zaki Yamani said Feb. 6 that the Qatar proposal was rejected because "it conflicts with Saudi Arabia's political and economic interests." Saudi Arabia had refused to go along with the OPEC majority in raising prices by 10% partly because it feared such an increase could slow or halt the world economic recovery, adversely affecting oil consumers and producers alike. (As press reports noted Jan. 17, Saudi Arabia had substantial amounts of money invested in Western companies' securities and government notes, whose value could be threatened by political instability or eroded by inflation.)

But the Saudi stand on prices also was intended as a political gesture to the Western world. The Saudis expected this gesture to be reciprocated by movement toward a settlement of the Israeli-Arab conflict and by some progress in the North-South conference between rich and poor nations.

The Middle East Economic Survey Feb. 14 said that Sheik Yamani reportedly had proposed a compromise plan of his own that involved a uniform 7% price increase for the entire year in 1977. This percentage corresponded with the figure Saudi Arabia apparently had been prepared to accept at the OPEC conference in December 1976, the newsletter said.

Movement toward a price compromise had been increasing steadily because some of the nations that had raised prices 10% were experiencing sharply reduced demand for their crude. The state-owned National Iranian Oil Co. (NIOC) disclosed Feb. 9 that its January oil exports had fallen nearly two million barrels a day from the December 1976 level of six million barrels a day, with a resultant loss in revenue of some $20 million a day. NIOC production was down only about 215,000 barrels a day from January 1976 levels.

Kuwait confirmed Jan. 26 that its oil output had slumped by 33% to date in 1977. The Wall Street Journal Jan. 20 had reported that major buyers of Kuwaiti crude, such as British Petroleum Co. (BP) and the Royal Dutch/Shell Group, were negotiating further reductions in their purchases.

Iraq, too, was believed to be suffering reduced demand for its oil, it was reported Jan. 26.

Iran, Iraq and Kuwait were generally considered to be the particular targets of the Saudi move to increase its own production in an attempt to force prices down. But Sheik Yamani said Feb. 9 the export slumps in Iran and Kuwait could not be directly attributed to a Saudi production increase because, in fact, bad weather in the Persian Gulf had interfered with shipping to such an extent in January that Saudi output had actually declined to just under 8.3 million barrels a day.

According to industry and press reports in late January and early February, Saudi Arabia had chosen four special European customers to receive some of the additional oil produced. The new customers, reported to be the Royal Dutch/Shell Group, British Petroleum, France's Cie. Francaise des Petroles and Italy's Ente Nazionale Idrocarburi, were each paired for supply purposes with one of the four U.S. companies that currently owned the Arabian American Oil Co. (Aramco).

(The four U.S. companies were Exxon Corp., Texaco Inc., Mobil Corp. and Standard Oil Co. of California. Aramco produced and sold the bulk of Saudi Arabia's oil and was soon to be taken over completely by the Saudi government.)

None of the four European companies, except the French concern, had been a major customer for Saudi crude previously; however, the French company had been buying large amounts of Iranian oil, as well as Saudi oil.

The existence of these arrangements had been confirmed by three of the U.S. companies and by the Royal Dutch/Shell Group, which had said it would receive about 200,000 barrels a day of Saudi crude through Mobil, it was reported Feb. 16.

But Yamani, in an interview Feb. 15, said the widely reported story was a "pure falsification by the press." He said the extra volume of oil to be produced would be handled by Aramco companies in accordance with their historical sales pattern. Thus, if a country had taken 15% of Saudi Arabia's oil in 1976, it would be entitled to 15% of the current increased output, according to Yamani.

Yamani noted that the four Aramco companies had signed an agreement aimed at assuring that savings on the cheaper crude would be passed on to customers. The agreement stipulated that the companies could sell the additional oil to buyer-users only, not to brokers.

Weather influences OPEC price war. Increased demand for oil due to the frigid U.S. winter, and restraints on Saudi Arabian production because of wind storms in the Persian Gulf helped two Middle East nations partially recover in February 1977 from the initial ill-effects of the OPEC price dispute.

The National Iranian Oil Co. announced that Iran's total oil exports had reached nearly 5.7 million barrels a day in February, some 30% higher than the levels of the previous month and of February 1976, according to the Associated Press March 2. Jamshid Amouzegar, Iranian minister of state, noted in an interview March 8 that Iran had not had to blacklist any company for failing to honor purchase agreements because of the high price of Iranian oil.

In Kuwait, oil officials had said daily average production was back up to 1.7 million barrels a day, after having dropped to 800,000 barrels a day in January, the AP reported March 24.

After the two-tiered price increase went into effect in early January, with the majority of OPEC members raising prices 10% and Saudi Arabia and the United Arab Emirates 5%, crude-oil sales by Iran and Kuwait had dropped more sharply than sales by other nations posting 10% increases. The resurgence in production levels in Iran and Kuwait, according to knowledgeable observers, was due in part to the wind storms in the Persian Gulf, which had prevented Saudi Arabia from loading as much oil as had been planned. Saudi Arabia thus had not been able to increase production as quickly as expected in its effort to mitigate the effects of the OPEC majority's 10% price hike.

In addition, the unexpectedly high demand for oil early in the year, especially in the U.S., had been greater than could be met by Saudi oil alone. After increased demand had wiped out the supply of oil stockpiled on the world market in late 1976, buyers had been forced to turn once again to Iran and Kuwait to meet their needs.

The return of good weather to the Persian Gulf in mid-February, however, helped Saudi Arabia increase petroleum exports in February to about 8.8 million barrels a day, up by about 500,000 barrels a day from January, according to Saudi officials March 10.

Venezuelan president on OPEC tour. President Carlos Andres Perez visited seven oil-producing nations April 21–May 5, 1977 in an apparently unsuccessful effort to negotiate an end to the OPEC price dispute.

Perez visited Qatar April 21–23; Kuwait April 23–25; Saudi Arabia April 25–27; Iran April 27–30; Iraq April 30–May 3, and Algeria May 4–5. He paid a six-hour visit to Abu Dhabi May 1, and stopped in Austria May 3–4 on his way to Algiers.

Perez had said April 19, before leaving Venezuela, that the purpose of his trip was to strengthen OPEC. He said Venezuela had been consulting OPEC members on possible compromises to end the two-tier price system currently used by the cartel.

Perez received his first setback in Saudi Arabia, where he met with Crown Prince Fahd in the absence of King Khalid, who was in London for treatment of a leg injury. Although Fahd characterized his talks with Perez as "useful and constructive," the official Saudi press agency reported April 28, after Perez' departure, that Saudi Arabia would not raise its prices to conform with those of other OPEC members. Sources quoted by the Washington Post May 1 said the Saudis were sticking to their price policy to reap political benefits during Fahd's planned summer visit to Washington.

(In a surprise move, Perez signed a joint communique with the Saudi Arabian government calling on Israel to withdraw from occupied Arab territories and to recognize the national aspirations of the Palestinians, it was reported May 6.)

After Perez conferred with Shah Pahlevi of Iran in Teheran, Venezuela and Iran issued a joint appeal to Saudi Arabia to raise its oil price in exchange for the cancellation of a further 5% price increase planned by 13 OPEC states for July 1, it was reported May 12. This plan, proposed earlier by Qatar, had been rejected by Saudi Arabia in February.

Perez' disappointment was evident in Vienna May 4, at a press conference at OPEC headquarters. "The unity of OPEC is unbreakable," the Venezuelan leader said, "but I can't tell you exactly when the expression of this solidarity will be resumed."

Perez made a new appeal for uniform OPEC prices May 9, after returning to Caracas. In a nationwide radio and television address, he warned that while the current price dispute had not "endangered the unity of the organization," continuation of the price split would engender new disputes that would harm OPEC.

Oil Aid

Two international finance plans were adopted to help nations economically harmed by the quadrupling of oil prices. The International Monetary Fund created an oil facility to make loans to IMF member nations with serious balance-of-payments deficits caused by the higher prices. And OPEC set up a fund to provide loans to developing nations.

Purchases under the 1975 oil facility. The International Monetary Fund announced Nov. 12, 1975 that 23 nations had purchased the equivalent of 1.897 billion special drawing rights under the IMF's oil facility for 1975. The special credits, which were financed by IMF borrowings from oil-exporting countries and other nations, were generally for four to seven years and carried an average annual interest rate of 7.75%. The 1975 oil facility became operational June 10 with the conclusion of borrowing agreements between the IMF and 11 lending nations.

The IMF program was funded with loans from 16 nations, which contributed 3.047 billion special drawing rights (SDRs) to the 1974 facility and SDR 3.856 billion to the 1975 facility.

The IMF paid lenders to the 1974 oil facility an annual interest rate of 7%—the same rate it charged nations seeking the special oil credits. Loans made to the IMF for the 1975 oil facility carried an annual interest rate of 7.25%.

An interest subsidy account also was established with $120 million in contributions, chiefly from oil-exporting nations, to ease the repayment problem by reducing interest rates for countries with the most serious financing difficulties.

The first transactions for 1975 were reported June 19 when Finland purchased the equivalent of SDR 71.25 million ($89.06 million) and Uruguay purchased SDR 21.56 million ($26.95 million).

Other purchases, expressed in SDRs and dollar equivalents:

Haiti SDR 1.59 million ($1.96 million); South Korea SDR 58.38 million ($71.8 million); announced July 9.

Greece SDR 51.75 million ($61.58 million); Iceland SDR 8.4 million ($9.99 million); Yemen SDR 4.6 million ($5.47 million); announced Aug. 5.

India SDR 201.34 million ($241.2 million); New Zealand SDR 49.46 million ($59.3 million); Turkey SDR 56.62 million ($67.9 million); announced Aug. 12.

Costa Rica SDR 7.2 million ($8.56 million); announced Aug. 20.

Costa Rica SDR 4.8 million ($6 million); Italy SDR 780.24 million ($925 million); Pakistan SDR 76.36 million ($92 million); Tanzania SDR 20.61 million ($25 million); announced Sept. 10.

Central African Republic SDR 1.69 million ($1.97 million); Philippines SDR 96.87 million ($112.38 million); Senegal

Cumulative Purchases Effected or Approved Under IMF Oil Facility

(millions of SDRs)

Member	1974	1975
All Countries:	**2,582.851**	**4,319.577**
Industrial Countries:	**675.000**	**1,780.240**
Italy	675.000	780.240
United Kingdom	—	1,000.000
Other Developed Countries:	**794.600**	**1,113.770**
Finland	—	186.360
Greece	103.500	51.750
Iceland	17.200	21.970
New Zealand	109.300	129.370
Portugal	—	114.760
Spain	296.200	275.930
Turkey	113.200	148.110
Yugoslavia	155.200	185.520
Developing Countries:	**1,113.251**	**1,425.567**
Argentina	—	76.090
Bangladesh	51.500	40.470
Burundi	1.200	—
Cameroon	4.620	11.790
Central African Rep.	3.300	2.660
Chad	2.205	—
Chile	118.500	125.220
Costa Rica	18.837	18.830
Cyprus	8.100	21.970
Egypt	—	31.680
El Salvador	17.890	—
Fiji	0.340	—
Ghana	38.600	—
Grenada	—	0.490
Guinea	3.510	—
Haiti	4.800	4.140
Honduras	16.785	—
India	200.000	201.340
Israel	62.000	81.250
Ivory Coast	11.170	10.350
Jamaica	—	29.200
Kenya	36.000	27.930
Korea	100.000	152.690
Malagasy Republic	14.300	—
Malawi	—	3.730
Mali	5.000	3.990
Mauritania	—	5.320
Morocco	—	18.000
Nicaragua	15.500	—
Pakistan	125.000	111.010
Panama	7.370	17.250
Papua New Guinea	—	14.800
Peru	—	52.660
Philippines	—	152.030
Senegal	15.525	9.910
Sierra Leone	4.914	4.970
Sri Lanka	43.500	34.130
Sudan	28.710	18.300
Tanzania	31.500	20.610
Uganda	19.200	—
Uruguay	46.575	48.070
Western Samoa	—	0.417
Southern Yemen	11.800	12.020
Zaire	45.000	32.530
Zambia	—	29.720

Nations Contributing to the Oil Facility for 1974 and 1975

Lender	1974	1975
Abu Dhabi	100	—
Austria	—	100
Belgium	—	200
Canada	246.9	—
West Germany	—	600
Iran	580	410
Kuwait	400	285
Netherlands	150	200
Nigeria	100	200
Norway	—	100
Oman	20	0.5
Saudi Arabia	1,000	1,250
Sweden	—	50
Switzerland	—	250
Trinidad & Tobago	—	10
Venezuela	450	200
Total	3,046.9	3,855.5*

*Purchases equivalent to SDR 460.977 million under the 1975 oil facility were financed from borrowing agreements for 1974.

SDR 9.91 million ($11.49 million); announced Sept. 29.

Grenada SDR 310,000 ($362,700); Kenya SDR 24.83 million ($29 million); Uruguay SDR 14.37 million ($16.8 million); announced Oct. 15.

Bangla Desh SDR 25.78 million ($30.4 million); Israel SDR 81.25 million ($95.9 million); South Korea SDR 38.91 million ($45.9 million); announced Oct. 31.

Chile SDR 79.79 million ($94.1 million); New Zealand SDR 32.97 million ($38.9 million); announced Nov. 6.

Argentina SDR 76.09 million ($89.78 million); announced Nov. 12.

1975 oil-facility program ended—The IMF April 1, 1976 concluded the final round of borrowing operations for the oil facility for 1975.

Since its adoption in September 1974, the oil facility had made the equivalent of nearly 7 billion special drawing rights (SDRs), or about $8 billion, available to 55 nations under a formula related to the increased cost of oil imports since 1973 and to a member's need for balance-of-payments financing.

Under the 1974 program, 40 countries made drawings from the oil facility amounting to SDR 2.583 billion. The portion that remained unused from the 1974 oil facility, SDR 460.977 million (about $556 million), was made available for the 1975 oil facility along with additional contributions from lenders. Forty-five member countries made purchases totaling SDR 4.320 billion under the 1975 oil facility.

The IMF eased eligibility rules for borrowing nations March 24, enabling them to return to the oil facility for a final round of credits. The IMF directors set a limit of total purchases from the oil facility at 78.46% of the member's line of credit based on the previously cited formula involving the rise in oil prices and the member's payments deficit. The previous limit had been 50%.

These final purchases from the 1975 program concluded the IMF's oil-facility transaction:

Announced Jan. 7: Cyprus, SDR 14 million ($16.4 million) and Mauritania, SDR 3.39 million ($4 million).

Announced Jan. 13: Ivory Coast, SDR 10.35 million ($12.1 million) and Portugal, SDR 73.12 million ($85.6 million).

Announced Feb. 23: Spain, SDR 75.93 million ($88.8 million) and Yugoslavia, SDR 129.37 million ($151.4 million).

Announced March 5: Egypt, SDR 20.19 million ($23.6 million), and Sierra Leone, SDR 3.17 million ($3.6 million).

Announced April 1: Egypt, the equivalent of SDR 11.49 million, or $13.3 million; Finland, SDR 115.11 million ($133.5 million); Mali, SDR 3.99 million ($4.6 million); Mauritania, SDR 1.93 million ($2.2 million); Morocco, SDR 18 million ($20.8 million); Southern Yemen, SDR 7.42 million ($8.6 million); Philippines, SDR 55.16 million ($64 million); Turkey, SDR 91.49 million ($106 million); Yugoslavia, SDR 56.15 million ($65 million); and Zambia, SDR 10.79 million ($12.5 million).

OPEC sets aid plan for poor nations. The Organization of Petroleum Exporting Countries agreed at a meeting in Vienna Nov. 17–18, 1975 to establish a $1 billion-a-year fund to help developing nations meet the rising oil prices.

OPEC said that the plan provided for interest-free, long-term loans to the underdeveloped countries to which all OPEC members would contribute "for balance of payment purposes and for development programs and projects."

Asked how the aid program would be financed, Iranian Oil Minister Jamshid Amouzegar said, "you will note that $1 billion is the equivalent of 10 cents a barrel on the amount of oil exported in a year." Iran had previously proposed a 10-cent tax on each barrel of oil exported by OPEC.

At a conference in Paris Jan. 26–28, 1976, however, a 1976 fund of only $800 million was set up. The cut was made after Ecuador and Indonesia said they could not contribute because of domestic financial difficulties.

The Paris meeting rejected an Iranian and Venezuelan proposal to establish a five-year program for the poor nations in favor of operating the $800 million fund for 1976 only. Contributions by individual nations were: Iran, $210 million; Saudi Arabia, $202 million; Venezuela, $112 million; Kuwait, $72 million; Iraq and Libya, $40 million each; United Arab Emirates, $33 million; Algeria, $20 million; Qatar, $18 million; and Gabon, $1 million.

The finance ministers in Paris May 11 voted to make available $400 million for the new International Fund for Agricultural Development, to be run by the World Food Council, a United Nations agency. The money was to come from the $800 million OPEC Special Fund. The ministers named Ibrahim Shihata, senior adviser to the Kuwait Fund for Arab Economic Development, to be director general of the OPEC fund. They said each member state would appoint a director to the board.

OPEC's Special Fund signed agreements in Vienna Dec. 23 providing $42.7 million in interest-free 25-year loans to six developing nations to help their balance of payments. The loans were: Sudan, $7.5 million; Western Samoa, $1.6 million; Sri Lanka, $8.1 million; Guinea, $2.35 million; Pakistan, $21.45 million; and Central African Empire, $1.75 million.

OPEC said Jan. 10, 1977 that it had allotted $111.7 million from the fund for balance-of-payments support to 24 developing nations. Among the largest loans scheduled were $21.8 million to India; $14.5 million to Egypt; and $13.9 million to Bangladesh.

The OPEC finance ministers Feb. 28 allotted $1 billion to the Special Fund for 1977.

OPEC aid questioned—At an International Monetary Fund meeting in Jamaica in January 1976, dispute arose over the poor nations by the oil-exporting countries.

The Group of 24, representing developing nations at the Jamaica conference, issued a communique at the meeting acknowledging "with satisfaction" the "substantial" increase in aid given by the members of OPEC. OPEC aid had "multiplied five-fold between 1973 and 1974 to a figure approaching $15 billion," evenly divided between bilateral and multilateral aid, according to the communique.

The group also said that on an annual basis, OPEC's aid to developing nations ran "well in excess of $21 billion" during the first half of 1975.

The World Bank also had praised OPEC's aid efforts, contending that OPEC transfers of economic assistance, measured against gross national product, were greater in percentage terms than the aid given by the 17 richest industrial nations, the Washington Post reported Jan. 12.

The Post cited a letter sent Aug. 30, 1975 by World Bank President Robert McNamara to Rep. David Obey (D, Wis.), in which McNamara claimed that the $11 billion in additional oil costs paid by developing countries during 1975 was "more than offset" by capital flows to the poor countries from OPEC and the Organization for Economic Cooperation and Development (OECD), representing industrial nations of the West.

These measures of OPEC generosity were disputed by Maurice J. Williams of the OECD's Development Assistance Committee.

Writing in the January issue of Foreign Affairs, Williams contended that the bulk of OPEC aid had gone to a few Arab and other Moslem nations. Williams also claimed that only half of OPEC's aid given in 1974 and 1975 was made as short-term transfers of funds—the kind of assistance needed to correct severe payments imbalances. Williams said a large part of OPEC's aid was extended as loans bearing a commercial rate of interest.

According to Williams, OPEC's subsidized aid for 1974 amounted to only $2.4 billion, or 16% of the total, compared with $11.3 billion, or 76% of the total given by Western nations.

Williams said he did not expect OPEC nations to increase or broaden the scope of their aid, noting a statement made in early 1974 by Abdel Rahman al-Atiki, Kuwait's finance minister:

"Nobody looked at the Arabs before. Why does everybody expect us to be the godfather? This part of the world has been neglected for centuries and its wealth has been carried away by foreigners without giving it a hand for development. The major part of our international financial aid will be put at the service of Arab countries, and to assist other Moslem countries, particularly in Africa."

New IMF lending facility sought. A key committee of the International Monetary Fund April 29, 1977 agreed in principle to establish a "temporary" and "supplementary" lending facility to assist countries with balance-of-payments deficits.

The aim of the new lending program was to recycle the petrodollars being accumulated at a rapid rate by oil-producing countries to oil-importing nations that had been unable to finance the five-fold rise in oil prices since 1973.

The IMF's Interim Committee also decided that the proposed facility would borrow and lend at "market-related interest rates." To qualify for loans from the program, countries would have to meet stringent economic conditions set by the IMF to insure an improvement in their payments position.

The worldwide financing crisis caused by the drastic shift in capital flow had provided the IMF with a new and important function—that of managing the oil-importing nations' mounting debt and enforcing strict disciplinary measures on countries that drew on the new facility.

Business Week, in its March 28 issue, noted that the re-emergence of the IMF at the center of the world monetary system was sudden and unexpected.

In January 1976, when the decision was made to adopt floating currency-exchange rates, the IMF had been deprived of its crucial role as manager of the foreign exchange system created at the Bretton Woods conference in 1944.

Stripped of its responsibilities in a revamped monetary system, the IMF was given the job of providing aid to developing countries, "only to see itself shoved aside by the great private banks of the U.S. and Europe," Business Week said.

The banks were flush with cash and searching for new customers because the recession had dampened corporate demand at home for loans. As late as December 1976, the banks were contending they could manage the enormous financing requirements of the oil-importing industrial countries, such as Great Britain and Italy, as well as those of the developing countries.

However, a price increase imposed in December 1976 by the Organization of Petroleum Exporting Countries, and the possibility of another in July 1977, revealed the strains on the private banking system.

"The private banks are already stretched to their lending limits with no end to the deficits of oil-importing nations in sight," Business Week said. "It is becoming increasingly clear that the hemorrhaging of the oil-importing nations will not soon end."

Experts estimated that the OPEC nations had accumulated a $45-billion surplus, while the non-OPEC nations had run up an overall payments deficit of $50 billion. Many analysts at first had expected that the petrodollar problem would prove only temporary. The drain on oil-importing nations' wealth would end, the analysts had said, when OPEC began to invest its excess revenues and spend its money on imports.

Analysts now believed that the temporary phenomenon had become permanent, Business Week said. The world had "divided into two camps: many nations with chronic balance of payments problems and a dozen oil-producing countries with never-ending surpluses."

OPEC's demand for imports was not expected to increase beyond current levels, and, "until oil-importing nations cut back their growth, reduce their annual payments deficits and risk political instability because of higher unemployment, or until the price of oil is cut sharply, international debt will keep on building," the magazine said.

Arthur F. Burns, chairman of the U.S. Federal Reserve Board, concluded March 10, "Our banks simply can't continue lending on the scale they are." Burns called on OPEC "to play a major role. They, too, have to become bankers."

Private banks saw the "institutionalization" of their oil-import loans as the only solution to the problem. The IMF, which already had stepped in to assist Great Britain and Italy, as well as the developing countries, was the logical institution to meet these rising financing needs.

The IMF also provided a channel for drawing surplus OPEC funds into the Western monetary system. OPEC nations had been unwilling to participate directly in a lending program with the stronger industrial countries. The IMF's proposed facility would permit them to make financing available to their export customers in an indirect way.

Saudi Arabia's participation was vital to the success of the plan. It was the richest of the OPEC nations, and its revenues were growing so fast that the 80 banks worldwide that held Saudi deposits were unable to absorb any more funds.

Negotiations on the Saudis' contribution to the IMF's proposed lending facility were continuing. In Washington May 3, after the Interim Committee's approval of the plan, Saudi Arabia's finance minister said his country would participate in the plan.

On May 4, however, Finance Minister Muhammad Ali Abdul-Khail said the contribution would be more modest than the talked-about $4-billion figure. (IMF managing director H. Johannes Witteveen had hoped to raise a total of $20 billion, but he lowered the target April 29 to $16 billion.)

Ali Abdul-Khail contended that Saudi Arabia was "poor" because it had "only" money, not the wealth derived from a developed economy. He also claimed that as Saudi Arabia increased its spending on domestic development projects, its revenue surplus would disappear after "a number of years." (Saudi Arabia's surplus was estimated at $30 billion.)

Gulf states set up investment bank. The establishment of an investment bank by five Persian Gulf oil-producing states to channel their wealth into development projects was announced in Kuwait Nov. 18, 1975. The capital was $110 million, with participation equally divided among Kuwait, Saudi Arabia, Bahrain, Oman and the United Arab Emirates.

OAPEC to aid oil prospecting. The Organization of Arab Petroleum Exporting Countries agreed at a meeting in Riyadh, Saudi Arabia Nov. 22–23, 1975 to form a $300 million petroleum services company to assist Arab countries in oil prospecting, processing and marketing. OAPEC states initially would contribute $45 million, with the remainder to be furnished in phases. The company would have its headquarters in Tripoli, Libya.

U.S.S.R. Exports Energy

Soviets distribute Iran's gas. Under a five-nation accord signed in Teheran Nov. 30, 1975, Iran agreed to supply 13.4 billion cubic meters of gas a year through a new pipeline to the U.S.S.R., which would ship an identical amount of the fuel westward for sale in West Germany, Austria and France. Czechoslovakia would join the venture later. The sale was expected to bring Iran $400 million annually during the 20 years of the pact.

The new 1,000-mile pipeline to be built from Iran to the Soviet Union was expected to be completed by 1981 at an estimated cost of $2–$3 billion, with Iran reportedly planning to put up 30% and to raise the remaining funds in Western Europe.

French deal on gas transit. The French state-run utility Gaz de France announced Nov. 3, 1975 that it had concluded an agreement with West German and Austrian firms for piping Soviet natural gas into France.

The gas would be piped, at a rate of 80,000 cubic meters annually beginning in 1980, from Waidhaus on the West German-Czech border to Gernsheim, the present West German terminal. A new pipeline would be built from Gernsheim to Austria and also to the French-German frontier.

The pipeline across Austria would also be used for the transportation of Soviet gas to Austria.

Austrian gas, trade accords signed. The Soviet Union and Austria signed their third contract providing for Soviet gas deliveries in exchange for steel pipes provided by the Austrian state-owned concern Voest-Alpine, the New York Journal of Commerce reported Sept. 4, 1975.

Under the agreement, the U.S.S.R. would deliver 500 million cubic meters of natural gas annually from 1978 to the end of the century in exchange for 300,000 tons of steel pipes. The Austrian Kontrollbank would grant credit at an interest rate of 6.5%, the report said.

Energy exports up. Soviet exports of crude oil to the West increased 40% in 1975 to 480 million barrels a day, according to the New York Times May 24, 1976. Exports to Communist countries increased by 8% to 1.6 billion barrels a day.

The rise in oil exports resulted from a 7% increase in production, to 9.7 million barrels daily, and a reduction in the growth of domestic consumption, which rose 5% in 1975, compared with 7%–8% rises in previous years.

The Financial Times of London reported Jan. 12, 1976 that the Greek Public Power Corp. had contracted with the Soviet trade firm Energomachexport for two 300-megawatt units for the lignite-fired thermoelectric station at Ptolemais in northern Greece.

The Journal of Commerce said Jan. 19 that Greece had purchased 500,000 tons of crude oil from the Soviet Union for a total of about $42 million or $12.72 a barrel. Half the value of the oil would be paid in convertible currency and the rest in the barter of agricultural goods.

Soviets cut Czech oil. Soviet deliveries of crude oil to Czechoslovakia would be sharply reduced in 1976–80, a Czech Presidium member said Oct. 4, 1976. The extent of the cuts and the cause were not disclosed. He said that the lack of energy resources had affected power production and would hinder national economic growth. He blamed power plant builders, engineers and miners for the problem.

Czechoslovakia had taken drastic steps to conserve energy resources, the Wall Street Journal reported Oct. 18. Street lighting in Prague suburbs was reduced, floodlighting of monuments was ended, shops were ordered to dim display lights and factories received periodic reductions in power supplies during working hours.

According to the report, the breakdown of aging power stations and the neglect of developing Czech coal reserves had contributed to the crisis. The increase in Soviet oil prices and the lack of foreign exchange to buy more expensive non-Soviet oil were other factors.

Oil prices raised. The Hungarian government radio Jan. 6, 1977 reported a price increase of about 22.5% on Soviet oil scheduled for delivery in 1977. The price per barrel was raised to $8.90 from $7.15. The increase was expected to cost Hungary an additional $2 million.

A similar increase was expected to affect other Eastern European oil importers, notably Poland, Czechoslovakia and East Germany, which each imported an average of twice as much Soviet oil as Hungary.

The 1976 price for Soviet oil had been approximately 7.6% higher than the 1975 price. The 1977 price hike exceeded the 5%–10% rise charged Jan. 1 by the Organization of Petroleum Exporting Countries (OPEC) but was still well below the OPEC charges of $12–$14 a barrel.

The abrupt 1977 Soviet price increase was believed to reflect the sharp OPEC price rises since 1974. International trade prices for members of the Council for Mutual Economic Assistance (COMECON) were based on average world prices over a three-to-five-year period preceding the year in which a COMECON export agreement was signed. A member purchasing in excess of the quota set by the agreement was charged the current world price for the excess.

Soviets seek to form oil cartel. The Soviet Union had approached a number of Arab members of OPEC about forming a separate oil cartel with Moscow and East European countries, the Kuwait Times said Jan. 24, 1977. According to Abu Dhabi press reports Jan. 29, Iraq had refused to discuss the idea out of concern for the unity of OPEC and other countries were resisting the proposal for similar reasons.

India oil-trade deal. India and the Soviet Union Feb. 19, 1977 agreed on the barter of Soviet oil for Indian goods. The Soviet Union was to furnish 1 million metric tons of crude oil in exchange for Indian goods of the same value, mainly pig iron and steel construction components. The deal was part of a larger agreement reached in principle for the Soviet Union to ship 5.5 million metric tons of crude oil to India from 1977–80.

Yakutsk deal. The Soviet Union concluded an agreement March 31, 1976 with the Bank of America and Japanese banks providing the U.S.S.R. with $50 million for exploration and development of gas reserves in the Yakutsk region of Siberia.

North Sea Oil & Gas

British progress continues. Queen Elizabeth II pushed a button at Dyce, Scotland, near Aberdeen, Nov. 3, 1975 to formally inaugurate the flow of oil between British Petroleum's Forties field in the North Sea and the refinery at Grangemouth.

The government had invested some $1.5 billion in drilling rigs and pipelines for an operation expected finally to reach a peak production of 400,000 barrels a day and to make Britain self-sufficient in oil by 1980.

The Halibut group, a combine headed by Britain's Burmah Oil Co., had announced the discovery of a new oil well in the North Sea Sept. 5. The well, located 7½ miles northwest of the Thistle field, tested at 3,600 barrels a day.

The Conoco oil group, in which the National Coal Board had a one-third interest, announced the discovery of a natural gas well in the North Sea Sept. 17. The well, located 125 miles northeast of Aberdeen, tested at 19.55 million cubic feet of gas a day.

The Financial Times July 9, 1976 reported three new North Sea finds, including one believed to contain major reserves. The largest find, announced by the Phillips Group, was linked to the important Brae Field, east of the Orkney Islands. Reserves in the Brae Field were estimated at over one billion barrels of oil and between two and three trillion cubic feet of gas. A second find, announced by Burmah Oil, was seven miles north of the Thistle Field, which was northeast of the Shetland Islands. The third, made by the Transocean Group, was just south of the Brae Field. (The Financial Times May 28 had reported a find some 110 miles southeast of the Shetland Islands on Block 9/19. It had been announced by Conoco North Sea, operator for a group involving Continental Oil, Gulf Oil and British National Oil Corp.)

Mesa Petroleum of Amarillo, Tex. announced Aug. 19 that it had made an important oil discovery 12 miles off the coast of Scotland.

The company said that the commercial significance of the find was not known, but that core samples indicated the presence of 300 feet of oil-bearing sands between 6,060 and 6,870 below sea level.

The firm said that the find was important because the sea depth of only 150 feet and the proximity to the coast would cut drilling and transport costs. Other recent finds had been made in waters from 300 to 500 feet deep, more than 100 miles from shore.

Mesa held a 25% interest in the find. The company's five partners in the venture were Kerr-McGee, 25%; Hunt Oil, 15%, P & O Petroleum, 15%; Cresslenn UK, 15%, and Exploration Holdings Corp., 5%.

A transaction described as the first major third-party sale of petroleum products from the North Sea for the U.S. was announced April 7, 1976. The deal, valued at $1 billion in propane and butane gases, was concluded by the Royal Dutch/Shell Group and Northern Natural Gas Co., a Nebraska-based utility.

The first shipment of North Sea oil was exported from Britain April 27 when 350,000 barrels of crude left the Forties Field by tanker for Hamburg, West Germany.

John Cunningham, undersecretary of state for energy, said Oct. 3, 1976 that oil output from the five operating North Sea fields totaled more than one million tons per month. In addition, he said, the goal for self-sufficiency in oil by 1980 was "on target."

Cunningham, speaking at the opening of the World Offshore Exhibition in London, said that the British success ratio of one find in three attempts compared with a worldwide offshore average of one in 20.

By the 1980s, he said, the United Kingdom would be one of the top 10 oil producers in the world. "From 100,000 tons of landward oil in 1970 to 100 million tons of offshore oil in 1980 is a remarkable successs story and one which is perhaps still not given due weight in assessing this country's economic position," he said.

Loans & other financial transactions—A total of $315 million in Euroloans were being provided the Occidental Petroleum Corp. and Thomson Scottish Associates, a private company, to finance their shares of the Piper and Claymore North Sea oil fields, it was announced Feb. 4, 1976.

The International Energy Bank and the Republic National Bank of Dallas were managing the loans, it was reported Feb. 5. The terms of the financing permitted the Occidental-Thomson consortium to utilize earnings from the Piper field as security for the Claymore loan.

It was reported Jan. 13 that the government had loaned $81.3 million to Burmah Oil Co. in connection with its negotiations to obtain control of the oil company's 21% interest in the Ninian field, one of the largest in the North Sea. The government then announced March 10 that it had bought Burmah's interest for $180 million.

British Petroleum (BP) and its largest shareholder, the government, agreed in principle July 1 to an arrangement by which the state-owned British National Oil Corp. (BNOC) obtained a 51% interest in BP's North Sea oil fields.

Under terms of the agreement, BNOC would have, effective Jan. 1, 1977, an option to buy at market prices up to 51% of the oil produced at BP's North Sea fields. In 1977 and 1978, BNOC would sell back to BP 100% of the option oil taken. Between 1979 and 1981, BNOC would sell a major portion of its option oil back to BP provided BP made available to BNOC an equal amount from its production sources throughout the world. From 1982 to 1989 the two firms would exchange crude oil so that BP had adequate supplies.

The agreement, negotiated by Energy Secretary Anthony Wedgwood Benn, was signed by Benn and by the chairmen of BP and BNOC. Benn invited Exxon Corp. and the Royal Dutch/Shell Group to meet with him to discuss government participation in their North Sea ventures.

BNOC Jan. 5, 1977 signed with Royal Dutch/Shell and Exxon subsidiaries a preliminary pact giving BNOC 51% of oil produced in their joint-venture North Sea fields.

Benn negotiated the agreement with representatives of Shell U.K. Ltd. and the Esso Petroleum Co.

British-Norwegian cooperation. The first oil from Norway's Ekofisk (North Sea) field began flowing Oct. 15, 1975 through pipelines ending at Teesside, England, it was announced by a spokesman for Phillips Petroleum, one of the participating groups.

Norwegian Industry Minister Bjartmar Gjerde Sept. 10, 1976 announced an agreement with Britain for closer contacts on North Sea oil and gas operations.

The agreement, reached Sept. 10 in London by Gjerde and British Energy Secretary Anthony Wedgwood Benn, provided for the establishment of a Norwegian-British coordinating committee on offshore matters. Some Cabinet members of each country would serve on the committee, Gjerde said, and would consider such matters as joint commercial marketing of the oil and gas, schemes for collecting gas on both sides of the North Sea median separating the two countries' areas of exploitation and safety.

Norwegian developments. Arve Johnson, managing director of Norway's state-run Statoil Company, announced Jan. 6, 1976 that the projected yields from the Statfjord Field in the North Sea could reach levels of 900,000 barrels per day by 1984, which would make it the largest find in those waters. The Statfjord Field was said to contain about 520 million tons of oil worth approximately $63.6 billion at current prices.

Earlier, Statoil Jan. 5 had revealed development plans for the oil find that included an investment of about 14 billion krone (over $2.5 billion) for production facilities.

In a related development, Parliament Jan. 7 agreed to a government plan establishing a national petroleum company to manage the refining and distribution of petroleum products for domestic use and export. As part of the program, approved 78–74, Norway would buy a 95% interest in Norsk Braendselsolje AS, a subsidiary of the British Petroleum Co. Ltd. The new state-owned company, provisionally known as Norsk Braendselolje II, would also incorporate in whole or part two other oil distribution and marketing firms. The cost to the government was reported to be $160 million. The government revealed Aug. 7 that it had approved an agreement to take over Norsk Braendselolje, paying $115 million for BP's 50% interest in the company and the 18% owned by Norwegians.

Norway announced Aug. 6 that Statoil and BP Petroleum Development of Norway A/S, a British Petroleum subsidiary, had been granted oil exploration rights in two areas of Norway's North Sea sector.

The two companies would hold equal shares in block 6 in field 29 and block 4 in field 30, located between the Stratfjord and Friggfields, on the boundary with the British exploration sector; however, if oil were found, Statoil's share would be increased on a sliding scale relating to the size of the find, up to a ceiling of 70%.

BP would be the operator of the drilling program, but a stipulation in the agreement would allow Statoil to take over the operator function if a find were to stretch into the British sector.

Oil-well blowout capped—A Norwegian oil well in the North Sea was capped April 30, 1977 after it blew out and spouted uncurbed for eight days. A 12-man crew of U.S. and Norwegian workmen succeeded in halting the flow after four previous attempts had failed. An estimated 8.2 million gallons of crude oil spilled into the waters between Great Britain and Norway while the well blowout continued.

The well, known as Bravo 14 in the Ekofisk field, was located 180 miles west of Stavanger, Norway.

The successful capping was accomplished by a U.S. team, based in Houston and led by Paul "Red" Adair, a specialist in oil-well fires and blowouts. Earlier attempts had failed because the capping equipment was not strong enough to resist the pressure from the well, which had pushed oil into the air at 400 miles per hour.

A major problem was that the well's blowout protector had been installed upside down by the drilling subcontractor, Moran International. After a computer calculated that increased hydraulic pressure was needed to close the protector rams, the piston in the hydraulic pump was lengthened an eighth of an inch and the rams closed off the runaway well. Mud was pumped into the oil and gas formation in the sea bed under the well to prevent further blowouts at the site.

During the blowout, a firefighting ship maintained a continuous water spray on the platform to prevent it from over-heating. There was constant danger of an explosion from static electricity or from sparks caused by falling steel debris.

The Norwegian government decided not to use chemical dispersants on the oil slick because of concern about the effect of its use on marine life. Attempts to skim the oil from the surface with long booms was almost a total failure, with less than 600 tons reported recovered May 2 from the effort.

The oil slick from the blowout well was reported April 23 to have covered about 200 square miles of the area in discontinuous patches. Wave action, and the

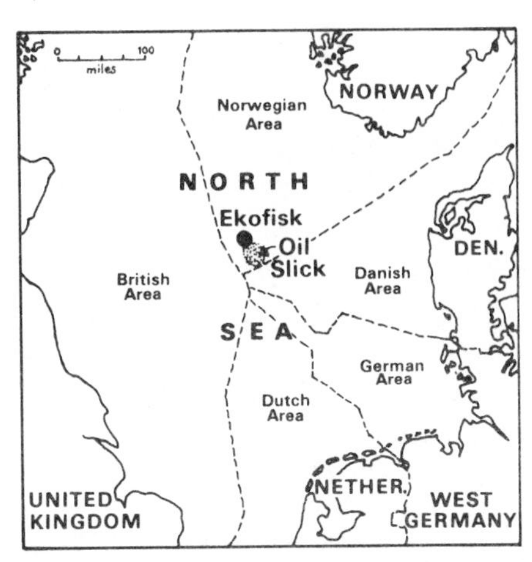

A blowout in Norway's Ekofisk well poured oil into the North Sea for eight days before being capped. No oil reached shore.

tendency of heavier oil to sink and lighter oil to evaporate, had prevented the formation of a single huge slick. A spokesman for Phillips Petroleum Co., which operated the well, May 2 said the slick from the Ekofisk well "had virtually disappeared," and that only scattered portions of the spill remained.

The danger to marine life, particularly fish and sea birds, generally came from the concentration of offshore oil development in areas where wildlife was most abundant. The fishing banks where half the world's fish were caught were concentrated in 1% of the ocean's seabeds. These were also the areas where most oil drilling occurred.

The major threat caused by the spill came from its proximity to mackerel spawning grounds. The well reportedly was contained before the oil could do significant damage.

Premier Odvar Nordli's Labor government April 28 suspended production at all Ekofisk field wells. The move was a reaction to pressure from left-wing Socialists, who supported the government in the Storting (parliament), where it had a one-vote majority. While the Conservative Party would have voted with the government to continue offshore oil production, the upcoming September elections in Norway made the government anxious to moderate the backlash from the electorate over the oil spill.

Offshore development had been a major issue in Norway before the Ekofisk blowout. In addition to having environmental concerns, many Norwegians believed that oil companies, such as Phillips, had contracts that were too generous and of too little benefit to Norwegians. Although Phillips paid large royalties and was limited in its production, the state had no direct ownership in the Ekofisk field.

The Ekofisk spill was expected to speed up negotiations between European states for cooperation on oil spills. The Organization for Economic Cooperation and Development (OECO) May 1 recommended that international regulations be drawn up to govern the safety of offshore oil wells.

Norwegian authorities reopened the Ekofisk fields May 2 and oil flow resumed in the pipelines between the wells and Great Britain.

Holland increases role. The Dutch government sent new regulations to Parliament increasing its participation in and share of profits from future North Sea oil and gas ventures, the Economy Ministry reported March 25, 1976. According to the new directive, the government would increase its holdings in new concessions from 40% to 50% and would up its share of the balance of profits from 50% to 70%.

Other Events

IEA agrees on energy plan. The governing board of the 18-nation International Energy Agency agreed at the conclusion of a meeting in Paris Jan. 30-31, 1976 to a long-term program of energy development to make its member states less dependent on the Organization of Petroleum Exporting Countries.

The agreement included a minimum price of $7 a barrel of oil to protect new investment; the removal of obstacles that might impede development of IEA energy resources; the bolstering of energy conservation measures; more joint research projects; and an exchange of information on conservation techniques in member countries.

The American representative at the conference, Assistant Secretary of State for Economic Affairs Thomas O. Enders, said the purpose of the accord was to provide "energy self-sufficiency in a reasonable time period," restore balance in the market and give consumer countries a role in price-fixing.

Canadian Deputy Under Secretary of External Affairs Peter Towe had argued during the meetings that his country could not commit itself to providing other IEA members free access to its oil. This position was challenged by Japan, West Germany and Denmark, who already had threatened withdrawal of earlier approval of a minimum price unless Canadian energy sources were made available to them. Despite Canada's refusal to make this commitment, it did provide "satisfactory assurances" that it would work toward removing restrictions, Enders disclosed.

Coal development cited in UN report. The United Nations Economic Commission for Europe released a study that said the major Western European governments had begun to place more emphasis on coal as a future source of energy, it was reported Feb. 10, 1976. The commission predicted that by 1985 coal would account for 21% of Western Europe's energy needs as opposed to earlier expectations of 11% to 12%. In addition to the European reaction to the oil crisis of 1973-74 when Arab oil production was reduced, the report said that general disappointment with the pace and safety of nuclear power development had contributed to the reassessment of the importance of coal.

Belgian firm buys Algerian gas. Sonatrach, Algeria's state petroleum corporation, announced Nov. 27, 1975 that it had signed a contract with the Belgian firm Cie. Distrigaz to deliver 2.471 trillion cubic feet of liquefied natural gas over 20 years, beginning in 1979. According to the French newspaper Le Monde Nov. 29, this was the first contract to have been signed individually by a member of the European consortium Eurogaz.

Tension over Aegean oil. Reports June 27, 1976 of "significant" hydrocarbon deposits off the Greek coast aggravated tension between Greece and Turkey. Greece warned Turkey against conducting any surveys in Greek waters off islands near Turkey, it was reported July 1.

Greece claimed continental-shelf rights for each of its islands off the Turkish coast and threatened military action if a Turkish seismic survey ship explored for oil near those islands. Turkey had claimed the islands were part of its continental shelf, and that Turkish territorial waters consequently extended to the deepest point of the Aegean, well beyond the Greek coastal islands.

The U.S. was said to have pressured Turkey to postpone its exploration operations, according to a Turkish newspaper report July 13. The newspaper Cumhuriyet indicated that Prime Minister Suleyman Demirel had agreed to put off the survey for two weeks.

Uganda scores Kenya over oil shortage. Ugandan President Idi Amin accused Kenya July 16, 1976 of embargoing oil sent to Uganda. He asked neighboring Zaire, Sudan and Rwanda to appeal to Kenya to resume normal shipments.

In separate meetings with ambassadors of the three countries, which also depended on Kenyan oil shipments, Amin said that over 200 tankers and other supply vehicles were held up in Kenya in a deliberate attempt to disrupt Uganda's economy. British subjects who arrived in Nairobi from Uganda July 16 reported that gasoline was scarce in Kampala and that many people had to walk to work, causing problems for Uganda's economy and its civil service operations.

Kenya denied prohibiting oil shipments, but admitted that the Kenya-Uganda railway was operating at only 60% capacity, allegedly because of a lack of spare parts. There were reports July 16 that Kenyan tanker drivers were refusing to enter Uganda because Ugandan soldiers had seized 30 Kenyan tankers bound for Rwanda and Burundi earlier in the week.

In a veiled warning to Kenya, a Ugandan military spokesman said in a July 17 radio broadcast that Ugandan warplanes were capable of flying to Kenya's Indian Ocean port of Mombasa and back without refueling.

Kenyan Vice President Daniel Arap Moi repeated Kenya's denial of a deliberate economic blockade July 18. There were reports that day that a train carrying fuel had crossed into Uganda after a foreign-currency deposit had been made in a Nairobi bank to cover transport costs.

In a sudden change of mood, President Amin announced in a broadcast reported July 20 that Uganda had no intention of attacking Kenya. He called Ugandans and Kenyans "blood brothers." Amin also commended British Prime Minister James Callaghan for his role during the Uganda-Kenya crisis.

Amin's abrupt change was interpreted by some observers as a reaction to reported mutinies in the Ugandan army.

The thaw in the Uganda-Kenya dispute was short-lived, as Uganda's fuel crisis showed no apparent signs of abatement. Uganda announced July 22 a ban on all sales of gasoline for private use, reserving the supply for government vehicles, industry, doctors and schools. Foreign airline flights were suspended effective July 25.

Amin resumed his threats to Kenya July 24. He cut off electricity supplies to Kenya from the Owens Falls power station for lack of fuel, and again threatened to attack Kenya unless fuel shipments came through. He accused Kenya of holding up 600 tanker trucks with oil that had already been paid for. Kenya charged that Amin had not paid for the shipment.

Radio Uganda reported July 25 that Amin had sent telegrams to the United Nations and to the Organization of African Unity (OAU) to appeal for help in the dispute with Kenya. He said that Uganda had fuel supplies to last only 5 more days and he called for fact-finding commissions from both organizations to study the situation.

An agreement designed to end the state of tension between Kenya and Uganda was signed by Kenyatta and Amin in Nairobi Aug. 7. Amin Aug. 10 freed 72 Kenyans who had been held since July 19 in "protective custody." Amin originally had promised to release 99 Kenyans, including 24 truck drivers, but there was no report on why they were not at the border crossing point with the others.

In related developments, electric power supplies to Kenya from Uganda's Owen Falls station were partially restored, the London Times reported Aug. 23. Kenyan oil deliveries began to reach Kampala and Ugandan troops were withdrawn from the Kenyan border, according to Washington Post reports of Aug. 17 and Aug. 28.

OECD stresses conservation need. The Organization for Economic Cooperation & Development warned Jan. 27, 1977 that its 24 member states—the nations of Western Europe, North America and Japan—would be unable to satisfy their demand for oil by 1985 if they maintained their current energy policies. The warning was contained in an unusually tough-worded report on the world energy outlook.

The report predicted that OECD members' demand for imported oil could rise to 35 million barrels a day by 1985, compared with 23 million a day in 1975. The forecast was based on three assumptions: unchanged energy policies, an average economic growth rate of 4.2% a year between the end of 1974 and 1985 and an oil price of $11.51 a barrel (the price of light Arabian crude before the latest increase).

Total world demand for oil from the Organization of Petroleum Exporting Countries (OPEC) would probably exceed 39 million barrels a day by 1978, the report said. It estimated that OPEC's oil-producing capacity would reach no more than 45 million barrels a day by 1985, however. The margin between supply and demand in the coming years thus would be very small, the report noted, even without taking into consideration the possibility that OPEC nations might hold production below capacity for conservation, political or other reasons.

In the face of the economic and political consequences of increasing demand for imported oil, the report said, the OECD countries had only one realistic course to follow: expand domestic energy sources and cut consumption of energy. To achieve the former, the report suggested elimination of price controls on crude oil and natural gas in the U.S., an increase in exploration efforts and a weakening of environmental protection-inspired restraints on the development and use of coal and nuclear power. The report suggested imposing selected mandatory controls, such as speed limits, to help cut energy use. It further proposed that member countries build larger buffer stockpiles of fuel sources to offset the effect of any deliberate supply reductions.

Such action would have to be taken within the next year or two, the report said, if its effects were to be felt within the next decade.

Noting the special role of the U.S. in the success of these efforts, the report said the U.S. was capable of reducing its oil imports by as much as all the other OECD members put together. It said, too, that U.S. consumers currently used as much energy as the rest of the world put together.

The report represented an expansion and revision of an earlier OECD study released in January 1975. Speaking of the revisions, an OECD spokesman said that if conservation measures were not taken, OECD net oil imports in 1985 would be 69% higher than had been forecast in the earlier study.

In other projections, the report said actual energy consumption by OECD countries would grow at a rate of 3.6% a year between 1974 and 1985, compared with a rate of 5.1% a year from 1960–73. (This prediction assumed an economic growth rate of 4.2% a year.) Domestic energy supplies in the OECD area were projected to grow by 3.5% a year from the end of 1974 to 1985, compared with 2.8% a year from 1960–74.

U.N. experts urge oil conservation. A preliminary summary of a United Nations conference of petroleum experts, reported April 26, 1977, indicated that oil and natural gas resources probably would exist at substantially higher than current costs for as long as 100 years. However, this conclusion was based on the assumptions that strict energy conservation measures would be taken promptly; that there would be great improvements in the technology of recovering conventional sources of petroleum and gas, and that it later would be possible to tap unconventional oil and gas sources not currently commercially exploitable.

The conference, sponsored by the U.N. Institute for Training and Research and the International Institute for Applied Systems Analysis, had been held in Austria in July 1976.

A U.N. official said that despite this forecast, the view of most participants at the conference seemed to agree with U.S. President Jimmy Carter's recent conclusion that petroleum supplies would be tight in the 1980s and 1990s.

Saudis to avoid oil embargo. Crown Prince Fahd of Saudi Arabia assured President Carter during talks in Washington May 24–25, 1977 that his country would not embargo oil to the U.S. to force concessions from Israel in order to achieve a Middle East peace settlement. Reports of a possible renewal of such action (previously employed in 1973) had begun to circulate in the Middle East May 22 when Egyptian Foreign Minister Ismail Fahmy was quoted as having said that if the forthcoming right-wing dominated Israeli government made no concessions, "the oil weapon will be used again automatically."

Carter, commenting on his talks with Fahd, said the Saudi leader had told him that the statement attributed to Fahmy "was a completely false report." Carter added, "there's no threat of an oil embargo at all."

Worldwide Domestic Developments

Angola

Oil output cited. Gulf Oil Corp., the largest foreign oil producer in Angola, reported that its production level would be maintained at about 120,000 barrels a day, the "maximum efficiency rate" of the wells, according to a Gulf spokesman quoted by the Wall Street Journal Aug. 19, 1976. Figures for oil output before the civil war were given at 140,000 barrels a day.

Gulf had announced Dec. 22, 1975 that it had suspended oil-drilling operations in Angola and withdrawn its personnel because "border warfare and continued civil war in Angola have made it impossible for the movement of personnel, supplies and equipment for the maintenance of operations there." The statement added that the U.S. state department Dec. 19 had asked the firm to withdraw its employes and their dependents.

Gulf also said it had established a "special interest-bearing account for the benefit of the state of Angola" in which it was placing in escrow some $95 million in taxes and $30 million in royalties due the Angolan government from Gulf's oil production offshore the Cabinda enclave.

Gulf alleged that the contending factions in Angola had made conflicting demands on the firm for the tax and royalty payments. The funds in the special account would be paid to the state of Angola "when it has a government that is in control of the territory and population and this government has been generally recognized by the world community," Gulf said.

(The government proclaimed by the Soviet-supported Popular Movement for the Liberation of Angola [MPLA] had been recognized by a number of African and socialist countries. The rival government established by the Western-backed Front for the Liberation of Angola [FNLA] and the National Union for the Total Independence of Angola [Unita] had not been accorded formal recognition by any nation.)

The Washington Post reported Dec. 23, however, that Gulf had been under pressure from the State Department to withhold the payments which would, in fact, have been made to the MPLA, which controlled Cabinda. Payment of the quarterly taxes to the MPLA would have created an ironic situation in that the U.S. government was supporting the FNLA and Unita with some $60 million in aid, while the U.S. oil company would simultaneously be paying more than twice that amount to the Soviet-backed group the two others were fighting.

Gulf, however, had begun negotiations with the MPLA government March 25, 1976, and agreement to resume oil production in Cabinda was reached by early April. An Angolan spokesman had said that his country was losing more than

$1.5 million each day that Cabinda operations remained suspended. The government had announced March 11 that $102 million of the money Gulf had placed in escrow had been turned over to the Luanda regime.

Australia

New oil prices in effect. A new sliding price system for oil, announced by Prime Minister Gough William Sept. 14, 1975, went into effect Sept. 18.

Hailed by the Australian Petroleum Exploration Association Sept. 15 as an incentive necessary to spur exploration in Australia, the raise would give the producer about $A6.90 ($US 9.17) a barrel (not counting a $A2 government excise tax), a price comparable to the price paid for imported crude. Domestic oil had been price-controlled at $A2 a barrel. Prices for oil from existing wells would rise between A23 cents and A85 cents, depending on the costs of production.

In a related development, national crude oil production rose by 6.3% in 1975 to approximately 150 million barrels, it was reported March 23, 1976.

Gas exploration halts. Santos Ltd. said Sept. 22, 1975 that it had halted natural gas exploration in South Australia's Cooper Basin. The company said the fields were unprofitable. Natural gas, which provided 27% of South Australia's primary energy, sold for one-third the price of liquid petroleum, although it cost twice as much to develop and market.

Oil discovery. The Australia High Commission announced Dec. 11, 1975 that a significant oil deposit had been discovered on the northwest shelf of Australia by the Woodside-Burmah oil and gas consortium. An exploraton well had produced 3,300 barrels per day, the company revealed.

Tenneco sells out. The Tenneco Oil Co. announced Dec. 19, 1975 that it had ended its operations in Australia and sold its remaining interest in International Oil Ltd., which operated the large, but increasingly depleted, onshore Moonie oilfield in Queensland. Tenneco, which had given notice of its intention to quit Australia three years ago, divested itself of 1.3 million shares of stock.

Northwest shelf opened for oil development. Western Australia State Mines Minister Andrew Mensaros said March 22, 1976 that the northwest coastal shelf area had been opened for petroleum exploration and development.

Coal-mining project planned. A consortium of Australian, British and Japanese firms announced plans for a coal-mining project at Warkworth in New South Wales, the Wall Street Journal reported July 7, 1976. The prospecting, which would cost about $124 million, was expected to be finished late in 1979. William Evans, director of Costain Australian Ltd., a British-controlled venture, said that coal reserves at Walkworth were estimated at 500 million tons and could be doubled with more drilling. The four firms involved were Costain, with a 30% share; C. Sleigh Ltd. of Australia, 45%; Australian Resources Development Bank, 15%; and Mitsubishi Development Pty. Ltd., 10%.

The consortium concern, Warkworth Mining Pty. Ltd., expected to apply by Jan. 1, 1978 for a permit to develop the mine.

'76 coal exports set record. The Joint Coal Board said coal exports had risen to a record 34.1 million tons in 1976, it was reported Jan. 25, 1977. This was an increase of 4.2 million tons over the 1975 sales total. Shipments to Japan in 1976 totaled 11.2 million tons, close to the record 11.3 million tons shipped in 1973.

Coal production in the Australian state of New South Wales rose 11% to a record 44.7 million tons. Exports rose 3% to 14.9 million tons. Consumption there increased to 20.3 million tons in 1976 from 19.3 million tons in 1975.

Coal exports from Queensland, the other Australian state with large reserves, rose to a record 19.2 million tons in 1976 from 15.4 million tons in 1975.

Coal usage by the Australian steel industry in 1976 was about the same as in the previous year, but the use of coal in power stations rose to 10.3 million tons in 1976 from 9.1 million tons in 1975.

The Joint Coal Board warned Jan. 25 that coal production in New South Wales would be cut back unless markets expanded beyond the 3% growth in export sales registered in 1976. Coal stockpiles held by companies at the end of the year were 60% higher than at the end of 1975.

Brazil

Foreign oil exploration approved. President Ernesto Geisel announced Oct. 9, 1975 that the state oil firm Petrobras would be authorized to sign "risk contracts" with foreign petroleum companies for oil exploration in Brazil and off its shores.

The contracts, under which the companies would be paid, in oil, only if they found new deposits, were opposed within the cabinet by Industry Minister Severo Gomes, who argued they were unconstitutional, and by Foreign Minister Antonio Azeredo da Silveira. They were also opposed by nationalist military officers and civilians. However, Geisel rallied important supporters to his side, including Gen. Adhemar de Queiroz, a former Petrobras president.

Geisel explained that "foreign capital is needed because there are immense areas to explore, beyond the possibilities of the country." He denied that the contracts would affect Petrobras' oil monopoly.

Brazil produced only 20% of its oil, importing the rest from the Middle East. It imported $2.8 billion worth of oil in 1974, more than half the value of all Latin American oil imports. Brazilian oil production at the end of July was 39.8 million barrels, .3% below the output a year earlier, it was reported Sept. 2. Petrobras announced that the offshore deposits at Campos contained no more than 200 million barrels of oil reserves, lower than earlier estimates, it was reported Oct. 3.

Petrobras' profits in January–June had risen to $300 million, 48% above the same period of 1974. Sales had increased by 41% to $3.2 billion, and investments by 132% to $750 million, the government reported Aug. 8. Petrobras received a $100 million revolving credit facility from a consortium of foreign banks led by Libra Bank of London, it was reported Nov. 28.

The high cost of oil imports contributed to Brazil's growing trade deficit, estimated for 1975 at $3.7–$3.8 billion, it was reported Dec. 3.

Oilfield found. Mines Minister Shigeaki Ueki said technicians had found a major new oilfield on the continental shelf off Rio de Janeiro (reported Jan. 6, 1976). The field was estimated to contain at least 240 million barrels of oil, and further testing would be conducted to determine whether the area could contain as many as 800 million barrels, as some early tests had indicated, Ueki said.

Canada

Gas exports to U.S. to be reduced. Exports of Canadian natural gas to the U.S. would be reduced because of an impending shortage, Donald Macdonald, minister for energy, mines and resources, said July 16, 1975.

"Natural gas supplies won't be adequate in the near term to meet both projected increases in domestic demand and existing export commitments," he said, commenting on a report compiled by the National Energy Board. The growth of domestic demand would also be restrained until further supplies of gas became available, Macdonald told the House of Commons.

(Canada produced 2.4 trillion cubic feet of natural gas a year, about one trillion cubic feet of which was exported to the U.S. Though this represented only 4.5% of the total gas consumed by the U.S., some areas, notably the Northwest, were especially dependent on Canadian sources of fuel.)

Further cuts in oil exports to U.S. Canada Nov. 20, 1975 made new cuts in its scheduled estimate of reduced oil exports to the U.S. Ottawa cut the 1976 export rate to an average of about 500,000 barrels a day, effective Jan. 1. This was 50,000 barrels below the level previously assigned. The government said all petroleum exports would be phased out by the end of 1981, or two years earlier than had been planned. The program to phase out oil exports to the U.S. had been announced in 1974.

According to the revised schedule, the 1976 export rate would be set at 510,000 barrels a day for the first months of the year, but would be gradually reduced to 250,000 barrels a day as the oil pipeline from Sarnia, Ontario to Montreal went into operation.

National Energy Board (NEB) Chairman Marshall Crowe Nov. 23, 1976 announced a fresh cut in Canadian oil exports to the U.S. to about 305,000 barrels a day, effective Jan. 1, 1977.

The daily export ceiling had been lowered Nov. 1 to 400,000 barrels from 435,000 barrels and the December maximum had been set at 385,000 barrels, the Oct. 13 Toronto Globe and Mail said. The monthly export cutbacks were part of a Canadian energy self-sufficiency program. As pipeline construction enabled more Canadian oil to reach eastern urban centers from western fields, the amount available for export from those fields was diminished.

The NEB announced Nov. 10 that duties on oil shipped to the U.S. would rise as of Jan. 1, 1977. The levy on light and medium crudes had been lowered Nov. 1 to $3.75 a barrel from $3.80 and had been raised to $3.35 a barrel from $3 on heavy crudes, other than Lloydminster crude, which remained at $2.90. (Lloydminster was a heavy crude used largely during the summer months for asphalt production.) The new charges announced by the NEB would be $4.60 a barrel for light and medium crudes, $3.75 for Lloydminster and $4.20 for other heavy crudes.

Alberta pact on petrochemical project. The Alberta Gas Ethylene Co., Ltd., the Alberta Gas Trunk Line Co., Ltd., Dow Chemical of Canada, Ltd. and Dome Petroleum, Ltd. said Sept. 21, 1975 that they had reached an agreement with the government of Alberta for the first phase of a petrochemical manufacturing project in the province.

The first phase, expected to involve capital outlays of $1.5 billion, would produce ethylene-based products which would help to improve Canada's balance of payments position by approximately $750 million annually, beginning in 1979, the companies said. (Ethylene was derived from ethane, which was extracted from natural gas. Alberta produced more than 80% of Canada's natural gas.)

The project was reportedly "fully consistent" with the provincial government's policies on petrochemical development and no subsidy would be required, the companies said.

Ottawa, Alberta pact raises gas price. The federal government and the province of Alberta reached final agreement on domestic natural gas pricing and the method of returning gas export earnings to producers, most of them within Alberta, Donald Getty, provincial minister of energy and resources, said Oct. 20, 1975. The pact served to resolve legal tangles that had threatened the gas pricing accord reached Sept. 23 and scheduled to start Nov. 1.

The problems were resolved, Getty said, when Ottawa agreed to Alberta's demand that proceeds from higher export prices be extended to all Alberta natural gas producers; previously, the federal government had wanted the higher export price to benefit only those pipeline concerns or producers which sold for export.

Under the agreement, the export price of natural gas rose from $1.40 to $1.60 a thousand cubic feet. Canadian domestic prices would rise to a lesser extent, with a Toronto city gate, or wholesale, price of $1.25 a thousand cubic feet and a new Alberta border "value" on gas leaving for extraprovincial markets, to be computed on the basis of the Toronto city gate price and the ruling wholesale domestic gas price.

Ottawa to allow oil-price increase. The federal government said May 18, 1976 that it would allow domestic crude oil prices to rise $1.75 a barrel to $9.75 by Jan. 1, 1977. A $1.05 increase—the first step in the two-stage hike—would take effect July 1 when the present $8 price level expired.

The decision reflected Canada's aim to move gradually toward world prices while trying to control high inflation rates, Energy Minister Alastair Gillespie said. It also took into account "the capacity of

Canadian industry to compete in world markets," he said.

The government action was taken after Prime Minister Pierre Elliott Trudeau and the provincial premiers May 6 discussed oil and natural gas price increases but failed to reach agreement.

In the inconclusive talks, the federal government had broadly endorsed a demand by Alberta and Saskatchewan, the two oil-producing provinces in western Canada, for a flat $2 oil-price increase. The oil-importing provinces of eastern Canada had opposed large increases; Ontario pressed for a "blended price formula" that would have raised the price only 20¢.

Trudeau told newsmen May 6 after the talks adjourned that if a consensus of premiers or at least "overwhelming support" were not forthcoming, the Petroleum Administration Act allowed Ottawa to set the price with the concurrence of Alberta and Saskatchewan. Were such agreement not secured, the act provided for the federal government to act unilaterally, Trudeau said.

Petro-Canada buys Arco unit. Petro-Canada, the Canadian state oil company, settled with the Atlantic Richfield Co. June 7, 1976 on the purchase of the Los Angeles-based firm's Canadian subsidiary, Atlantic Richfield Canada, for $335 million. The purchase agreement covered the Arco subsidiary's oil and natural gas production and 11 million acres of undeveloped oil and gas properties in the provinces of Alberta and British Columbia. It excluded the company's six disputed Athabasca tar-sand leases in Alberta.

Sarnia pipeline opened. The first continuous flow of crude oil from western Canada arrived in Montreal June 15, 1976 through a new 520-mile pipeline extension from Sarnia, Ontario. The pipeline had been completed in March and the first oil to pass through the extension had arrived in Montreal June 6, though heavily mixed with water.

The opening of the Sarnia pipeline meant that eastern Canada, especially Quebec, would be less dependent on imported oil. Previously, the area east of the Ottawa Valley had relied on foreign oil exclusively, while oil from western Canada was used in the rest of the country. The $232-million Sarnia-Montreal extension, which connected with the pipeline beginning in Edmonton, Alberta, was constructed so that oil could flow both toward and from Montreal.

Oil subsidies to end. The Canadian government changed a law so that it could end imported-oil compensation payments to Montreal refiners who refused to use their share of domestic crude oil, it was reported Aug. 25, 1976.

The change in the Petroleum Administration Act was meant to insure use of oil sent from western Canada through the recently completed Montreal-Sarnia pipeline. Subsidies on imported crude were $3.30 a barrel to compensate for the difference in the price of domestic oil, which was controlled below world price levels. The difference between the import subsidy and revenue from export charges made it profitable for some eastern Canadian companies to continue importing foreign oil, while trying to maximize exports to the U.S.

U.S., Canadian firms to exchange oil. The Canadian National Energy Board and the U.S. Federal Energy Adminstration approved three oil exchanges by private companies involving an average 10,000 barrels of crude oil daily, it was reported Sept. 2, 1976.

Under terms of the exchange, Canadian and U.S. domestic oil would be traded on a strict barrel-to-barrel basis.

The trade was set up to aid U.S. refineries that had been cut off from their traditional supplies of Canadian crude as a result of Canada's policy of phasing out oil exports.

Natural-gas surplus registered. A large buildup of natural gas supplies in Alberta forced buyers to stop taking deliveries of fuel, the New York Times reported Sept. 3, 1976. TransCanada Pipelines, Ltd., main buyer of Alberta fuel, announced that it would not accept deliveries of new supplies until November 1977, according to the report.

The Alberta Gas Trunk Line Co. of Calgary, which transported fuel in the province, estimated the surplus at nine trillion cubic feet of gas.

The surplus resulted from a mild winter, which reduced consumption, and expectations of fuel shortages, which encouraged conservation measures. Rising wellhead prices of gas led to increased exploration and also prompted consumers to switch to fuel oil, increasing supply and reducing demand.

Provinces vote oil-price rises. A majority of the provincial energy ministers attending a federal-provincial energy conference May 12, 1977 agreed to raise prices of oil and natural gas by C$1 a barrel every six months, starting in July.

The result to consumers would be an increase of 7¢ (Canadian) for a gallon of gas and 6¢ (Canadian) for a gallon of heating oil by March 1978, according to Energy Minister Alastair Gillespie. He said seven of the 10 provinces had agreed to the price increases for at least one year.

Energy ministers from Ontario, Manitoba and Nova Scotia continued to oppose the increases, saying they would aggravate inflation and unemployment. They also had requested assurances that if the increases came into effect, the extra income would be used to expand development of new oil deposits. Gillespie said a formal agreement would have to be concluded by Ottawa with each province separately because of the opposition, but he added that the 1977 price increases were definite.

Canada, U.S. sign pipeline pact. After more than two years of negotiations, the U.S. and Canada Jan. 28, 1977 signed a treaty guaranteeing oil and gas movement through pipelines in each other's territory. The treaty had major implications for a proposed $8.5-billion trans-Canada pipeline to carry gas from Alaska to the U.S. Midwest and southern states.

Report asks pipeline delay. After three years of inquiry, the Berger Commission May 9, 1977 released a study proposing a delay of at least 10 years in the construction of a natural gas pipeline through the Mackenzie Valley of the Northwest Territories.

Headed by Judge Thomas Berger of the British Columbia Supreme Court, the commission said in its report, entitled Northern Frontier, Northern Homeland, that all land claims by Indian, Inuit (Eskimo) and Metis (mixed-blood) peoples should be settled and their economic security assured before Canadian Arctic Gas Pipeline Ltd., a consortium of U.S. and Canadian firms, constructed the pipeline to bring Alaskan gas through Canada into the U.S. The commission further recommended abandoning a pipeline route through the northern Yukon Territory that would have linked the northern Alaskan gas fields to the proposed Mackenzie Valley pipeline because of the potential threat to the environment.

The report gave qualified approval to an alternate route for a pipeline through the southern Yukon. The route, called Alcan, posed less of a threat to the environment and to the welfare of the native peoples, according to the report, because it passed through fewer populated areas and followed the Alaska highway instead of penetrating into unspoiled land. However, the Alcan route was being investigated by a separate commission that had not yet presented its final report.

Oil industry and business representatives expressed disappointment with the Berger report, while most native groups welcomed its conclusions. Some native groups had supported the pipeline because they anticipated an increase in jobs and economic growth as a result, while other native groups called the Berger report a victory for native self-determination. In Parliament, Progressive Conservative leader Joe Clark and New Democratic Party leader Edward Broadbent endorsed Berger's recommendation of a 10-year moratorium.

There was no official comment from government ministers. Parliamentary debate on the Berger report was expected to begin in two months. A final government decision was expected by the fall.

Mackenzie drilling suspended—Two of Canada's major oil companies said they would suspend drilling in the Mackenzie River delta until the Canadian government approved a pipeline route through the Northwest Territories, it was reported June 4.

Gulf Oil Canada Ltd., a subsidiary of Gulf Oil Corp., and Mobil Oil Canada

Ltd., a subsidiary of Mobil Corp., had already spent $125 million on joint exploration and initial development of a $400-million natural gas processing facility. They said they would complete two wells currently under way but would halt further exploration until the question of a pipeline to the U.S. was settled. They added that exploration would be held up until the federal government issued new land-use regulations to replace the ones withdrawn in 1971.

The Gulf-Mobil team had found gas in 13 wells offshore and had found oil in one. Gulf said it would continue offshore drilling in the Beaufort Sea in partnership with Dome Petroleum Ltd. in spite of government environmental restrictions in that area.

The Gulf-Mobil pullout left Imperial Oil Ltd. as the only major operator in the delta. Earlier, Shell Canada Resources Ltd. had suspended operations as a result of uncertainty over land-use regulations.

Beaufort Sea drilling. The federal government June 1, 1977 authorized Dome Petroleum Ltd. of Calgary to conduct exploratory drilling for oil and natural gas in the Beaufort Sea, north of the Yukon Territory, until 1979. Dome Petroleum had estimated that offshore reserves in the area amounted to about 40 billion barrels of oil and 320 trillion cubic feet of natural gas.

The project was placed under strict environmental controls by the federal government, including suspension of all drilling during the last month of the short Arctic summer season. The suspension would give the government and the oil company time to clean up any oil spills in the event of an accident.

The drilling also would be subject to a review at the end of each year before permission to continue the next year's drilling was issued. The project had been started in 1976, but permission for 1977 drilling had been held up pending a study of the effects of an oil-rig blowout in the North Sea in April.

China

China claims oil self-sufficiency. Hsinhua, the official press agency, said in a report monitored Dec. 1, 1975 in Hong Kong that China had achieved self-sufficiency in oil. "After a quarter of a century's efforts," the news agency said, "China has built up an oil industry embracing such branches as prospecting, drilling, oilfield construction, oil refining and scientific research." China could now export oil, Hsinhua claimed.

A Nov. 30 Hsinhua report said that Chinese petroleum-loading terminals at unspecified locations could accomodate 25,000-ton, 50,000-ton or 100,000-ton oil tankers. The report did not disclose the number of terminals completed. Hsinhua said that wharves for 10,000-ton ships had also been finished, capable of handling coal, mineral ore and other goods.

The Karamai oilfields in the Sinkiang Uighur Autonomous Region of China's northwest had tripled its production capacity and output of crude oil as compared with 1965 levels, Peking Review reported Nov. 7. The oilfields, it said, had fulfilled many 1975 objectives by June 1974, including oil extraction, refining, maintenance and well construction.

Negotiations were reported in progress for the Chinese purchase of sophisticated American oilfield apparatus, the New York Times reported Nov. 28. Houston-based equipment companies dealing with the Chinese said that China apparently had given petroleum development major priority in the proposed new five-year plan.

'76 oil production up 13%. China had produced 13% more oil in 1976 than in 1975, Hsinhua, the Chinese press agency, reported Jan. 6, 1977. It gave no total, saying only that production had been 7.6 times that of 1965. Foreign analysts cited by the New York Times believed China's output totaled 625.8 million barrels in 1976.

Natural gas production had risen 11% in 1976 over 1975, Hsinhua said.

The agency also confirmed that China had recently opened an oilfield in North China. No precise location was given, but the field was believed to be in Hopeh Province near the Pohai Gulf or the city of Tientsin. Production at the country's leading oil field, Taching in northeast China, had increased 8.7% from 1975.

Hsinhua also said the petroleum industry turned over to the state 8% more funds than in 1975. Petroleum sales provided China with the largest single source of foreign exchange. No official estimates were available, but exports were believed to have dropped to 74.5 million barrels in 1976 from 89.4 million barrels in 1976.

The Communist Party newspaper Jenmin Jih Pao, however, called March 11 for strict conservation of energy resources.

A campaign of "rigid economy" in the consumption of coal, oil and electricity had been urged. "We must send every ounce of coal, every watt of electricity and every drop of oil we have to places it is most needed in revolution and production," the party newspaper said.

Dubai

Oil nationalized. Dubai July 10, 1975 announced the nationalization of the five foreign oil companies that produced and marketed the country's 250,000 barrels of oil daily. The firms were to receive $110 million in compensation for their lost assets and were to continue their operations as usual under the new management. The companies involved were the Dubai Petroleum Co., Dubai Marine Areas, Ltd., Deutsche Texaco, the Dubai Sun Oil Co. and the Defzee Dubai Petroleum Co.

Ecuador

Oil export prices revised. Ecuador cut the prices of certain petroleum exports July 9, 1975 by reducing the income tax on oil-producing companies.

The tax cut, retroactive to July 1, was ordered to help increase petroleum production and government revenues. It applied only to oil exported to the Caribbean and the west coast of Central and North America, and not to Chile and Peru, which would take 50% of Ecuador's oil exports in 1975.

Natural Resources Minister Adm. Luis Salazar said the oil price would be reduced by 43¢ a barrel from the old price of $10.84 per barrel. The tax rate was cut from 58.5% to 53.1%, and the royalty rate was maintained at 16.67%.

The government had been under considerable pressure to reduce the tax from the Texaco-Gulf consortium, the largest oil producer in Ecuador. Texaco-Gulf had considerably reduced production in recent months, claiming the government's large take of oil income made production only marginally profitable. The government had charged in return that the U.S.-based consortium was attacking Ecuador as the weakest link in the Organization of Petroleum Exporting Countries (OPEC), and that it was "strangling [the] country's fiscal programs," the New York Times reported Feb. 19.

Oil production had fallen to 165,000 barrels a day from a potential of 210,000 barrels per day, according to the Wall Street Journal July 10. As a result Ecuador had become the first OPEC nation to show trade and budget deficits and plummeting foreign reserves.

Barely four months later the government raised the price of crude oil exports Nov. 17 to $11.45 per barrel, an increase of $1.04 per barrel, and it raised the tax on oil exports Nov. 18 to $10.57 per barrel, an increase of 40¢ per barrel.

Oil income in 1975 was $378 million up to Oct. 15, 30% below the level for the same period of 1974, it was reported Nov. 21.

Anglo yields oil installations. Anglo-Ecuadorean Oilfields Ltd. handed over its Santa Elena concession to the government Jan. 16, 1976 after its contract ended and the concession reverted to the state oil company CEPE.

The British firm had operated in Ecuador for 56 years, dominating the oil industry until the Texaco-Gulf consortium of the U.S. began exporting oil from the eastern Amazon region in 1972. Production from Anglo's 600 wells had declined steadily over the past 10 years, standing currently at 2,000 barrels per day.

Government buys Gulf assets. The government announced Jan. 1, 1977 that it had purchased the local assets of Gulf Oil Corp. of the U.S., effective the previous day.

The purchase gave the government a 62.5% share in Ecuador's petroleum operations. The other 37.5% was held by

Texaco Inc. of the U.S., which had begun a joint venture with Gulf in Ecuador in 1965. The two companies had developed the Oriente oilfields, which currently produced about 200,000 barrels per day, and had built the trans-Ecuadorean pipeline which carried the petroleum to the port city of Esmeraldas.

A spokesman for Gulf estimated the company's unamortized investment in Ecuador at $120 million. He said Ecuador had agreed to pay Gulf $82 million in cash immediately, and the balance on completion of an audit by an international firm.

Gulf had said in mid-1976 that it had "chosen to withdraw from Ecuador because of continued inability to achieve an understanding with the Ecuadorean government which would allow Gulf to receive a reasonable return on its invested capital along with being unable to reach acceptable solutions to other problems."

The purchase followed a series of disputes between the government and Gulf. The disputes culminated in a government threat to expropriate Gulf's assets without compensation unless the company paid $53 million it allegedly owed the government. Gulf paid the money Sept. 30, 1976, hours before the government's deadline, but announced that it no longer wished to operate in Ecuador.

The money, according to the government, represented taxes and other obligations on Gulf exports since February. In withholding the funds, Gulf had charged that the government had taken crude oil "in excess of entitlements"; had withheld income taxes, profit-sharing and other taxes on Gulf in excess of legal requirements; had required Gulf to satisfy local demand for crude oil in excess of obligations and at a price resulting in substantial losses; had not made full payments on CEPE's purchase of equity participation in Gulf's producing and pipeline operations in Ecuador, and had not repaid other debts by CEPE.

Texaco had withheld about $11 million in payments to the government, claiming the government had underestimated production costs in Oriente jungle oilfields when it assessed Texaco's taxes. Texaco paid up Aug. 20, 1976 after Col. Rene Vargas, the national resources minister, had warned that the government might cancel Texaco's oil concessions.

France

New oil company. A new French oil company was formed July 9, 1976 by merger of the state-owned Elf-Erap and the Societe Nationale des Petroles d'Aquitaine (SNPA), which was owned 54% by Elf-Erap and 45% by private investors. The new company was called the Societe Nationale Elf-Aquitaine (SNEA).

SNEA's total assets would be nearly $2 billion and it would have a stock market capitalization of $1 billion. As a result of the move, Elf-Erap's ownership of SNPA increased to 70% and SNPA received most of Elf-Erap's petroleum industry assets.

The merger was criticized by Gaullists on the grounds that it had not received legislative approval. Socialists and Communists claimed it would lead to denationalization, which also required legislative approval.

Solar power plant opened. The government opened its first operational solar generating plant Jan. 25, 1977 at Odeillo in the Pyrenees mountains in southwestern France. The plant was connected with the national electricity grid, through which its potential output of 64 kilowatts could power about 60 household electric hand irons.

France used the opening to declare its lead in the race to export solar energy technology.

The Odeillo plant used huge parabolic mirrors to reflect the sun's rays against a boiler that absorbed their heat. Inside the boiler, a fluid was heated to a gas that was used like steam to drive a turbo-alternator to produce electricity.

Iran

Revenue drop causes deficit. The government reported Feb. 3, 1976 that a sharp reduction in oil revenue had slowed the country's growth rate by 60% since the spring of 1975, and it announced a budget with a record $2.4 billion deficit for the next fiscal year, starting March 21. Expenditures were estimated at $45 billion and receipts at $42.6 billion, including $20.3 billion in oil and gas revenues.

Oil revenues in the current fiscal year were nearly $3 billion less than the almost $22 billion that the government had anticipated, State Planning and Budget Minister Abdol Majid Majidi said in submitting the economic report.

Reporting on the proposed budget, Premier Amir Abbas Hoveida informed the National Assembly Feb. 3 that the $2.4 billion deficit would be financed partly by borrowing $1.1 billion abroad. Hoveida accused Western oil companies of having committed a "flagrant breach" of pledges to Iran by not making expected investments and by failing to maintain adequate levels of oil production and export. The firms' drop in output totaled an average 750,000 barrels a day, he said. Iran's oil production had dropped more than 11% in 1975 to an average 5.4 million barrels a day, largely because a consortium of 14 Western oil companies took less of Iran's petroleum.

Iran, Occidental cancel deal. Iran and the U.S.-based Occidental Petroleum Corp. announced Aug. 27, 1976 the cancellation of a $125 million deal to develop Caspian Sea oil and to process and market other Iranian petroleum.

The deal collapsed because both sides "had been unable to agree on the definitive terms of the previously announced letter of intent [June 20] and had therefore decided to terminate further negotiations," the Iranian government said.

U.S. firms, Iran hold barter talks. U.S. Federal Energy Administrator Frank G. Zarb said in Teheran May 9, 1976 that three U.S. aircraft companies were negotiating with the Iranian government about bartering their planes for Iranian oil. The firms were later identified as Boeing Co., Northrop Corp. and General Dynamics Corp.

Boeing and Northrop May 11 confirmed that they were discussing sale of their aircraft to Iran in exchange for oil. General Dynamics said it was considering the sale of its F-16 fighter-planes to Iran, but declined to say whether the payment would be in oil or cash.

Zarb, who was concluding a 10-day tour of Saudi Arabia and Iran, said the proposed barter arrangement was one of the topics he had discussed with Finance Minister Hushang Ansari, Interior Minister Jamshid Amouzegar and other Iranian officials.

Zarb said he had told the Iranians and the Saudi Arabians that the U.S. favored an absolute freeze on the price of oil for the remainder of 1976. Since the world economic recovery was in "a tender state," Zarb said, a new price rise by the Organization of Petroleum Exporting Countries would run "counter to the rate of speed or very momentum of that recovery."

BAC trades arms for Iranian oil. The British Aircraft Corp. (BAC) signed a $640 million arms-for-oil agreement with Iran Nov. 18, 1976 to provide Iran with the Rapier surface-to-air guided missile. In return for supplying Iran with the tracked Rapier, BAC was to receive the cost of the weapons system in crude oil. The oil was to be marketed for the BAC by the Anglo Dutch Shell Oil Company outside the U.K. in order to avoid problems created by the fluctuations in the value of the pound. This arrangement would give the BAC maximum foreign revenue from its weapons sales.

The precise number of missiles and quantities of oil involved in the exchange were not specified. The National Iranian Oil Company agreed to supply the crude oil at commercial rates over an eight-year period. The delivery was to be at the rate of about 16,000 barrels a day with the first delivery expected in December.

Japanese loan. Under an agreement signed Aug. 27, 1976 in Tokyo, Japan was to lend Iran $296 million for construction of a petrochemical complex at Bandar Shahpur on the Persian Gulf. Of the total, $96 million would be repayable over a 20-year period, including a five-year grace period, at an annual interest rate of 4%. The remaining $200 million would come from Japan's Export-Import Bank, repayable in 11 years, including a five-year grace period, at an annual interest rate of 7.5%.

Ireland

Rockall oil rights dispute. The Irish government Feb. 3, 1977 claimed rights to economic resources within 200 miles of the uninhabited British island of Rockall in the North Atlantic. The action was in response to the British grant of oil exploration permits to two British firms in the area. The Irish claim to economic rights paralleled a Danish claim to fishing rights within the area.

The Irish claim was based upon the legality of using uninhabited islands to claim an economic zone. Neither the Irish nor the Danes disputed British sovereignty over the island itself.

The continental shelf surrounding Britain extended to Rockall, according to the British.

Italy

Oil, gas finds. The state-owned oil Corporation Ente Nazionale Idrocarburi (ENI) announced a fourth methane gas strike in the Ionian Sea off Crotone in southern Italy (reported July 23, 1975). The find was located about six miles south of the first field, Luna, which was expected to begin production soon. Experts speculated that the discoveries could lie in a geological formation extending across the Mediterranean Sea east of Malta to the Libyan oilfields.

The French oil firm ELF-ERAP had announced June 4 discovery of oil in the Adriatic Sea off the coast of the Italian port city of Vasto. ELF was drilling operator for an Italian-French consortium.

Kuwait

Gulf & BP sell out. The Gulf Oil Corp. and the British Petroleum Co. Dec. 1, 1975 signed an agreement with Kuwait to sell the remaining 40% of their concessions in the Persian Gulf state to the Kuwaiti government at a cost of at least $50.5 million. The accord, retroactive to March 5, also would guarantee the two firms 950,000 barrels of oil a day for the first five years of a new purchase contract at a reported 15-cent-a-barrel discount. Gulf and BP had been withdrawing 1.3 million barrels a day out of Kuwait's total output of 2 million barrels a day.

Libya

Libya, Occidental settle dispute. Occidental Petroleum Corp. and Libya had settled a dispute over oil concessions, with the firm resuming operations in the North African country, company officials announced Dec. 4, 1975.

Major provisions of the accord: Occidental would produce no less than 300,000 barrels a day for the next three years from two concessions in which it had a 49% interest and might continue at that rate for two more years. After 1980 the daily output rate would be reduced gradually to 250,000 barrels daily until the end of 1984.

Occidental would have the right to buy Libya's 51% of production at rates that would not result in a financial loss to the company. Occidental in turn would abandon its demand for arbitration of the dispute and pay $440 million it currently owed the Tripoli government in cash or notes, with interest on any part of the principal that was deferred.

Libya would compensate Occidental for the crude oil sold to other companies and shipped from Libya during their dispute.

At the height of the dispute, Occidental Petroleum had seized two shipments of questioned Libyan oil being unloaded from tankers in Louisiana ports Nov. 3 and 9 after obtaining court orders in the state approving such action. The seizures involved a 49% take-over of shipments of 390,000 barrels and 354,351 barrels.

An Occidental statement said the Libyan National Oil Co. had "purported to sell the cargo to an affiliate of the Coastal States Gas Corp. in violation of Occidental's concessionary rights to an undivided 49% of the oil from Concessions 102 and 103" in Libya.

Meanwhile, Occidental announced Nov. 7 that Libya had lifted its ban on preventing the firm's non-Libyan employes from leaving the country provided they were replaced.

Mexico

New oil finds. The state oil company Pemex announced the discovery of "important" new offshore oil deposits in the

Gulf of Campeche, off the state of Campeche in southeastern Mexico, it was reported Aug. 13, 1975.

Pemex gave few details of the find but said the Campeche deposits justified immediate commercial exploitation. The discovery of new oilfields in the Cotaxtla region of Veracruz State had been announced with similar reticence March 18.

Oilfields discovered in Chiapas and Tabasco States in 1974 had produced a total of 100 million barrels of crude in the last 12 months, amounting to more than 40% of national petroleum production, it was reported Aug. 4. Mexico's crude oil exports averaged 89,000 barrels per day, according to a report June 6.

Rise planned in oil production, export. The government had announced that in the next six years Mexican petroleum production would rise to 2.2 million barrels a day and exports of crude and refined oil would reach 1.1 million barrels daily, it was reported Jan. 14, 1977.

Mexican oil production for 1977 was estimated at 953,000 barrels a day, and exports were estimated at 153,000 barrels daily. The sharp increase planned by 1982 was dictated by the need to increase export revenues to pay back foreign creditors, according to the Financial Times (London) Jan. 25. At current prices, oil could bring Mexico $5 billion in annual export revenues by 1982, or 60% more than the value of the country's total visible exports in 1976.

The increased revenues would also help Mexico expand its petroleum refining installations and increase the proportion of refined exports to crude exports.

The rise in production and exports would be facilitated by the recent discovery of new oil deposits in four regions of the country: offshore near the port of Tampico, on the Gulf of Mexico; at Cotaxla near Veracruz, on the gulf further to the south; near Nuevo Laredo on the U.S. border and at Sebastian Vizcaino in Baja California Sur.

The planned increase in exports would intensify pressure on Mexico to join the Organization of Petroleum Exporting Countries (OPEC), the Financial Times reported Jan. 25. Mexico heretofore had resisted joining OPEC because membership in the cartel would exclude Mexico from trade preferences offered by the 1974 U.S. Trade Act and would bring tensions to Mexico's relations with Washington, according to the Times.

Nevertheless, Mexico had followed OPEC prices. In January it raised its crude sales price to $12.65 per barrel, 10% higher than OPEC's previous base of $11.50 per barrel but only 2.8% higher than the $12.30 per barrel Mexico had been charging, it was reported Jan. 14.

Reserves exceed 60 billion bbl. Pemex March 18, 1977 estimated Mexico's total petroleum reserves at more than 60 billion barrels.

Pemex Director General Jorge Diaz Serrano said the country's proven reserves were only 11.1 billion barrels. However, he said this figure did not include known reserves on the continental shelf off the state of Campeche, nor huge oilfields not yet in production.

The 60-billion figure, reported in the press in February, had been attributed to U.S. government officials and oil industry executives. The estimate made Mexico's petroleum reserves comparable to Iran's and Kuwait's and six times greater than the reserves in Alaska's North Slope.

Until Diaz Serrano's announcement, Mexican officials purposely had underestimated the country's oil reserves, apparently to minimize conflicting pressures from the U.S. and the Organization of Petroleum Exporting Countries. There also had been considerable disagreement among Mexican officials over whether the oil should be exploited quickly for short-term economic advantage or should be saved "for future generations," according to the New York Times Feb. 18.

Now, however, President Jose Lopez Portillo had embarked on a program of accelerated exploitation and exports of crude oil, the Times reported. Lopez Portillo had acted not only because of Mexico's economic crisis and its desperate need for foreign exchange, but because of his confidence in the size of the reserves, the newspaper said.

Nevertheless, Mexico faced major financial and technological problems in extracting the oil. Pemex was particularly hampered by the enormous cost of extraction, delays in obtaining necessary equipment from abroad, and a shortage of skilled Mexican engineers and technicians, the Times reported.

In a related development, the Mexican government announced March 14 that it would issue 2 billion pesos—about $90 million—worth of "petrobonds" that would have a maturity value pegged to the world price of oil. The bonds were expected to stimulate domestic savings since the Mexican peso was floating downward against the dollar and the dollar price of oil was rising.

The Netherlands

Gas reserves decline. A State Geological Service report, released March 28, 1977 by the Ministry of Economics, showed that the nation's overall reserves of natural gas had dropped 3% in 1976 to 2,396.9 billion cubic meters. The decline, estimated at 80 billion cubic meters, was attributed to the large production of natural gas within Holland in 1976.

The reports' estimate of gas reserves included onshore and offshore deposits, and proven and unproven reserves.

Since the first massive discoveries of natural gas in the Netherlands in the late 1960s, further exploration had continued to increase the estimated reserves.

The natural gas reserves enabled the Netherlands to enjoy the only balance of payments surplus in the European Community, except for West Germany. About half of Dutch gas production was exported to other West European nations.

Peru

Pipeline opens. A trans-Andean pipeline began operations May 25, 1977 with crude petroleum flowing from Peru's eastern Amazon jungle over the Andes mountains to the port of Bayovar on the Pacific Ocean.

Officials said the pipeline, which initially carried 30,000 barrels of oil per day, would enable Peru to reduce its daily oil imports to 10,000 barrels by July. Peru currently imported 50,000 barrels per day from Ecuador and Venezuela.

A feeder line to an oilfield operated by the U.S. firm Occidental Petroleum Corp. north of the pipeline was expected to be completed by the end of 1978. Officials said it would carry 80,000 barrels a day and make Peru self-sufficient in oil.

The state oil company, Petroperu, said it had spent about $997 million in the Amazon region for prospecting, exploration, exploitation and pumping operations, it was reported March 14. Of the total, $207 million went to exploration and production, $670 million to the pipeline and $120 million to the feeder line to the Occidental field.

Foreign firms wooed. Petroperu March 3, 1977 announced better terms for exploration and production contracts with foreign oil companies. It did so in an attempt to bring back to Peru many of the concerns that had left in recent years after claiming that operations in the country were unprofitable.

The new contracts would be more flexible, dropping provisions of the "Peruvian model" contracts, in effect since 1970, under which Petroperu took more than 50% of a company's production and paid it off with the rest. Under the new pacts Petroperu would continue to take a large share of production, but companies could choose to be paid in cash, and payments and terms would conform to changing world market conditions.

Qatar

Takeover of Shell group completed. The Qatar government completed the takeover of the Royal Dutch-Shell group's oil assets in the country under an agreement signed Feb. 9, 1977. The accord gave the state's Qatar Petroleum Producing Authority the remaining 40% stake in the oil facilities owned jointly by Shell and the Italian ENI-Agip firm. The government agreed to pay Shell an undisclosed amount for the takeover and fees for continued operations by a new firm called Qatar Shell Services Co.

The agreement placed all foreign-held operations in Qatar in government hands.

Shell had produced 245,000 barrels of crude oil daily in 1976, about half of Qatar's total output.

Rhodesia

Mobil denies violating sanctions. Mobil Oil Corp. said Aug. 27, 1976 that efforts to investigate charges that its South African affiliate was supplying Rhodesia with petroleum had been blocked by South African law.

In a letter to the U.S.-based United Church of Christ, which had made the charges, Mobil said that the South African official secrets act prohibited any dissemination of information relating to storage, distribution and sale of petroleum products. The company also wrote that attempts to get information from its Rhodesian affiliate had met with no response because of similar legal restrictions.

The South African affiliate, according to Mobil, had not supplied Rhodesia with petroleum since 1966, but the affiliate said it was powerless to prevent its customers from selling there.

The United Church Board for World Missionaries, the overseas arm of the United Church of Christ, said in an Aug. 30 Wall Street Journal report that it had notified the Securities and Exchange Commission Aug. 27 that it was filing a shareholder resolution calling upon Mobil to "take every measure immediately" to insure that petroleum supplies did not reach Rhodesia, including refusal to make bulk sales to buyers unable to verify that the products were not destined for Rhodesia. (According to a church official, the church and its affiliates owned about 85,000 shares of Mobil.)

The charges had originally been made by the church group in a June 21 report of its Center for Social Action. The Rev. Larold K. Schulz, the center's executive director, released the document at a joint press conference with the Peoples Commission, a radical left-wing organization. Schulz explained that because previous disclosures about Rhodesia had been ignored by the news media, a joint announcement seemed the best way "to get the kind of impact necessary." He said that the information had come originally from a group called Okhela, which was composed of white South Africans who gave clandestine support to black nationalists.

As outlined by Schulz, the scheme involved the sale of petroleum products by Mobil (South Africa) to Genta, a purchasing agency owned by the Rhodesian government, which then resold these goods to all the oil firms in the country, including Mobil (Rhodesia). Afterwards, in order to "make it look as if Mobil (South Africa) was not involved in any trade with Rhodesia," Schulz said, a "paper chase system" was devised "whereby sales and payments would be passed through various South African companies which acted as intermediaries." Schulz claimed that "similar procedures" had been worked out by the Royal Dutch/Shell Group, British Petroleum, Caltex and Total. He said that three U.S. citizens had acted simultaneously as directors of Mobil (South Africa) and executives of Mobil (U.S.A.).

A spokesman for Mobil denied the charge June 21, noting that the "management of our international division has gone to considerable effort to make sure that all of our affiliates, particularly those in southern Africa, have been informed of U.S. law," which forbade the sale to Rhodesia of most products.

Saudi Arabia

Mobil to build Saudi pipeline. A Mobil Oil Corp. subsidiary would build a pipeline across Saudi Arabia under an agreement signed with Petromin, the state-owned firm (announced Feb. 8, 1977). The facility would reduce shipping distances from the country's eastern oilfields to European and North American ports by more than 3,000 miles.

The 750-mile line, to be constructed at a cost of $1.55 billion, would stretch from the Ghawar fields near the Persian Gulf across the peninsula to the west coast port of Yenbu on the Red Sea. From there the oil would be shipped either through the Suez Canal or to the Sumed Pipeline, which crossed Egypt to the Mediterranean coast. Construction was expected to be completed by the end of 1981.

The line would be totally financed by Saudi Arabia and wholly owned by Petromin. It would supplement the Trans-Arabian pipeline, which shipped oil to the Mediterranean from Saudi Arabia through Jordan, Lebanon and Syria.

Saudis to increase oil output. Saudi Arabia planned to increase its capacity to produce oil by nearly 50% between 1977 and 1982, partly to help keep world oil prices down and also to meet mounting demand, an official of the Arabian American Oil Co. (Aramco) said May 11, 1977.

Company chairman Frank Jungers foresaw "16 million barrels a day of capacity by 1982," adding that "this is the planning," although "it could change." He noted that this would exceed a production target of 8 million to 8.5 billion barrels a day that had been set in 1976.

(Saudi Petroleum Minister Sheik Ahmed Zaki Yamani had indicated in an interview May 10 that production in May would likely be "close to April," which had totaled 10.1 million barrels a day, or "it might be a little more.")

Jungers said Aramco's production capacity was currently "slightly over 11 million barrels a day." The company had announed earlier in the week that its 1976 reserves had increased 2.3 billion barrels to 110.2 billion. Aramco had discovered three new oilfields in 1977; in 1976 only about 700 of its 1,800 wells were in operation.

Fire halts major Saudi pipeline—The flow of two-thirds of Saudi oil production to the Persian Gulf for shipment was shut down May 11 by a fire that broke out in Aramco's major pipeline at the Abqaiq field, about 35 miles southwest of Dhahran. The blaze, which killed one workman and injured 13, was brought under control May 12 and extinguished May 13, an Aramco spokesman said May 14.

The blaze was suspected to have been caused by sabotage, according to Arab businessmen who arrived in Beirut May 12 from Riyadh. They said Saudi authorities had received a warning the previous week from unidentified sources who said that Arab and African "Communist agents" had entered the country earlier in the month for the purpose of damaging the pipelines. The Saudis were said to have discounted the information at the time.

Aramco sources May 15 said six persons had been killed in the fire. They ruled out sabotage.

The company May 19 estimated that losses from the blaze would total $100 million in destroyed oil and damaged equipment. As a result, Aramco's four U.S. owners—Exxon Corp., Texaco Inc., Mobil Corp. and Standard Oil Co. of California—had acted to reduce their sales of Saudi Arabian crude oil. Exxon and Mobil had informed their customers, including Japan, that deliveries would be cut by up to 30% in May.

Soviet Union

U.S.S.R. growth contrasted to U.S. decline. An editorial in the April 30, 1976 issue of Science, the magazine of the American Association for the Advancement of Science, contrasted the Soviet Union's growth in energy with the U.S.' "drift into dependence." It said:

> Proved Soviet oil reserves are 80 billion barrels (11 billion metric tons) in contrast to 33 billion barrels for the United States. Soviet production is about 9.8 million barrels a day (b/d) and rising, that of the United States is 8.1 million b/d and dropping. The Soviet Union is a net exporter of oil. . . .
>
> The proved reserves of natural gas in the Soviet Union (800 trillion cubic feet or 22.4 trillion cubic meters) are by far the largest in the world; those of the United States are 228 trillion cubic feet. During 1975, production in the United States (20.1 trillion cubic feet) exceeded that in the Soviet Union (10.2 trillion cubic feet). But U.S. reserves are declining rapidly. If present trends continue, which is likely, Soviet production will be exceeding that of the United States in about 4 years. Already the Soviet Union is an exporter of natural gas. . . .

New oil minister named. Nikolai A. Maltsev, a former deputy minister of the petroleum industry, was promoted to minister (reported April 5, 1977). He replaced Valentin D. Shashin, who had died March 22.

Sri Lanka

Sun, wind, manure to produce power. Energy from the sun, wind and bio-gas (produced by fermenting animal manure) would be used to generate electricity for an entire community in southern Sri Lanka by the end of 1977, according to the United Nations Educational, Scientific and Cultural Organization (UNESCO) March 30, 1977. A rural energy center, sponsored jointly by the U.N. Environment Program and the government of Sri Lanka, would be the first anywhere to use the three sources in combination so that even small amounts of one could add to the others to produce the energy needed for development, UNESCO said.

The success of the center, according to UNESCO, would be a milestone in the development of renewable energy sources to supply the needs of Third World nations, particularly in remote areas. By comparison with hydroelectric, fossil-fuel and nuclear power plants, the new system would provide energy from cheap ingredients, would be nonpolluting and could be efficiently tapped, UNESCO asserted.

Turkey

BP, Mobil accept oil import prices. The British Petroleum Co. (BP) and the Mobil Oil Corp. averted the take-over of their refining assets in Turkey by agreeing to import crude oil into the country at prices imposed by the Turkish government, the Financial Times of London reported May 1, 1975.

The companies had halted oil imports after the government, in December 1974, decreed that imports could not cost more than $10.09 a barrel. The firms said the lowest crude price they could find was $11.20. Ankara Feb. 7 had given the companies three months to resume crude imports or face revocation of their refining licenses. The new import prices ranged from $10.21 to $10.50.

Venezuela

Oil nationalization law. A bill to nationalize Venezuela's petroleum industry was signed into law Aug. 29, 1975 by President Carlos Andres Perez and his entire cabinet.

Perez declared that with the nationalization, Venezuela would "assume the most demanding responsibilities on the path toward the liberation of Latin America." He asserted that "a new international economic order has been proclaimed," and lamented that "the industrialized countries . . . are experiencing a crisis of misunderstanding that prevents them from accepting the fact that the exploited countries are assuming the active defense of their own interests and making their own decisions."

Under the new law, control of the oil industry was shifted Jan. 1, 1976 to a new state corporation, Petroleos Venezolanos (Petroven), which was headed by retired Gen. Rafael Alfonso Ravard, who was sworn in Sept. 1, 1975 with the other members of Petroven's board of directors.

The major foreign (mostly U.S.) companies that were nationalized were allowed to sign two-year, renewable technical assistance agreements with Petroven to continue providing essential aid to the oil industry. They would also continue to buy most of Venezuela's oil output, according to a report Sept. 5.

The nationalization would not affect the supply of Venezuelan oil to the U.S. or the cost of petroleum products to U.S. consumers, according to a statement Aug. 29 by Frank G. Zarb, administrator of the U.S. Federal Energy Administration.

The bill, passed by the Chamber of Deputies July 28 and by the Senate Aug. 18, received final approval from Congress Aug. 21. It had been pushed through Congress by Perez' Democratic Action Party (AD).

Although the bill was approved by all political factions, six of the eight opposition parties in Congress had bitterly opposed its fifth article, which allowed Petroven to undertake joint projects with private firms in certain areas of the nationalized industry, including supply of technology, transportation of oil, foreign marketing and management.

Ex-President Rafael Caldera, a senator and leader of the opposition Social Christian Party (Copei), denounced the fifth article in the Senate Aug. 5, asserting the foreign (mostly U.S.) firms that Petroven might sign contracts with "have never supported any increase in our sovereignty and, far from being good partners, they have always placed obstacles in our path."

President Perez had argued July 5 that the fifth article would enable him to "conciliate the interests of those who possess technology but need our oil and those of the country that possesses oil but not the advanced technology." He noted that under the article, mixed enterprises could be undertaken only in the national interest and only with the approval of both houses of Congress.

The bill provided for compensation to the nationalized companies not to exceed the net book value of their assets, which the government put at $1.16 billion but the companies estimated at $5 billion. If the companies rejected the government's compensation offer, they would be expropriated.

Perez had said July 5 that in practice, Petroven would supervise and control the operations of the nationalized companies, "conserving their structure and organization as much as possible." (Venezuelans would be named to the companies' top management positions, but other foreign employes presumably would be allowed to keep their jobs, according to press reports.)

Petroven, to have an initial operating budget of $470 million, would be directed by a nine-member council initially appointed by Perez for a six-month term and subsequently elected by the state company's membership.

The government expressed confidence that the transfer of ownership would proceed smoothly, but doubts were voiced by members of the opposition, the oil industry, private enterprise and the press. Marcel Padron, a columnist for the conservative newspaper El Universal, asked in an article reported July 16: "Who will pay everyone? Who will be the executives? What will happen to the workers' loans? . . . How will the new state enterprise [Petroven] be protected from the political contamination that has happened to all the others?"

Other critics asserted the government lacked legal instruments to perform all the complex tasks heretofore handled by the foreign oil firms and to consolidate the score of foreign companies into one coherent operation. "There simply hasn't been enough basic carpentry done. They have underestimated the complexity of running the [industry]," said one oil executive quoted by the Washington Post July 16.

Despite their misgivings, the foreign companies—including Exxon Corp. and Gulf Oil Corp. of the U.S. and Royal Dutch/Shell of Great Britain and the Netherlands had cooperated with the nationalization planners. During this 1975 period, petroleum output had fallen to an average 2.5 million barrels per day, the government announced Aug. 6. The figure was nearly 19% below that of the same period of 1974. Finance Minister Hector Hurtado had said May 29 that production would be cut to 2 million–2.2 million barrels per day in 1976 to conserve Venezuela's oil resources.

Average oil production for 1975 had dropped to 2.46 million barrels per day by the end of September, it was reported Oct. 6. The government said Oct. 15 that it would reduce oil exports in 1976 to an average of 1.94 million barrels per day, compared with the current 2.2 million barrels a day. Despite this cut the government expected to earn $8.6 billion from oil exports in 1976, compared with the $7 billion earned in 1970 when exports were at a peak of 3.5 million barrels per day.

Meanwhile, the government announced Oct. 22 that it was increasing its oil prices by an average of 75¢ per barrel retroactive to Oct. 1, the date set for a 10% price increase by the Organization of Petroleum Exporting Countries (OPEC), to which Venezuela belonged.

Companies accept compensation offers—Most foreign oil companies in Venezuela agreed Oct. 28, 1975 to accept government offers totaling more than $1 billion in compensation for their nationalized assets. Venezuela agreed to pay partly in cash on the turnover of operations to Petroven and partly in interest-bearing bonds.

The largest of the offers, $512 million, went to Exxon Corp. for its wholly-owned subsidiary, Creole Petroleum Corp. Other major offers were accepted by Shell International Petroleum Co. and Gulf Oil Corp., whose subsidiaries, along with Creole, produced 83% of Venezuela's daily oil output. More than 30 other firms also accepted compensation offers, according to government officials.

The payment for Creole was decided in negotiations between the government and Exxon that also determined the terms under which Exxon would continue to provide technical aid to Petroven, according to a report Oct. 12.

Oil nationalized. Venezuela's oil industry was formally nationalized Jan. 1, 1976 in a ceremony at Lake Maracaibo attended by top government officials and representatives of foreign nations including members of the Organization of Petroleum Exporting Countries (OPEC).

President Carlos Andres Perez raised the Venezuelan flag over the nation's first productive well, Zumaque No. 1, drilled in 1914 by Shell Caribbean Oil Co. He noted that the nationalization process, though "difficult," had been carried out "in a climate of friendship and peace."

"We are not nationalizing because we will earn more money," Perez declared. "We are nationalizing because oil is the nation's basic industry . . . and it is neither convenient nor acceptable that [it] be in foreign hands." He said OPEC members should use oil not as a weapon but as an instrument to correct injustices and further the dialogue between rich and poor countries.

Perez added at a news conference Jan. 2 that the nationalization would facilitate creation of a Venezuelan "social democracy" which would "halt the concentration of wealth in the hands of a few" and provide "a genuine distribution of goods, above all to the least powerful classes."

Through oil, Venezuela had earned $7.6 billion in 1975, or more than 80% of the nation's income.

Under the nationalization, all oil companies were absorbed by the state monopoly Petroleos Venezolanos (Petroven), which thus became the largest single supplier of crude oil to the U.S. Venezuela sold the U.S. more than 1 million barrels of oil daily, or more than one-third of the U.S.' oil imports.

Some foreign companies would remain in Venezuela to provide technical assistance and advice to subsidiaries of Petroven, and most of the 500 foreigners employed by the foreign firms would continue to work for the new Venezuelan companies. Exxon Corp. of the U.S. announced Jan. 6 that it had signed a contract to purchase an annual average of 965,000 barrels of oil per day from Petroven and to provide a wide range of services to Lagoven, the Petroven subsidiary operating the assets formerly owned by Creole Petroleum Corp.

Meanwhile, oil production was reported dropping sharply, having fallen to 1.75 million barrels per day in mid-December 1975 from the average of 2.4 million barrels a day in the first 11 months of the year. Production in 1976 might average 1.5 million barrels per day or less, the Wall Street Journal reported Jan. 7.

The production drop was generally attributed to declining demand and full storage tanks, though the government ascribed the fall to its desire to take over the industry with low stocks of oil, the London newsletter Latin America reported Jan. 2. The foreign oil companies had told the government they could not sell more than 1.4 million barrels per day at current Venezuelan prices because of the reduced world demand, but they had offered to buy as much as 2.36 million barrels a day from Venezuela at lower prices, Latin America noted.

In another oil development, Occidental Petroleum Corp. of the U.S. reported Jan. 1 that it was writing off its entire $73.4 million investment in Venezuela. The government had not offered to compensate Occidental for its nationalized assets because of reports that Occidental had bribed government officials.

Oil income & exports. Venezuela's nationalized oil industry had earned a total income of $9.9 billion in 1976, the Journal of Commerce reported Jan. 11, 1977. Government earnings reached a record $9.12 a barrel. Petroven, the state monopoly, scheduled $320 million in new investments in 1977, with emphasis on reactivating exploration, the Journal reported. Nearly $150 million would be spent to drill 55 wildcat wells and survey 7.7 million acres in traditional oil areas. Other surveys would be made in adjacent areas and off the northeastern coast near Trinidad, where foreign companies had made major commercial oil and gas strikes.

Venezuela currently was selling abroad 2 million barrels per day, with 1.1 million going to the U.S. The Exxon Corp. said Jan. 6 that although it would try to buy more low-priced Saudi Arabian oil, it would have to keep buying Venezuelan oil at current levels because demand for heating fuel in the U.S. was so high. Exxon said Venezuela had raised its oil prices by 6.1% to 9.8% since Dec. 17, 1976, when OPEC had approved price increases.

President Carlos Andres Perez said Jan. 1 that Venezuela would grant credits to Latin American nations to help them meet OPEC's new prices.

Venezuelan exports, mainly crude and refined oil, fell to $10.3 billion in 1976 from $10.9 billion the previous year.

Other oil developments. Among other developments involving the Venezuelan oil industry:

- The National Association of Public Accountants said July 23, 1975 that U.S. oil firms in Venezuela had "secretly extracted" $1.1 billion worth of petroleum outside their legal concessions, and they therefore owed the government that sum. Foreign oil executives admitted their firms had produced some oil from underground areas outside their concessions, asserting oil "migrated" underground, but they questioned how the accountants could reach the precise $1.1 billion figure, which was roughly equivalent to the net book value of the foreign companies' assets as estimated by the government.
- The government had completed planning for its long-awaited petrochemical development project, the Latin America Economic Report said (June 11, 1976). The project, to cost $2.32 billion, comprised four main petrochemical complexes, two in western and two in eastern Venezuela. All were expected to be in operation by 1981.
- The Venezuelan Scientific Research Institute said Sept. 9, 1976 that it had developed a new way to transform sulfur-laden "heavy" crude oil into more marketable, sulfur-free "light" petroleum. The discovery would make it easier for the state oil company to develop the huge Orinoco tar belt which stretched across eastern Venezuela.

Canadian utility nationalized. The government nationalized the Canadian International Co., the main supplier of electricity to western Venezuela, which had four subsidiaries in the country's major oil-producing regions, it was reported Nov. 12, 1976. The firm received $100 million in compensation.

Vietnam

Foreign oil firms negotiate. U.S., French, British, Japanese and Canadian oil companies had been negotiating with North and South Vietnamese representatives in Paris since the summer of 1975 to resume offshore oil exploration in the South China Sea (reported April 24, 1976). Drilling had been suspended by the firms when South Vietnam fell to Communist forces in April 1975.

The American oil companies that had been involved in the drilling prior to the Communist victory were Union Texas, Skelly, Marathon, Mobil, Shell (U.S.), Cities Service and Exxon. All but Exxon had informed the U.S. State Department that they would like to negotiate a return to Vietnam. Although American firms were still officially banned from trading with the Vietnamese under a Congressional embargo act, the oil negotiations were said to have received unofficial endorsement by the State Department.

The first talks in Paris were said to have been held in July 1975 with former Viet Cong officials. Later discussions had been conducted with North Vietnamese representatives.

Foreign Minister Nguyen Thi Binh said in the French newspaper Le Monde May 7, 1976 that the Vietnamese government was "ready to cooperate with foreign countries and oil companies including those that operated in Vietnam before the liberation of our country, such as [Anglo-Dutch] Shell and Gulf Oil." She insisted, however, that "such cooperation will be made on the basis of mutual interest and in full respect for our sovereignty and independence."

West Germany

Coal rationed. Measures to reduce consumption of coal, which had been put into effect Sept. 8, 1976, were officially announced Nov. 17. They included limiting the amount of coal available in 1976–77 to 1.8 tons for all two-room apartments, down from an allotment of 2–2.5 tons, depending upon the number of occupants. Coal exports, a chief source of Western currency, remained at the 1975 level of 38 million tons. Transportation difficulties, rising demand and inefficient energy use were cited as reasons for the shortage.

In a related development, West Germany granted Poland $270 million in credits for equipment and research for a coal gasification project, the Dec. 8 Journal of Commerce reported.

Yugoslavia

Dow in record $700 million oil deal. The Dow Chemical Co. of Midland, Minn. signed an agreement March 26, 1976 in Belgrade with Yugoslavia's largest oil and petrochemical company, Industrija Nafte (INA), for the joint construction and operation of a $700 million petrochemical complex on the northern Adriatic island of Krk, near the port city of Rijeka.

U.S. officials said the venture marked the largest U.S. investment ever made in Yugoslavia. Negotiation on the project had begun in May 1974 and a letter of intent was signed in January 1975.

When completed in 1982, the complex would produce 1.8 million tons of petrochemicals annually with a value of $550 million. The project would consist of manufacturing units for a variety of plastics, monomers and hydrocarbons, plus an ethylene plant.

The venture entailed major transfers of U.S. technology to Yugoslavia and a commitment by Dow to continually update the project's technology and provide training facilities for key Yugoslav employes.

As required by Yugoslav law, the domestic partner, INA, would hold a 51% interest in the project, although management would be apportioned co-equally between the two partners. Dow would be able to repatriate its profits.

Zaire

Offshore oil flow begins. President Mobutu Sese Seko Nov. 28, 1975 officiated at rites inaugurating the flow of Zaire's first oilfield, about 10 miles offshore from the town of Moanda at the mouth of the Zaire (formerly Congo) River.

Under the terms of the $20 million international venture, the government of Zaire would receive a 20% share of the oilfield's profits. Gulf Oil of Zaire, a subsidiary of Gulf Oil Corp., held a 50% share, and private Japanese and Belgian firms were lesser partners.

The flow from the field, whose potential was estimated at 25,000 barrels a day, was scheduled for export, while Gulf continued to import a waxier crude more suitable for Zaire's existing refinery, the New York Times reported Nov. 28.

Nuclear Energy

U.S. Policy Controversies

U.S. nuclear energy policy engendered controversy for many reasons. There was disagreement over the extent of the hazards that atomic plants posed to people specifically and to the environment in general. There was dispute over the costs of nuclear installations and over whether or not nuclear power was really necessary. And there was fear that nuclear weapons would become available to additional nations and even to terrorists as nuclear fuels and technology became increasingly common.

Ford's private-industry plan criticized. A Ford Administration plan to increase private industry's role in nuclear power was criticized by the General Accounting Office Nov. 1, 1975. In a report requested by Chairman John O. Pastore (D, R.I.) of the Joint Committee on Atomic Energy, the GAO contended that a preferable way to increase use of enriched uranium in the nuclear power field would be to expand existing gaseous diffusion plants rather than have private industry build a new facility.

The GAO report was focused largely on a proposal of Uranium Enrichment Associates, a partnership of Bechtel Corp. and Goodyear Tire & Rubber Co., for a $2.75 billion gas-diffusion process project.

The GAO considered the Bechtel-Goodyear project more costly than adding on to current plants and probably subject to more project delay. "Its fundamental shortcoming," however, the GAO said, was that the industry plan assured a good profit to private investors while "shifting most of the risk during construction and proving the plant can operate to the government."

It considered the Bechtel-Goodyear project "not acceptable."

Other projects proposed in response to President Ford's plan were a $700 million project by Exxon Corp., a $900 million plant by a Signal Cos. subsidiary and a $1 billion joint venture by Atlantic Richfield Co. and Electro-Nucleonics Corp., the Wall Street Journal reported Oct. 2. All these would employ the gas-centrifuge process.

Carter address at U.N. While still a candidate for the Democratic presidential nomination, Jimmy Carter expressed his opinion on the major nuclear energy issues May 13, 1976 in an address at the U.N. on "Nuclear Energy and World Order." He said:

U.S. dependence on nuclear power should be kept to the minimum necessary to meet our needs. We should apply much stronger safety standards as we regulate its use. And we must be honest with our people concerning its problems and dangers.

I recognize that many other countries of the world do not have the fossil fuel reserves of the United States. With the four-fold increase in the price of oil, many countries have concluded that they have no immediate alternative except to concentrate on nuclear power.

But all of us must recognize that the widespread use of nuclear power brings many risks. Power reactors may malfunction and cause widespread radiological damage, unless stringent safety requirements are met. Radioactive wastes may be a menace to future generations and civilizations, unless they are effectively isolated within the biosphere forever. And terrorists or other criminals may steal plutonium and make weapons to threaten society or its political leaders with nuclear violence, unless strict security measures are developed and implemented to prevent nuclear theft.

Beyond these dangers, there is the fearsome prospect that the spread of nuclear reactors will mean the spread of nuclear weapons to many nations. By 1990, the developing nations alone will produce enough plutonium in their reactors to build 3,000 Hiroshima-size bombs a year, and by the year 2000, worldwide plutonium production may be over 1 million pounds a year—the equivalent of 100,000 bombs a year—about half of it outside of the United States.

This prospect of a nuclear future will be particularly alarming if a large number of nations develop their own national plutonium reprocessing facilities with the capacity to extract plutonium from the spent fuel. Even if such facilities are subject to inspection by the International Atomic Energy Agency and even if the countries controlling them are parties to the Non-Proliferation Treaty, plutonium stockpiles can be converted to atomic weapons at a time of crisis, without fear of effective sanction by the international community.

The reality of this danger was highlighted by the Indian nuclear explosion of May, 1974, which provided a dramatic demonstration that the development of nuclear power gives any country possessing a reprocessing plant a nuclear weapons option. Furthermore, with the maturing of nuclear power in advanced countries, intense competition has developed in the sale of power reactors, which has also included the sale of the most highly sensitive technologies, including reprocessing plants. With the spread of such capabilities, normal events of history—revolutions, terrorist attacks, regional disputes, and dictators—all could take on a nuclear dimension. . . .

Nuclear energy must be at the very top of the list of global challenges that call for new forms of international action. . . .

I would not presume to anticipate the outcome of your expert deliberations. But I suggest that new lines of international action should be considered in three main areas:

(1) action to meet the energy needs of all countries while limiting reliance in nuclear energy;

(2) action to limit the spread of nuclear weapons; and

(3) action to make the spread of peaceful nuclear power less dangerous.

1. We need new international action to help meet the energy needs of all countries while limiting reliance on nuclear energy.

In recent years, we have had major United Nations conferences on environment, population, food, the oceans and the role of women—with habitat, water, deserts, and science and technology on the schedule for the months and years immediately ahead. These are tentative first steps to deal with global problems on a global basis.

Critics have been disappointed with the lack of immediate results. But they miss an important point: a new world agenda is emerging from this process—an agenda of priority problems on which nations must cooperate or abdicate the right to plan a future for the human condition.

The time has come to put the world energy problem on that new agenda. Let us hold a World Energy Conference under the auspices of the United Nations to help all nations cope with common energy problems—eliminating energy waste and increasing energy efficiency; reconciling energy needs with environmental quality goals; and shifting away from almost total reliance upon dwindling sources of non-renewable energy to the greatest feasible reliance on renewable sources. In other words, we must move from living off our limited energy capital to living within our energy income. . . .

Existing international ventures of energy cooperation are not global in scope. The International Energy Agency in Paris includes only some developed non-Communist countries. The Energy Commission of the Conference on International Economic Cooperation does not include countries such as the Soviet Union and China, two great producers and consumers of energy. And the International Energy Institute now under study does not call for a substantial research and development effort.

A World Energy Conference should not simply be a dramatic meeting to highlight a problem which is then forgotten. Rather, it should lead to the creation of new or strengthened institutions to perform the following tasks:

Improving the collection and analysis of worldwide energy information;

Stimulating and coordinating a network of worldwide energy research centers;

Advising countries, particularly in the developing world, on the development of sound national energy policies;

Providing technical assistance to train energy planners and badly needed energy technicians;

Increasing the flow of investment capital from private and public sources into new energy development; and

Accelerating research and information exchange on energy conservation.

An international energy effort would also be the occasion to examine seriously and in depth this fundamental question:

Is it really necessary to the welfare of our countries to become dependent upon a nuclear energy economy and if so, how dependent and for what purposes? Surely, there is a moral imperative that demands a worldwide effort to assure that if we travel down the nuclear road we do so with our eyes wide open.

Such a worldwide effort must also provide practical alternatives to the nuclear option. Many countries, particularly in the developing world, are being forced into a premature nuclear commitment because they do not have the knowledge and the means to explore other possibilities. The world's research and development efforts are now focused either on nuclear energy or on the development of a diminishing supply of fossil fuels.

More should be done to help the developing countries develop their oil, gas, and coal resources. But a special effort should be made in the development of small-scale technology that can use renewable sources of energy that are abundant in the developing world—solar heating and cooling, wind energy, and "biconversion"—an indirect form of solar energy that harnesses the sunlight captured by living plants. Using local labor and materials, developing countries can be helped to produce usable fuel from human and animal wastes, otherwise wasted wood, fast growing plants, and even ocean kelp and algae. . . .

The exact institutional formula for coping with energy effectively on a world level will require the most careful consideration. The IAEA is neither equipped nor staffed to be an adviser on energy across the board; nor would it be desirable to add additional functions that might interfere with its vitally important work on nuclear safeguards and safety.

One possibility to be considered at a World Energy Conference would be the creation of a new World Energy Agency to work side by side with the International Atomic Energy Agency in Vienna. A strengthened International Atomic Energy Agency could focus on assistance and safeguards for nuclear energy; the agency on research and development of non-nuclear, particularly renewable, sources.

2. We need new international action to limit the spread of nuclear weapons.

In the past, public attention has been focused on the problem of controlling the escalation of the strategic nuclear arms race among the superpowers. Far less attention has been given to that of controlling the proliferation of nuclear weapons capabilities among an increasing number of nations.

And yet the danger to world peace may be as great, if not greater, if this second effort of control should fail. The more countries that possess nuclear weapons, the greater the risk that nuclear warfare might erupt in local conflicts, and the greater the danger that these could trigger a major nuclear war.

To date, the principal instrument of control has been the Non-Proliferation Treaty which entered into force in 1970. By 1976 ninety-five non-weapons states had ratified the Treaty, including the advanced industrial states of Western Europe, and prospectively of Japan. In so doing, these nations agreed not to develop nuclear weapons or explosives. In addition they agreed to accept international safeguards on all their peaceful nuclear activities, developed by themselves or with outside assistance, under agreements negotiated with the International Atomic Energy Agency—a little appreciated, but an unprecedented step forward, in the development of international law. . . .

The NPT was not conceived of as a one-way street. Under the Treaty, in return for the commitments of the non-weapons states, a major undertaking of the nuclear weapons states (and other nuclear suppliers in a position to do so) was to provide special nuclear power benefits to treaty members, particularly to developing countries.

The advanced countries have not done nearly enough in providing such peaceful benefits to convince the member states that they are better off inside the Treaty than outside.

In fact, recent commercial transactions by some of the supplier countries have conferred special benefits on non-treaty members, thereby largely removing any incentive for such recipients to join the Treaty. They consider themselves better off outside. Furthermore, while individual facilities in these non-treaty countries may be subject to international safeguards, others may not be. . . .

We Americans must be honest about the problems of proliferation of nuclear weapons. Our nuclear deterrent remains an essential element of world order in this era. Nevertheless, by enjoining sovereign nations to forego nuclear weapons, we are asking for a form of self-denial that we have not been able to accept ourselves.

I believe we have little right to ask others to deny themselves such weapons for the indefinite future unless we demonstrate meaningful progress toward the goal of control, then reduction, and ultimately, elimination of nuclear arsenals. . . .

3. We need new international action to make the spread of peaceful nuclear power less dangerous.

The danger is not so much in the spread of nuclear reactors themselves, for nuclear reactor fuel is not suitable for use directly in the production of nuclear weapons. The far greater danger lies in the spread of facilities for the enrichment of uranium and the reprocessing of spent reactor fuel—because highly enriched uranium can be used to produce weapons; and because plutonium, when separated from the remainder of the spent fuel, can also be used to produce nuclear weapons. Even at the present early stage in the development of the nuclear power industry, enough materials are produced for at least a thousand bombs each year.

Under present international arrangements, peaceful nuclear facilities are sought to be safeguarded against division and theft of nuclear materials by the International Atomic Energy Agency in Vienna. As far as reactors are concerned, the international safeguards—which include materials accountancy, surveillance and inspection—provide some assurance that the diversion of a significant amount of fissionable material would be detected, and therefore help to deter diversion.

Of course, as the civilian nuclear power industry expands around the globe, there will be a corresponding need to expand and improve the personnel and facilities of the international safeguards system. The United States should fulfill its decade-old promise to put its peaceful nuclear facilities under international safeguards to demonstrate that we too are prepared to accept the same arrangements as the non-weapon states.

That would place substantial additional demands on the safeguards system of the IAEA, and the United States should bear its fair share of the costs of this expansion. It is a price we cannot afford *not* to pay.

But in the field of enrichment and reprocessing, where the primary danger lies, the present international safeguards system cannot provide adequate assurance against the possibility that national enrichment and reprocessing facilities will be misused for military purposes.

The fact is that a reprocessing plant separating the plutonium from spent fuel literally provides a country with direct access to nuclear explosive material.

It has therefore been the consistent policy of the United States over the course of several administrations, not to authorize the sale of either enrichment or reprocessing plants, even with safeguards. Recently, however, some of the other principal suppliers of nuclear equipment have begun to make such sales.

In my judgment, it is absolutely essential to halt the sale of such plants.

Considerations of commercial profit cannot be allowed to prevail over the paramount objective of limiting the spread of nuclear weapons. The heads of government of all the principal supplier nations hopefully will recognize this danger and share this view.

I am not seeking to place any restrictions on the sale of nuclear power reactors which sell for as much as $1 billion per reactor. I believe that all supplier countries are entitled to a fair share of the reactor market. What we must prevent, however, is the sale of small pilot reprocessing plants which sell for only a few million dollars, have no commercial use at present, and can only spread nuclear explosives around the world.

The International Atomic Energy Agency itself, pursuant to the recommendations of the Non-Proliferation Treaty review conference of 1975, is currently engaged in an intensive feasibility study of multinational fuel centers as one way of promoting the safe development of nuclear power by the nations of the world, with enhanced control resulting from multinational participation.

The Agency is also considering other ways to strengthen the protection of explosive material involved in the nuclear fuel cycle. This includes use of the Agency's hitherto unused authority under its charter to establish highly secure repositories for the separated plutonium from non-military facilities, following reprocessing and pending its fabrication into mixed oxide fuel elements as supplementary fuel.

Until such studies are completed, I call on all nations of the world to adopt a voluntary moratorium on the national purchase or sale of enrichment or reprocessing plants....

A-power gets low priority—Carter told a luncheon gathering of the Public Citizen Forum in Washington Aug. 9, 1976 that atomic power would have the lowest energy priority of his administration. He would emphasize efforts to conserve exing energy sources, substitute coal for oil and increase research on and development of solar power.

Carter promised to make drastic budget cutbacks in the nuclear breeder reactor that produced its own radioactive fuel. (The program had been at the top of the Ford Administration's energy priorities.) Instead, Carter said, the U.S. should examine the working breeders used by France and Great Britain.

Carter also said he opposed legislation pending in Congress that would allow private industry to get into the business of enriching uranium for fueling nuclear power. He refused, however, to endorse antitrust legislation of Ralph Nader that would "break up" the vertically integrated major oil companies whose operations ranged from exploration to the retail sale of gasoline.

A-plant growth rate down 63%. The Energy Research & Development Administration said Jan. 22, 1976 that plans for nuclear power plant construction were down 63% in 1975 from the 1974 figure, that plans for only 11 new reactors had been announced by power companies in 1975 compared with the 30 announced in 1974.

In addition, utilities canceled previously announced orders for 13 reactors and delayed commercial operation of 72 other plants.

The cutback was attributed to a lower than anticipated demand for electricity and to "problems related to construction, financing and regulatory procedures on local and federal levels."

There were 58 nuclear power reactors authorized to operate in the U.S., capable of producing a total of 39,595 megawatts of electricity; 87 more plants were under construction and 93 others in the planning stage.

Growth rate 'essentially zero'—Nuclear Exchange Corp. reported that the net number of orders for new nuclear power units by U.S. utilities had been "essentially zero" since 1974, the Journal of Commerce reported Aug. 1, 1977. Cancellations of ordered units had virtually balanced new orders over that period.

The nuclear firm cited several factors in explaining the lack of growth:

—Lowered projections for growth in power demand.

—Problems and delays in licensing and other regulatory matters.

—Increased resistance to nuclear plants on environmental grounds.

Nuclear Exchange Corp. said manufacturers of nuclear equipment still retained their production capability, but that capability was "likely to decline soon without new business."

NRC neutral on A-centers—In a report Jan. 22, 1976, the Nuclear Regulatory Commission took a neutral stand on creation of nuclear-energy centers, or a cluster of up to 20 commercial nuclear-power plants on one site. While such plans could be made "feasible and practical," the agency said, there was not any "great or unequivocal advantage or compelling need" for them. In any case, the agency said, it was not in its province to render the decision on the matter, which should be left to Congress or non-regulatory agencies.

Vepco halts work on A-plants—Virginia Electric & Power Co. (Vepco) March 18, 1977 canceled plans to complete two nuclear power units at its Surry, Va. generating station. The utility said losses could amount to $146 million, which would be charged to customers over a 10-year period.

The two 900,000-kilowatt units were to have been completed in 1986 and 1987 at a projected cost of $1.8 billion. A Vepco statement said the change in plans had been caused by reduced estimates of consumer need for electricity and "growing concern over the many uncertainties that face the nuclear industry at this time together with the increasing financial burdens these uncertainties impose. . . ." Delays caused by labor problems, intervention by environmentalists and the necessity of meeting tighter controls imposed by the Nuclear Regulatory Commission (NRC) all had made inflation a major factor in the cost of the plants.

An NRC spokesman noted March 18 that since June 1974, construction of 29 nuclear power plant units had been delayed or deferred, some indefinitely.

Vepco fined—Vepco Feb. 23 had paid a $31,900 fine imposed by the NRC for 30 violations discovered during an investigation of construction practices at the utility's North Anna nuclear power plant. The fine was the third largest ever levied in civil actions by the NRC or its predecessor, the Atomic Energy Commission. Vepco had received the two larger ones also.

Vepco had been fined $60,000 Sept. 11, 1975 for statements about the safety of a power plant site. The Nuclear Regulatory Commission's licensing board, which ordered the fine, considered the statements false in denying the existence of a geological fault at the plant site, in central Virginia.

Nuclear Fuel Problems

Congress probes fuel issue. The adequacy of the U.S.' supply of fissionable fuel for nuclear power plants was investigated by the Congressional Special Subcommittee to Review the Nation's Breeder Reactor Program, a unit of the Joint Committee on Atomic Energy. A federal fast-breeder demonstration project was under way at the Clinch River near Oak Ridge, Tenn. Such a plant would be able to produce more nuclear fuel than it used.

A subcommittee report, inserted in the Congressional Record Feb. 19, 1976 by Subcommittee Chairman Mike McCormack (D, Wash.), said in part:

Uranium is the primary fuel used by present day light water reactors. Its availability is a critical element in the debate over the need for breeder reactors and the timing of their introduction into commercial use by the electric utility industry.

In the material reviewed by the Subcommittee only ERDA and EPRI presented independently developed numerical projections. ERDA projects that 3.6 million tons of uranium will be available, at a production cost of $30 a pound or less. ERDA also projects there is another 13 million tons of uranium at economic and environmental costs which ERDA and others believe might well preclude the use of this material. ERDA engineers and geologists independently develop resource projections based on such inputs as industry data, field examinations and available geologic reports. The majority of those presenting information to the Subcommittee cited ERDA projection as the most reliable and the one that should be used to plan our energy programs.

If ERDA's recent estimates prove correct, the size and composition of the resource base have serious implications for the non-breeder reactor power program. Information provided to the Subcommittee indicates that more than the approximately 700,000 tons (620,000 of reserves and 90,000 of byproduct) of reasonably assured reserves will be needed over the lifetime of reactors presently operating, under construction, or on order. Plants that will be contracted for from now on will depend for fuel on "potential resources" which have not as yet been discovered or verified. Several respondents reported that at some time during the 1990's all of the "potential resources" will also have been committed to the lifetime needs of new reactors. If converter reactors are to be built after that time, they will depend for fuel on either uranium which has not yet been projected to exist as "potential resource", higher cost uranium from low grade ore deposits, or recycled plutonium.

In an analysis done for the Electric Power Research Institute (EPRI) Milton F. Searl projected that there is a 50 percent chance that there are more than 13.2 million tons and a 5 percent chance that there are more than 28.9 million tons of uranium in the United States. His forecast is based on an extrapolation from ERDA data to obtain the expectation of finding a given amount of uranium below a cost of $100/pound. Searl's forecast is mentioned in the NRDC publication "Bypassing the Breeder" as the forecast which NRDC subscribes to. "Bypassing the Breeder" is frequently cited by those opposed to the breeder as part of their basis for believing the breeder should be postponed or abandoned.

Several witnesses expressed their subjective view that uranium resources will exceed the amounts which ERDA projects. These estimates did not include numerical projects and did not appear to be based on independent analysis. . . .

The Subcommittee believes that the ERDA forecast of 3,600,000 tons of uranium at a cost of $30 or less per pound is the most prudent projection on which to base energy plans. It is recognized that these numbers may change as the findings of the National Uranium Resource Evaluation program become available.

The Subcommittee concludes that the uranium supply forecast by ERDA will be inadequate to provide for the nuclear power projected in this report beyond the mid 1990's. The utilization of the breeder concept would increase the energy potential of the presently projected 3.6 million tons of uranium such that it would become equivalent in energy output to about 126 million tons of low cost uranium, an amount of nuclear fuel sufficient to supply nuclear powerplants for centuries. . . .

In recognition of the likely inability of uranium supplies to economically support the future energy load projected for nuclear power, this Nation has had a breeder reactor development program underway for over 25 years. The type of breeder reactor that is generally considered the most advanced and to possess the greatest likelihood of commercial development is the Liquid Metal Fast Breeder Reactor (LMFBR). An LMFBR development program is in progress with the objective of establishing a broad technological base leading to a competitive commercial industry. An essential element of this program is the construction of a mid-sized demonstration plant, the Clinch River Breeder Reactor (CRBR).

In view of the concerns that have been expressed in Congress and by the public with respect to various aspects of the LMFBR program, the Subcommittee undertook to examine the need for the program, and its potential benefits and risks. In the public hearings and other information collection activities conducted by the Subcommittee, the overwhelming consensus was that the LMFBR is needed, and that this need is urgent. A minority group, generally known to oppose nuclear energy, asserted that the breeder was not needed.

The Subcommittee was impressed that those Government agencies with responsibilities for planning or providing for the Nation's energy needs supported the urgent development of the breeder, as did almost all industrial or utility organizations queried. The main reason offered in support of the need for a commercial breeder on a timely basis was its ability to provide sufficient fuel for future electrical generating requirements. In addition, many proponents of the breeder noted its attractiveness from a cost-benefit basis, i.e., future projected savings in fuel costs by the breeder are expected to far outweigh development costs.

With regard to uranium reserves, the basic argument is that the limited amount of "assured" and "potential" reserves (3,600,000 tons of U_3O_8 in the United States) will be fully committed to "burner" reactors, such as the light water reactor, by perhaps the mid-1990's depending on the energy growth rate, plant capacity, and other factors. No additional reactors of this type could be built after this date unless additional uranium resources are found or low grade ores with their accompanying higher costs and environmental impacts are used. A breeder reactor, on the other hand, would permit the extraction of up to 50 times as much energy from these uranium resources, thereby extending our nuclear fuel supplies from decades to centuries.

With regard to cost-benefit analysis, the basic argument is that the use of breeder reactors will avoid reliance on low grade, high cost uranium ores, with substantial resultant savings in fuel costs which will be passed on to consumers. These savings are estimated, under all but the most pessimistic conditions, to be considerably greater than development costs of the LMFBR program. The net savings would, according to these arguments, reach $150 billion by the year 2020, as well as substantially reduce the requirements for mining and enriching uranium.

The Subcommittee also heard testimony and received information to the effect that the need for the breeder had been overstated, and that the projected benefits were not to be had. In general, those groups or individuals opposing the breeder were found to do so based on the same arguments that proponents cite to favor its development, but the opponents generally place a different interpretation on the factors cited above. For example, the opponents suggest that development costs will outweigh savings in fuel costs, (i.e., that cost benefit analysis supports the abandonment of the breeder rather than its development), that AEC–ERDA estimates of uranium resources are too conservative, and that substantial quantities remain to be discovered at economical prices in this country, that energy demand will be less than projected by breeder proponents in future years, and that safety and environmental problems are beyond man's control.

A corollary question considered by the Subcommittee was, assuming a breeder reactor is needed, whether or not this country should continue to put its major effort on the LMFBR, or if more (or less) effort should be devoted to alternate breeder reactor concepts. Again, respondents were essentially unanimous in agreeing that the LMFBR should continue to be the focus of breeder efforts and receive top priority.

The Subcommittee also examined whether or not the overall LMFBR program objectives, content and approach are correct, and what steps can be taken to minimize the costs of the program, and improve performance with respect to program schedules. With regard to overall objectives and approach, the great majority of respondents was again enthusiastic about the program goals and the means proposed to achieve them. A few respondents who as noted before, are generally viewed as opposing nuclear energy, found the whole program premature and ill-advised. They recommended steps such as successful operation of the FFTF before a demonstration plant is built, if at all.

The placement of greater reliance on foreign technology was suggested as a means of improving program performance and reducing costs. While the surface advantages of this approach are evident, several disadvantages were also pointed out, such as the state of dependence the U.S. might be placed in (such as now exists on foreign oil), failure of the U.S. to develop its own industry, undesirable impact on our balance of payments situation, and the need for foreign designs to be modified to meet U.S. safety and licensing requirements.

Other questions examined by the Subcommittee included the total R. & D. costs for the LMFBR and their means of recovery, the predicted capital costs of commercial LMFBR's and the methods to provide that capital, and the overall issue of whether or not the LMFBR would be economically viable. The general consensus of information presented was that the $10.6 billion ERDA figure for the total LMFBR research, development and demonstration program was a reasonable estimate, although a few respondents voiced strong feelings that previous inabilities to meet cost estimates meant the figures would go much higher.

With regard to timing for the LMFBR, the consensus was that the breeder is needed by about the time it would become commercially available under current development plans, i.e., the early 1990's. . . .

There are, however, practical limits to the contribution which present light water reactors can make to ur energy needs. The present generation of light water reactors is fueled by Uranium-235, which makes up less than one percent of uranium as it occurs in nature. Known domestic reserves of uranium ore will be committed to light water reactors by the early 1980s. Potential undiscovered domestic resources will be committed to these reactors during the last decade of this century.

Because of its ability to convert the vastly more plentiful Uranium-238 into useable fuel, the breeder can extend our available uranium resources from decades to centuries. There is presently enough Uranium-238 left over from uranium enrichment operations for light water reactors to provide fuel material for breeders for centuries.

This material is already mined and processed and is currently stockpiled at gaseous diffusion plants without value in the present energy picture. With the breeder, however, this Uranium-238 can provide the equivalent energy of 400 years of oil or 700 years of coal. It has been calculated that at present day coal prices this energy resource is worth some $20 trillion.

The Liquid Metal Fast Breeder Reactor is the most technically advanced of the long-term energy sources . . ., and in light of the uncertainties for solar-electric and fusion power, the LMFBR must remain a high priority development project if this Nation is to have an assured source of electrical energy in the 1990s and beyond. Other nations have recognized the potential of the breeder reactor and are also actively pursuing its development.

The breeder is an attractive option from several standpoints. In the area of safety, the breeder will equal or exceed the impressive safety that has been built into light water reactors. Its design will follow the same defense-in-depth concept which provides for high quality construction and redundant systems to insure safety. One safety advantage of the breeder is that its sodium coolant operates at low pressure and ruptures or breaks in the system are more readily accommodated than such an incident in, for example, the high pressure systems of conventional water-cooled reactors.

The breeder is also attractive from an environmental standpoint. About 200 acres of land must be strip-mined each year to provide fuel for a 1000 MWe coal plant. About five to seven acres must be mined to provide uranium fuel for a conventional nuclear plant. Breeders require less than a single acre to yield fuel with the same potential. Because of the stockpile of Uranium-238 which exists today, breeder reactors would require no mining at all for at least 100 years and probably more. The breeder uses uranium so much more effectively, the uranium from one acre of land would fuel roughly 20 breeders for a year.

We believe the breeder reactor will be a safe and environmentally acceptable method of producing electrical power. The safety design of the Clinch River plant is backed by considerable experience in building and operating liquid sodium fast breeder reactors since 1951. It should be noted that the first electricity from nuclear power in the world was produced by an LMFBR—Experimental Breeder Reactor Number 1 in Idaho in December 1951.

Knowledge gained from this facility and others that have operated over the years will be incorporated into the design of the Clinch River plant, and assumptions regarding economy, reliability, safety, and environmental impact will be tested through on-line operations on the Tennessee Valley Authority system.

Briefing on breeder need. At a briefing by scientists at the Oak Ridge (Tenn.) National Laboratory Sept. 18, 1976, Michael L. Butler presented the case for "the need for the breeder," a nuclear reactor that transforms Uranium-238, which is not usable as a nuclear fuel, into plutonium, which can fuel nuclear power plants (but which can also be used in nuclear bombs).

Noting the U.S.' "increased reliance on coal and nuclear power" in at least "the near term," Butler said:

In some respects, the Clinch River Project is unique in the history of technology development. It is a partnership between the Federal Government and the electric utility industry, and a joint Government-industry Project office under ERDA management has been established to direct the Project.

Some 741 electric utility systems nationwide are contributing $257 million to the Project, the largest industry commitment ever made for a single energy research and development project. The remainder of the Project's cost of $1.950 billion is being provided by the Federal Government.

One of the chief goals of the Clinch River Project is to demonstrate that a breeder reactor can meet the rigid licensing standards imposed on nuclear plants by the Nuclear Regulatory Commission. Accordingly, the Clinch River plant will be licensed like all other nuclear plants in this country. . . .

Liquid metal fast breeder project—ERDA Administrator Robert C. Seamans Jr. had announced Jan. 2, 1976 a decision to proceed with a 10-year research and development program for a liquid metal fast breeder reactor.

Among the unresolved issues of the program were reactor safety and protection and waste management, major issues of the current reactor program. According to ERDA deputy administrator Robert Fri Jan. 2, the question was "do we stop the program altogether or continue it to resolve the problems. We've got to conduct the program to answer the questions."

Schlesinger opposes plutonium use. White House energy adviser James R. Schlesinger said March 25, 1977 that the Carter Administration opposed the development of plutonium fuel systems for nuclear power plants. "For the immediate future we will not be using plutonium recycling," Schlesinger said.

He addressed his remarks to 19 men and women participating in a White House energy conference. The 19 had been selected from among 20,000 respondents to a questionnaire on energy policy sent out by the White House March 2 to approximately 450,000 persons, 300,000 of whom had been chosen at random. The other 150,000 recipients included members of Congress and business and civic leaders.

Schlesinger indicated that the Administration would go along with construction of more conventional uranium-fueled reactors to help meet the country's electricity needs over the next 25 years. He said the Administration wanted to separate conventional reactors "from the plutonium economy" and "separate the use of nuclear power from the spread of nuclear weapons."

The nuclear-power industry and the Ford Administration had contended that existing uranium-fueled nuclear plants would need a new fuel by the end of the 20th century. They had argued that the nation's uranium resources would run so low that plutonium use would become necessary. They added that the plutonium-fueled fast-breeder reactor currently being developed would produce more plutonium than it consumed and thus would be an inexhaustible energy source.

But Carter Administration officials and some academic experts believed the nation's uranium resources had been underestimated and there was enough uranium to fuel existing types of plants well into the 21st century.

Plutonium licensing—The Nuclear Regulatory Commission had announced Nov. 12, 1975 that it would issue interim licenses for the processing and use of plutonium in the nuclear fuel cycle. Previously, the commission had announced in May a postponement for at least three years of a decision on approving widespread reactor use of plutonium.

Breeder reactor, plutonium use scored. A panel of 21 scientists and economists March 21, 1977 urged major changes in the nuclear energy policy of the U.S., including an end to the crash program to develop a commercial fast-breeder reactor and an indefinite postponement of plans to reprocess plutonium for use as a reactor fuel.

The panel, organized by the Mitre Corp. with a grant from the Ford Foundation, endorsed the continued use of current generation uranium-fueled nuclear power plants, but urged more emphasis on solving safety and radioactive-waste problems. The panelists said adequate electricity for the short term could be generated by a combination of nuclear and coal-fired plants. As for the 21st century, they said, "We believe that some mix of coal, solar and fusion energy, assisted by conservation, would be capable of supplying society's long-term energy needs."

The breeder reactor was envisioned as a replacement for current nuclear power plants, which used uranium. The breeder used plutonium and was controversial because plutonium was a basic ingredient of atomic bombs. Plutonium, which rarely occurred naturally, was also a by-product of existing uranium-fueled plants and could be reprocessed for use as a reactor fuel.

The panelists, agreeing with critics of the breeder reactor and plutonium reprocessing, objected primarily to the danger of the worldwide spread of nuclear weapons as a result of the availability of plutonium. They said federal research on the breeder should be continued, but in such a way that commercial use of the breeder would be delayed. (The federal government in recent years had poured billions of dollars into breeder-reactor research.) They said further that plutonium reprocessing had "little if any" economic value and should be "postponed indefinitely."

The report was based on two key assumptions: that conventional nuclear reactors, which currently generated about 10% of the electricity in the U.S., would be about as expensive and dangerous to operate as coal-fired plants; and that the current official estimates of uranium reserves and resources substantially underestimated the amount of uranium that would be available during the next two or three decades.

The Ford Foundation said the panelists, who had worked on the study for almost two years, had been chosen because of their middle-of-the-road views on nuclear energy. Among them were John Sawhill, former head of the Federal Energy Administration, and two members of the Carter Administration: Defense Secretary Harold Brown, who was president of the California Institute of Technology when the work on the report was done, and Joseph S. Nye Jr., deputy to the undersecretary of security assistance in the State Department.

The panel was known as the Nuclear Energy Policy Study Group, and its report was entitled "Nuclear Power Issues and Choices." In its section on the breeder issue, the report said:

> The priority and timing of the plutonium breeder is inevitably a central budget and policy issue since the commitment to this program currently dominates federal energy research and development activities. The plutonium breeder, which produces more plutonium than it consumes in operation, can in principle improve the utilization of uranium by a factor of as much as 100. When used in light-water reactors (LWRs), current estimated uranium reserves would provide only one-tenth the energy of coal reserves; in breeders, these same uranium reserves could in principle provide ten times the energy of coal reserves. The breeder thus opens up a vast additional energy resource and answers the criticism that nuclear power will price itself out of the market as soon as low-cost uranium is exhausted.
>
> The Liquid Metal Fast Breeder Reactor (LMFBR) has become the centerpiece in the U.S. energy research and development program. The LMFBR program is focused on the early commercialization of a power plant to compete with the current generation of LWRs. ERDA has estimated that this program will cost at least $12 billion to complete, assuming utilities will be able and willing to start buying breeders within ten years without government subsidies.

The plutonium breeder involves a full commitment to the plutonium fuel cycle and would introduce tremendous quantities of plutonium into national and international commerce. In these circumstances, the pressure for indigenous plutonium reprocessing facilities would grow rapidly and be difficult to oppose. The breeder would thus greatly complicate the proliferation problem and increase the possibility of theft or diversion of material suitable for weapons. The economics of the breeder have generally been considered so persuasive that this serious disadvantage has until recently been largely dismissed in government planning.

Past government policy on the LMFBR has been predicated on a belief that nuclear power would exhaust reserves of low-priced uranium in a few decades, making breeder introduction economically attractive by the early 1990s. Our analysis, however, indicates that the early economic potential of the breeder has been significantly overstated. The LMFBR, as presently envisaged, will have higher capital costs than the LWR and must therefore operate as a significantly lower fuel cycle cost to be economically competitive. There appears to be little prospect that these fuel cycle costs can be reduced to a point that would give the LMFBR a significant economic advantage over the LWR in this century or the early decades of the next century. The current assessment of uranium reserves probably substantially understates the supplies that will become available; uranium, at prices making light-water reactors competitive with breeders, will be available for a considerably longer time than previously estimated. New enrichment technologies may also extend these supplies. Moreover, coal available at roughly current costs will look increasingly attractive if the costs of nuclear power rise. Finally, demand projections on which breeder economic assessments have been made in the past were unrealistically high and have already been substantially reduced. These considerations lead us to the conclusion that the economic incentive to introduce breeders will develop much more slowly than previously assumed in government planning.

This conclusion applies to other countries, as well, provided that they have access to low-enriched uranium to meet their nuclear fuel requirements. Moreover, the contribution of breeders to energy independence is questionable for most countries since the complexity and scale of the breeder fuel cycle would make an autonomous breeder system too costly for all but the largest industrial economies. Therefore, the prospect of a large export market for breeders in this century is illusory.

Despite this negative assessment, we believe that a breeder program with restructured goals should be pursued as insurance against very high energy costs in the future. This situation could develop if additional uranium reserves do not become available, environmental problems place limits on the utilization of coal, and other alternative energy sources do not become commercially viable at reasonable prices in the first decades of the next century. The present U.S. program, directed at the early commercialization of the LMFBR, is not necessary to the development of the breeder as insurance. The ultimate success of the breeder may even be compromised by telescoping development stages to meet an early deadline, freezing technology prematurely. We believe therefore that the breeder program should deemphasize early commercialization and emphasize a more flexible approach to basic technology. In such a program, with a longer time horizon, the Clinch River project, a prototype demonstration reactor costing $2 billion, is unnecessary and could be canceled without harming the long-term prospects of breeders. In fact, premature demonstration of a clearly noncompetitive breeder could be detrimental to its ultimate prospects.

Although long lead times are required for a project as complex as the breeder, we believe that the decision on commercialization, now set for 1986, can safely be postponed beyond the end of the century. The cost, if any, of such postponement will be small, and there is a strong possibility that postponement will help in restraining large-scale, worldwide commerce in plutonium and buy time to develop institutions to deal with this problem. The option of bypassing the plutonium breeder altogether should not be prematurely foreclosed since there is at least a possibility that the plutonium breeder may never become necessary, or even economically competitive, compared to other energy sources that may become available in the next century.

Plutonium fuel system called safe—The Ford-Mitre study was strongly criticized by two nuclear researchers at the Electric Power Research Institute in Palo Alto, Calif., in interviews reported March 26. Milton Levenson and Chauncy Starr, officials of the institute, noted that a breeder reactor nuclear fuel-recycling system in which the fuel remained so radioactive that terrorists could not have stolen it had operated almost unnoticed for more than four years in the 1960s at what was currently known as the Idaho National Engineering Laboratory.

Starr said "major flaws" in the Ford report included "unverified and risky" assumptions about future uranium and coal supplies and prices and a failure to estimate the "economic and social penalties" if energy supplies turned out to be smaller than forecast.

Both men attacked the study's implicit assumption that fuel-reprocessing for future breeder reactors would be done the same way as at existing plants for recovering plutonium from spent uranium reactor fuel. Such plants dissolved the spent fuel completely so that near-pure plutonium could be separated from radioactive byproducts of the nuclear reaction.

Levenson said that in a fast-breeder reactor, such radioactive debris would not interfere with the nuclear reaction and so the plutonium fuel would not have to be pure. The fuel could remain highly radioactive and thus, he said, immune to "diversion" by terrorists.

Levenson said the process used in Idaho had been developed not to guard against terrorism but to speed up the fuel-reprocessing operation.

Carter proposes ban on plutonium use. President Carter April 7, 1977 proposed that the U.S. "defer indefinitely" the use of plutonium to fuel commercial nuclear power plants. Carter also indicated the government's fast-breeder reactor program would be slowed and reshaped to emphasize the development of fuels other than plutonium.

The President said concern about the spread of nuclear weapons internationally had been the main factor in his decision. (Plutonium was the key ingredient of atomic bombs.) In a phrase intended as a signal to other nations, as was much of his statement, Carter said, "We have concluded that a viable and economic nuclear power program can be sustained" without the reprocessing of plutonium.

In line with his general proposal, the President said a plant under construction at Barnwell, S.C. would "receive neither federal encouragement nor funding for its completion as a reprocessing facility." Allied-General Nuclear Services had already spent an estimated $250 million on the plant and had sought as much as $500 million in federal assistance to complete it as a fuel-reprocessing demonstration project. Allied-General was indirectly equally owned by Allied Chemical Corp. and General Atomic Co., itself a joint venture of Gulf Oil Corp. and the Royal Dutch/Shell Group.

Carter said his Administration would "restructure the U.S. breeder-reactor program to give greater priority to alternative designs . . . and to defer the date when breeder reactors will be put into commercial use." Under questioning from reporters, he said the government's fast-breeder demonstration project on the Clinch River near Oak Ridge, Tenn. would "not be terminated as such" but would receive a reduced amount of federal funds. Future government support of the facility, he said, would focus on experimental research and development rather than the possibility of commercial utilization.

Knowledgeable sources said that top environmental officials in the Administration had recommended that the Clinch River project be discontinued altogether.

The U.S. Energy Research and Development Administration (ERDA) estimated that $500 million had been spent on designing the Clinch River reactor and that additional outlays of $50 million to $100 million would be required just to complete the design effort.

Carter said that to make sure the U.S. did not need to use plutonium, he would propose an increase in the production of enriched uranium for nuclear power plant use. He also said he would propose legislation to allow the U.S. to guarantee supplies of uranium fuel to foreign countries so they would not have to rely on reprocessed plutonium either. (According to the Washington Post April 13, the U.S. was insisting that countries that purchased nuclear power plants from the U.S. waive their rights to the plutonium generated by burning U.S.-supplied uranium in those plants.)

The President said the U.S. would continue to embargo the export of technology or equipment for uranium enrichment or plutonium reprocessing. (Both processes could be used to make nuclear weapons.) He said the U.S. would continue discussions aimed at reducing the spread of atomic weapons with countries that supplied and used nuclear materials. But he said he would not ask countries that had plutonium-reprocessing capabilities, such as West Germany, France, Great Britain and Japan, to discontinue reprocessing.

"We are not trying to impose our will," the President said, noting that other industrial powers had a "special need" that the U.S. did not have for atomic energy because they lacked the domestic supplies of coal, oil and natural gas that the U.S. had. "But I hope," Carter added, "that by this unilateral action we can set a standard."

Britain and France had been negotiating large fuel-reprocessing contracts with Japan and other nations. (The precise implications of Carter's statement were especially important to Japan, which received all its enriched uranium from the U.S.)

Most of the nuclear power stations in the world used a U.S.-designed reactor that needed enriched uranium fuel—uranium with a 93% saturation of the isotope Uranium-235. Enriched uranium supplied by the U.S. was exported under special license from the federal Nuclear Regulatory Commission in compliance with the terms of the Nuclear Nonproliferation Treaty. Even after use, the nuclear material—which would then contain plutonium—could not be transferred to another country for storage or reprocessing without another license that complied with arms control regulations.

The Carter Administration, however, currently was holding up approval of 28 export licenses involving the shipment of enriched uranium fuel to 13 countries, including most of the nations of Western Europe, Japan, Brazil and Canada, according to a Washington Post report April 14. The Post April 13 had reported that in the case of Spain, the U.S. had said it would approve certain export licenses if Washington were given a veto over when and where spent fuel was reprocessed, if at all. The Post said the U.S. was asking for veto power over all spent fuel, no matter who had supplied it.

The French government April 8 noted that Carter had said he would not impose the U.S. will on other countries, which the French took to mean that he did not intend to prejudge the means other countries might use eventually to meet their energy needs. In that respect, French observers noted, Carter had taken into account the comments presented by France, West Germany, Great Britain and Japan during U.S.-initiated talks held prior to the President's speech.

The West German government declined to comment on the statement. Hours before Carter spoke, West Germany had issued its own nuclear-policy statement calling for curtailment of the spread of nuclear weapons through "multinational, nondiscriminatory and generally binding" agreements on safeguards and the peaceful uses of nuclear energy rather than through the restrictions on technology advocated by Carter. The German statement pointed out that the peaceful use of nuclear energy was for many countries necessary to secure their social and economic progress.

Carter plutonium ban assailed. A 41-nation conference on nuclear energy April 13 adopted a resolution opposing Carter's policy because it attempted to restrict the development of fast-breeder reactors. The resolution was approved at the end of a five-day conference in Persepolis, Iran attended by 500 scientists, government officials and representatives of the nuclear industry.

The conference had been called by the American Nuclear Society and similar organizations in Japan and Europe. Among the nations represented were the U.S., the Soviet Union, France, West Germany and Japan.

The resolution said that most countries looked to nuclear power as the only means to energy independence. For those countries without large uranium resources, this independence could come only through the use of breeder reactors, it said.

The resolution also contended that the Carter restrictions unilaterally abrogated the section of the Nuclear Nonproliferation Treaty that promoted the free flow of nuclear knowledge. This, the resolution said, weakened the confidence of other nations in U.S. promises to provide adequate uranium fuel supplies as alternatives to plutonium.

In other reaction, officials of the International Atomic Energy Agency (IAEA) in Vienna believed the Carter policy could not be successful, according to the Los Angeles Times April 17. They said that not a single exporter of nuclear technology had indicated support for it.

Sigvard Eklund, IAEA director general, said the policy could "disrupt the atmosphere of mutual trust and confidence that has been built up during the last 25 years." Another IAEA official said it "is a fallacy to suppose that because plutonium and reprocessing are unnecessary and uneconomical for the U.S., they are also unnecessary and uneconomical for the rest of the world."

On the U.S. domestic front, Robert E. Kirby, chairman of Westinghouse Electric

Corp., told the company's annual meeting April 27 that Carter's proposed curb on fast-breeder reactor research was "most unwise." He said breeder reactors promised energy resources equivalent to at least three times those represented by Middle East oil reserves.

Westinghouse, a diversified manufacturer of industrial and electrical products, had been designing the nation's first large-scale breeder research plant. Its construction on the Clinch River at Oak Ridge, Tenn. had been postponed by Carter.

Kirby said "the connection between the breeder program and possible nuclear weapons proliferation is so thin as to be insignificant." He warned that halting breeder research would "remove the U.S. from any voice in the development of international controls of this important new technology."

U.S. Sen. Frank Church (D, Ida.) echoed that theme May 2 when he called with reprocessing capability would help prevent other nations from acquiring it.

Reaction in U.S. mixed, cautious abroad—In the U.S., President Carter's statement drew criticism from the Atomic Industrial Forum, a trade group financed by reactor manufacturers and others with business interests in the nuclear energy field. A group spokesman said Carter was "in effect asking Americans to forego an essential element in their energy future in order to create a bargaining chip" for use in international negotiations. Other U.S. energy industry officials expressed relief that the President had not decided to completely cancel research and development of the breeder reactor.

A coalition of 12 environmental and other groups, including the Sierra Club and the National Council of Churches, lauded Carter's ban on plutonium use but said his "failure to take an equally strong and definitive position on the breeder-reactor program may seriously undermine the effect of this as a nonproliferation policy."

Abroad, the statement met generally polite but firm resistance. It was pointed out in many capitals that Carter had not made an outright declaration that U.S.-supplied fuels could not be reprocessed. Great Britain, France, the U.S.S.R., West Germany and Japan had all invested heavily in plutonium-fueled fast-breeder reactors as insurance against exhaustion of uranium supplies or dependence on foreign uranium suppliers. Moreover, the President's policy "a formula for nuclear isolationism" that would "reduce, not enhance, U.S. influence in shaping worldwide nuclear policy."

In a speech at the Massachusetts Institute of Technology, Church urged Carter to go ahead with the Clinch River breeder-reactor demonstration project and reopen the question of federal funding for the privately owned but unfinished nuclear reprocessing plant in Barnwell, S.C.

Church advocated effective IAEA supervision of all uranium enrichment and reprocessing plants, saying the "agency should own all the nuclear fuel in its global system, be responsible for the fuel's security and accountable for its use."

Westinghouse uranium case. U.S. District Judge Robert Mehrige Jr. in Richmond, Va. Feb. 3, 1976 approved an agreement between Westinghouse Electric Corp. and 13 domestic and three foreign utilities for division of 15 million pounds of uranium Westinghouse had in its inventory or on order.

The uranium was to be supplied under terms of contracts Westinghouse terminated in September 1975, or was seeking to terminate.

Westinghouse acted under the Uniform Commercial Code allowing companies to "excuse" themselves from contracts under certain economic conditions. The condition Westinghouse put forth was the price of uranium, which had risen from an average price in the contracts of $9.50 a pound to its current market price of more than $25 per pound.

The agreement did not touch upon the major issue in the court fight, the dispute over an additional 65 million pounds of uranium Westinghouse was obligated to provide the utilities over the next 20 years under the contracts it was seeking to cancel.

The dispute had begun Sept. 8, 1975 when Westinghouse announced it would not be able to meet its contractual obligations with 27 utility firms to supply them with about 65 million pounds of uranium oxide, or yellow-cake, because of a sharp rise in price.

The increase had made it "commercially impracticable" to honor the contracts, Westinghouse said.

In early 1977, Westinghouse and three Pittsburgh-area utilities reached an out-of-court settlement in a contract dispute involving Westinghouse's refusal to honor its agreements to supply the utilities with 770,000 pounds of uranium oxide. The tentative settlement with the three firms was announced March 3 by Judge I. Martin Wekselman of the Allegheny County Court of Common Pleas in Pittsburgh.

Westinghouse's proposed agreement with Duquesne Light Co., Ohio Edison Co. and Pennsylvania Power Co. would not directly affect the lawsuits filed by 24 other utilities against Westinghouse for backing out of their uranium-supply contracts.

Under the proposed agreement with the three utilities, Westinghouse would provide an "undisclosed amount of cash, plus certain valuable new equipment, technical services and engineering." (These would help offset higher prices the utilities would pay for uranium.)

Since Westinghouse also was negotiating an out-of-court settlement with the remaining 24 utilities, whose suits had been consolidated, the company agreed to make an upward revision in its settlement with the three utilities if the agreement compared unfavorably with any later settlement.

Westinghouse, which had contracted to sell the uranium for an average $9.50 a pound, said producers were charging $40 a pound. Fulfilling the fixed-price, long-term contracts at current prices would cost the company "in excess of $2.5 billion" over 18 years, Westinghouse said, and would have an "extremely adverse financial impact" on the company.

The proposed settlement also was related to an antitrust suit filed by Westinghouse in October 1976 against 29 foreign and domestic uranium producers and their agents. Westinghouse alleged that they were part of an international uranium cartel formed in Paris in 1972. The defendants were charged with illegally raising and fixing uranium prices and refusing to sell Westinghouse the amount needed to supply its utility customers.

In the agreement with the Pittsburgh utilities, Westinghouse agreed to provide them with "a portion of the net proceeds from any recovery" in its $7.5-billion suit against the uranium producers.

Westinghouse had built about 130 nuclear reactors around the world since 1957. The Pittsburgh utilities argued in their suit that Westinghouse had made extravagant supply promises to potential customers in a high-powered drive to sell the reactors. (The Pittsburgh-area utilities operated the Beaver Valley Nuclear Unit No. 1 in Shippingport, Penn.)

Westinghouse had contended during the nonjury trial, which ended in February, that uranium prices would have remained stable during the length of the contracts if oil-exporting countries had not quintupled the oil prices in 1973–74. This action drove up the prices of other types of energy, Westinghouse argued.

Delmarva plant canceled. The General Atomic Co. had canceled a contract to supply Delmarva Power & Light Co. with nuclear fuel for its $1.1 billion Delaware power station, the Wall Street Journal said Oct. 29, 1975.

General Atomic, an equal partnership of Gulf Oil Corp. and Scallop Nuclear Inc., agreed to pay Delmarva $125 million for costs incurred on the project. The cancelation was brought about by cost escalation and by General Atomic's problems with commercializing its high-temperature, helium-cooled reactor.

The Philadelphia Electric Co. had halted plans Sept. 16 to construct a nuclear power plant on the Susquehanna River in southern Pennsylvania after General Atomic announced it would not be able to provide the reactors. (The New York Times Sept. 9 reported the conclusions of a survey by the Edison Electric Institute, a trade association of investor-owned electric utilities, which said that five proposed nuclear reactors and eight conventional units had been canceled during the first half of 1975.)

Cartel dispute focuses on Gulf. Controversy had been aroused by charges of a price-fixing cartel of uranium-producing countries. Controversy focused on Gulf Oil Corp. April 25, 1977 when the Wall Street Journal reported that several officials of the firm "were at one time active" in the cartel "through a wholly owned Canadian subsidiary."

The oil company's participation in the cartel, through Gulf Minerals Canada Ltd., "began early in 1972 and continued for at least two years," the Journal said.

Court "documents show that Gulf executives were active in both the Canadian branch of the cartel and in its worldwide deliberations. In fact, a Gulf official helped draft price-fixing and bid-rigging rules for the cartel in March 1974," according to the Journal.

Gulf's role in the price-fixing scheme was under investigation by a number of U.S. groups. A grand jury had been empaneled in 1976 to consider findings from the Justice Department's probe, begun a year earlier, of possible antitrust violations in the uranium industry. (The Journal report was based in part on documents connected with the grand jury investigation and with other court cases involving the uranium industry.)

Charges that a uranium producers' cartel existed had been first reported in the press in August 1976. Secret documents, apparently stolen from an Australian uranium producer, Mary Kathleen Uranium Ltd., were made public by Friends of the Earth, an environmental group. The documents discussed the cartel's efforts to fix prices and divide the non-U.S. free-world market by setting quotas. (Uranium, which was selling for about $6 a pound in 1972, rose to $41 a pound in 1976.)

More charges surfaced later in 1976 when Westinghouse Electric Corp., a major supplier of uranium to U.S. utilities with nuclear generators, claimed the cartel had been responsible for the nearly seven-fold increase in uranium prices since 1972.

Westinghouse cited the swift rise in price as justification for canceling its long-term contracts to supply 27 utilities with fuel for their generators.

In a later action, Westinghouse brought suit against 29 foreign and domestic uranium producers, charging they had conspired to fix prices in violation of U.S. antitrust laws.

According to the Journal article, Gulf was among more than 20 companies, including Denison Mines Ltd. of Canada and Rio Tinto Zinc Corp. of Great Britain, that had formed the cartel under the sponsorship of Canada, France, and South Africa. (Australia later also was named as a backer of the cartel.)

At a meeting in Johannesburg in June 1972, the Journal said, the cartel agreed to set a floor price on uranium. "Within 20 months, the cartel price rose about 57%." Members also divided the non-U.S. free-world market among themselves, with one-third of the share going to Canada, as the largest exporter.

Canada withdrew its support from the cartel in early 1975, according to the Journal. "The marketing arrangement had been overtaken by market forces," Alastair Gillespie, Canadian minister for energy, mines and resources, said in September 1976. That same month, the Journal said, the government issued a regulation barring Canadian residents from making any comment about uranium unless compelled to do so by Canadian law or authorized to speak by the energy minister.

State and federal probes—After publication of the Journal article, the controversy shifted to investigations conducted by the New York State Assembly's Office of Legislative Oversight and Analysis and to a subcommittee of the U.S. House of Representatives. Both groups focused on Gulf's role in the price-fixing cartel and possible antitrust implications.

The New York investigators released a report June 9 charging that Gulf's participation in the cartel was more extensive than previously reported.

According to the New York report, Gulf officials had attended cartel meetings in Paris in March and April 1972, and four executives had been present in Johannesburg in June 1972 when the cartel established the floor for uranium prices. (The report said that at least 17 cartel meetings were held between February 1972 and May 1974.)

In a position paper released June 9, Gulf admitted that its chairman, Jerry McAfee, had been "advised" of the Canadian subsidiary's participation in the cartel meetings but said he had not been "involved" in any of them.

McAfee, who had replaced the ousted Bob R. Dorsey in the wake of Gulf's overseas bribery scandal, had been president of Gulf Oil Canada Ltd. from 1972 to 1974.

Gulf also contended that its Canadian subsidiary had been "compelled" to participate in the cartel by the Canadian government. The New York report acknowledged that the Canadians had pressured Gulf to join the cartel. But the report said Gulf officials had "forced the Canadian government to 'require' their [Gulf's] participation in the cartel."

Gulf claimed that at the time of its Canadian subsidiary's "required entry into the [cartel] arrangements, U.S. antitrust laws were reviewed carefully with a conclusion that such activities were legal and that there wouldn't be any impact on domestic or foreign commerce of the U.S."

Gulf also claimed that the "marketing arrangement was ineffective" and had not been "responsible for a dramatic increase" in uranium prices. The New York report disagreed, saying the "cartel was, in large measure, responsible for the 700% increase in the cost of uranium."

The New York report also charged that the Canadian government was refusing to provide information about the cartel and was "effectively pressuring the U.S. Department of State with threats of Canadian retaliation against . . . [a planned natural gas pipeline across Canada] if the uranium investigation surfaces as a criminal inquiry."

The House Commerce Committee's Subcommittee on Oversight & Investigations held public hearings June 16-17 to question Gulf officials on documents about Gulf's role in the cartel. The papers had been obtained under subpoena from Gulf's Canadian subsidiary.

Gulf's contention that the cartel had had "no discernible impact" on U.S. uranium prices was undercut June 16 by testimony from one of its officials, S. A. Zagnoli, president of the Canadian subsidiary.

Rep. Albert Gore (D, Tenn.) asked Zagnoli about the Tennessee Valley Authority's agreement with a West German supplier in 1974 to buy 1.5 million pounds of uranium at $12.50 a pound. Gore charged that the price, "fixed behind closed doors" by the cartel, had increased electricity costs for TVA customers.

Zagnoli conceded that "to the extent they [U.S. utilities] bought overseas, they probably paid a higher price." Zagnoli also differed with the parent company on the cartel's impact on world trade in uranium. Zagnoli said the cartel had been "moderately successful." It was "effective in allocating the market" and "did raise prices overseas," Zagnoli said.

The subcommittee June 16 released a Gulf memo on the subject of rising uranium prices. The memo said: "What appears to be happening is that the international producers are in effect setting the world price" by establishing a floor that was higher than what U.S. utilities were willing to pay. "The U.S. producers refuse to sell at any price that doesn't give them a substantial margin above the 'floor' being quoted by the non-U.S. producers. Thus, in essence, the international producers can stop any transactions by constantly nudging the floor upward. . . . In the interim, the U.S. buyer becomes increasingly frustrated, offers a higher price in order to get some response and the cycle starts over again."

The subcommittee also questioned Gulf officials about their assertion that the Canadian government had "compelled" Gulf to participate in the cartel. The heavily Democratic panel was openly skeptical about the claim, charging Gulf with using it as a shield against prosecution under U.S. antitrust laws.

"Your argument is that Gulf is some kind of corporate Patty Hearst, forced to act against your will," Gore said.

The subcommittee June 16 cited a letter by a Gulf lawyer discussing the value of the Canadian government's involvement in the cartel. "The more intricately involved the Canadian government . . . remains in this matter," the lawyer wrote, "the better the degree of protection for Gulf. The obvious reason for my concern is possible Gulf vulnerability under the American antitrust laws. . . . My inclination is that Gulf representatives should always record their strong objection and total disagreement with any predatory cartel action affecting or intended to affect American trade or commerce. . . . Gulf's recorded objections and disagreement will in all likelihood just be noted and overriden by other cartel members."

In testimony June 17, a former Gulf official said he had helped draft a letter for the Canadian energy minstry that requested his participation as the Gulf representative on the cartel's operating committee.

The ex-Gulf employe, Lawrence T. Gregg, also said he had attended the cartel's 1972 meeting in Johannesburg and had proposed that the group raise its uranium price. (A smaller increase than Gregg's proposal was adopted—the

$12.50 price paid by TVA. Gregg said the TVA contract was discussed at the meeting. He also contended that TVA would have paid a lower price if the cartel had not been in existence.)

Canadian reaction—Finance Minister Donald Macdonald June 16 charged that U.S. policies in the early 1970s had forced Canada and other producing nations to form a cartel that would stabilize world prices.

Macdonald noted that the U.S. had imposed an embargo on uranium imports in 1970 that was not lifted until 1976. Canada, the world's leading exporter of uranium, was excluded from the American market, which bought 70% of the world's uranium supply, Macdonald said.

At the same time, he added, U.S. companies, like Westinghouse, "with the American Administration supporting them, were engaging in loss-leader tactics."

"We acted to protect ourselves from these predatory American tactics," Macdonald said.

In a speech before the House of Commons June 17, Macdonald said he objected "that the U.S. would seek to apply its laws in Canada. . . . I do not regard that as a friendly act."

Nuclear Exports & Laxity Linked to A-Arms Proliferation

Lilienthal calls for nuclear export ban. David E. Lilienthal, the first chairman of the Atomic Energy Commission (forerunner of the Energy Research and Development Administration), Jan. 19, 1976 urged an embargo on U.S. exports of nuclear reactors, materials and technology to halt the "terrifying" spread of atomic bomb capabilities. He made the recommendation at a hearing of a Senate Government Operations Committee panel headed by Sen. John H. Glenn Jr. (D, Ohio).

Lilienthal, calling the U.S. "the atomic patsy of the world," said many citizens would be "shocked" if they realized the extent to which the U.S. "has been putting into the hands of our own commercial interests and of foreign countries quantities of bomb material."

His call for an embargo was endorsed by Nobel prize-winning physicist Hans Bethe, who worked on the U.S. atomic bomb project during World War II, and physicist Herbert York, who worked on the U.S. hydrogen bomb project later, but they said the embargo should be a temporary one until an accord were reached on security for nuclear trade.

Tighter security sought for A-material. The Nuclear Regulatory Commission was petitioned by the Natural Resources Defense Council Feb. 2, 1976 to take immediate emergency steps to tighten safeguards at the 16 research and processing plants licensed to use weapons-grade nuclear materials.

The council, a private environmental organization, considered current security inadequate. Among the council's proposals were use of federal marshals at the plants and possible repossession by the federal government of all weapons-grade nuclear materials.

Ford warns vs. A-arms spread. President Gerald Ford told Congress July 29, 1976 that a "world of many nuclear-weapons states could become extremely unstable and dangerous."

Presenting the annual report of the U.S. Arms Control and Disarmament Agency, Ford said that 20 additional countries had the technical competence and the material to build nuclear weapons. Six nations—the U.S., the Soviet Union, China, Britain, France and India—already had such knowledge and materials. Ford said that all countries supplying nuclear fuel, equipment and assistance should impose International Atomic Energy Agency standards on their sales. These standards included prohibition of all nuclear explosions, insistence upon measures to prevent sabotage and theft, and imposition of the same standards on the equipment if it were transferred to a third country.

Fred C. Ikle, director of the Arms Control & Disarmament Agency, had criticized U.S. nuclear aid July 23 as "too cavalier."

In an interview after he had testified before the Senate Foreign Relations Committee, Ikle said that the U.S. had sold reactors and fuel abroad with little regard for safety standards. While the situation had shown improvement in recent years, Ikle said, the U.S. had supplied reactors to countries which did not need them, such as South Vietnam and Zaire. He also charged that U.S. nuclear aid to Europe had made possible uncontrolled European sales of reactors and reprocessing equipment to less-developed countries.

Ford proposes tighter nuclear curbs. President Ford proposed tighter control over nuclear technology Oct. 28, 1976 in an effort to curb the worldwide spread of nuclear weapons. The new policy was announced at the White House and by Ford in Cincinnati, where he was campaigning. Ford noted there that the new policy included an earlier proposal to Congress for a $4.4-billion addition to a federal uranium fuel plant at Portsmouth, Ohio, which, Ford said, would mean "6,000 new jobs for southern Ohio."

A statement Oct. 28 from the headquarters of Jimmy Carter, then the Democratic presidential nominee, characterized Ford's new policy as "a shortsighted, campaign-inspired attempt to correct the timid record of the past." Carter's statement said that Ford's plan "only thinly disguises" his interest in a massive aid program for plutonium reprocessing "on a so-called evaluation basis."

Under Ford's new policy, domestic reprocessing of nuclear fuel would no longer be considered "inevitable" and would not be undertaken "unless there is sound reason to conclude that the world community can effectively overcome the associated risks of [weapons] proliferation." Reprocessing involved plutonium, a byproduct of nuclear-power reactors. The plutonium could be used as a fuel for power plants or as the key component in the manufacture of atomic bombs.

The new policy involved a shift from a large-scale "demonstration program" in this area to an "evaluation program" taking into account the weapons risk. The President called upon other nuclear nations to forsake, for at least three years, the spread of technology or facilities for reprocessing.

Also in the international area, Ford proposed:

- To offer "binding letters of intent" to assure countries that the U.S. would supply them with nuclear fuel if they accepted "responsible restraints" on their nuclear policies.
- To initiate diplomatic negotiations to strengthen international nuclear controls.
- To revise U.S. bilateral pacts toward anti-proliferation policy.
- To establish an international authority to store spent nuclear fuel of various nations and to accept, in the meantime and under certain conditions, the spent fuel of other nations for safekeeping.

A-fuel reported missing. A report by the General Accounting Office (GAO) cited Aug. 5, 1976 at a session of the House Small Business Committee's subcommittee on energy and the environment said that government nuclear facilities were unable to account for several tons of nuclear fuel that could be used for making weapons.

The U.S. government had maintained that most of the missing fuel could be accounted for by material embedded in reprocessing machinery and by crude statistical controls. However, the classified GAO report, which was disclosed during testimony, charged that lax government inventory controls and security measures were responsible for the loss of as much as 6,000 pounds of weapons-grade plutonium and uranium.

A similar report was made by the GAO to Congress May 2, 1977. The report said that although the material had not necessarily been stolen, "the fact that it is missing greatly detracts from the integrity of the safeguards system."

The GAO also said investigations at three facilities chosen as a sample had revealed flaws in the security systems designed to prevent unauthorized seizure of dangerous materials.

Officials said that 4,800 pounds of enriched uranium and 3,400 pounds of plutonium were missing or unaccounted for as of September 1976.

Officials of the Nuclear Regulatory Commission (NRC) and the Energy Research and Development Administration (ERDA) minimized the possibility that the bomb-grade material had been stolen.

The announcement marked the first time the government had made public detailed information on losses of fissionable material.

Rep. John Dingell (D, Mich.) said an audit by the General Accounting Office showed the losses to be higher than NRC and ERDA contended.

Dingell Aug. 8 chaired a hearing of the House Interstate and Foreign Commerce subcommittee on energy and power at which ERDA officials admitted there were another 16 tons of enriched uranium that could not be accounted for. The material was from the facilities at Oak Ridge, Tenn. and Portsmouth, Ohio, where atomic weapons were manufactured.

The 16-ton figure had not been included in the Aug. 4 report, ERDA Administrator Robert Fri said, because it was only a rough estimate. Fri said the weapons-grade material was not actually lost or stolen; instead, it was just the uranium that had condensed along the hundreds of miles of pipes at each facility. (The uranium was diffused through the pipes in gaseous form in order to separate various isotopes.)

Fri said about another 64 tons of non-weapons-grade uranium was believed to have accumulated in the pipes of the two facilities.

Fri also testified that investigators of every suspicious incident involving missing weapons-grade material had always concluded "there is no evidence that a significant amount of special nuclear material had been stolen or diverted."

After the hearing, however, subcommittee staff director Michael Ward told reporters that officials in the U.S. intelligence community had "strong suspicions that a diversion occurred."

A-plants told to tighten security. The U.S. Nuclear Regulatorv Commission Feb. 22, 1977 ordered the nation's 63 federally licensed nuclear power plants to take major new safety precautions against terrorist attacks.

The order required each facility to hire at least 10 guards or trained personnel armed with semiautomatic rifles. It required guards to respond to all threats with an equal degree of force, including shooting to kill if necessary. Most plants currently were protected by three or four guards with pistols.

The order also required implementation of elaborate procedures to limit employe access to certain vital areas of the plants and construction of a bullet-resistant alarm center to ensure that a call could be made for outside assistance in case of attack.

Most of the new measures were to be taken within 90 days, but any construction required would not have to be completed for 18 months. Plants not in compliance by the deadlines would be closed down. An NRC spokesman estimated the cost to each facility at about $1.5 million to $2 million a year, compared with the $500,-000–$750,000 most plants currently spent on safety.

The new regulations were designed primarily to protect against sabotage by a group of "several" outsiders cooperating with at least one employe of a plant. NRC officials said they did not know of any group in the country with the motivation and skill to attack a nuclear reactor and cause an accident that would release deadly radioactivity into the air. However, they said the number of threats against nuclear facilities had increased to 68 in 1976 from an average of about 16 a year through 1975.

The threat of thefts of plutonium, the raw material of atomic bombs, was considered remote at nuclear reactors, officials said, because the plutonium was contained only in burned fuel rods at the core of the reactor and was too hot to touch. Nevertheless, the NRC was considering other safety measures for the 13 facilities licensed to process the plutonium and enriched uranium that was used mostly for to fuel reactors powering Navy ships.

Carter for export curbs. President Jimmy Carter April 27, 1977 proposed legislation to establish tough new controls on U.S. exports of nuclear technology and materials to prevent the spread of nuclear weapons. In a message to Congress, Carter said the bill was "intended to reassure other nations that the U.S. will be a reliable supplier of nuclear fuel and equipment for those who genuinely share our desire for nonproliferation."

Under the President's bill, nuclear materials would not be exported until the executive branch notified the Nuclear Regulatory Commission that the proposed sale "will not be inimicable to the common defense and security."

The bill also provided interim rules on exports that would be applied while foreign governments were being pressured to renegotiate existing agreements with the U.S. to incorporate the tougher new rules. The interim rules were intended to discourage the use or transfer of nuclear materials for military purposes.

Any new international agreements would have to include the new rules, according to the proposed legislation. One of the new rules required that in countries currently without nuclear weapons, "all nuclear materials and equipment" be subject to safeguards of the International Atomic Energy Agency (IAEA), regardless of whether those materials had been supplied by the U.S. No U.S. nuclear exports would be permitted to countries that did not submit to international controls. Carter said this requirement should be viewed as an interim measure, as the first preference of the U.S. was that all nations sign the Nuclear Nonproliferation Treaty.

Among other rules to be included in new agreements were:

- A requirement for IAEA safeguards on all U.S.-supplied material and equipment for an indefinite period.
- Extension of current U.S. rights of approval on re-transfers and reprocessing to cover all special nuclear material produced through the use of U.S. equipment.
- Exemptions from the new rules could be granted by the president "if he considers this to be in our overall nonproliferation interest."
- U.S. nuclear supplies would be cut off to any nation that exploded a nuclear device or materially violated IAEA safeguards or any guarantee it had given under its agreement with the U.S.

In his message to Congress, Carter said that similar legislation previously introduced in Congress was too strict and did not allow the president enough flexibility. Some earlier congressional proposals, for example, would require the U.S. to stop exports immediately to foreign nations that did not promptly accept the new rules through renegotiation of existing agreements. The President's bill would allow exports to continue during negotiations about application of the rules.

The tough standards and penalties in the Carter legislation reflected the fact that the U.S. retained considerable technological leverage in the recently intensified international bargaining over nonproliferation of nuclear weapons, according to a New York Times news analysis April 28. Many nations currently using light-water nuclear power plants that were pioneered in the U.S. depended heavily on the U.S. for supplies of enriched uranium fuel for them.

The analysis noted, however, that the leading customers for U.S. nuclear equipment and material, including Britain, Japan, France and West Germany, were all investing heavily in efforts to break their dependence on the U.S. Their efforts included work on plutonium-fueled fast-breeder reactors, which used unenriched uranium to "breed" their own supplies of plutonium.

Nuclear exporters OK safeguard pact. Delegates at the Conference of Atomic Energy Suppliers in London Sept. 21, 1977 initialed a pact to require uniform safeguards on sales of nuclear energy and technology to other countries. The pact was aimed at containing the spread of nuclear weapons technology.

Efforts to halt the spread of nuclear weapons had been hampered because not all nuclear-exporting nations required their customers to observe international safeguards against making weapons. Under the terms of the London agreement, nuclear exporters would require customers to: pledge not to develop or detonate atomic weapons even for peaceful purposes; permit inspection of nuclear facilities by the International Atomic Energy Agency (IAEA); show that nuclear installations were protected against sabotage and theft; agree to IAEA controls on resales of nuclear material to third countries, and pledge not to duplicate nuclear facilities purchased under the London controls.

The agreement had been reached in principle in 1976 and had been reaffirmed earlier in 1977. The final pact clarified the language of the agreement and added provisions for dealing with violations. The text would be submitted for IAEA approval. The countries involved in the London conference were: Belgium, Canada, Czechoslovakia, East Germany, France, Great Britain, Italy, Japan, the Netherlands, Poland, Sweden, Switzerland, the U.S.S.R., the U.S. and West Germany.

A-Plant Safety

Scientists question A-plant program. A petition calling for a "drastic reduction" in the program for construction of nuclear power plants because of safety hazards was sent to the White House and Congress Aug. 6, 1975 by the 2,300-member Union of Concerned Scientists (which was said to include many non-scientists).

The petition urged a major research effort on reactor safety, plutonium safeguards and nuclear waste disposal. The "record to date," it said, "evidences many malfunctions of major equipment, operator errors and design defects as well as a continuing weakness in the quality control practices with which nuclear plants are constructed."

On the radioactive waste disposal problem, it said no "technically or economically feasible methods have yet been proven." Nuclear wastes, it added, could prove "a grim legacy . . . to future generations."

A third problem, cited by the petition, involved the potential use of plutonium from commercial reactors for making nuclear weapons. Various studies indicated, it said, "multiple weaknesses in safeguard procedures intended to prevent the theft or diversion of commercial reactor-produced plutonium for use in illicit nuclear explosives or radiological terror weapons."

In addition to calling for a slowdown in the nuclear power plant program, the petition urged a greater national effort to conserve energy and to develop alternate energy sources.

Atomic engineers resign over safety. Three high-level nuclear engineers for General Electric Co.'s San Jose, Calif. division resigned Feb. 3, 1976 to join a campaign against nuclear power, a California initiative then scheduled for a statewide ballot. The anti-nuclear proposal, bitterly fought Proposition 15, was defeated by a vote of 3,986,770 to 1,924,304 June 8. It would have prohibited construction of new nuclear generating plants and required the phasing out of existing plants unable to meet strict conditions. One of the conditions was that the current federal limit of $560 million of liability from any single nuclear accident be eliminated.

The three resigning engineers cited the safety problems as the overriding reason for their move. Aligned with this was their concern that safety problems were not fully reported by nuclear companies to the federal government, which had the regulatory responsibility.

Those resigning, who had from 16 years to 22 years of experience with GE, were Dale G. Bridenbaugh, 44, manager of performance evaluation and improvement at GE's nuclear energy division in San Jose; Richard B. Hubbard, 38, manager of quality assurance; and Gregory C. Minor, 38, manager of advanced control and instrumentation.

A statement from GE Feb. 3 pointed out that the company employed several thousand nuclear engineers and it "emphatically" disagreed with the viewpoint of the three resigning. "The overwhelming majority of the scientific and engineering community, including GE scientists and engineers," it said, "believes the benefits of nuclear power far outweigh the risk."

Bridenbaugh said in his letter of resignation:

> My reason for leaving is that I have become deeply concerned about the impact—environmentally, politically socially, and genetically—that nuclear power has made and potentially can make to life on earth. As we have discussed in the past, there is an inherent close intertie between commercial power and weapons technologies and capabilities. I am strongly opposed to the deployment of such capabilities and I fear the implications of a plutonium eceonomy. The risk involved in such a system is far too great for the short term benefit. I see no way for us to develop the ability to maintain the perfect human and technical control needed for the long periods of time necessarily involved with the highly toxic materials we are producing. This problem is not something I wish to pass on to my children and to succeeding generations to control. Contributing to the advancment of such proliferation now seem immoral and is no longer an acceptable occupation for me.
>
> Furthermore, in my recent assignment as the Project Manager of the Mark I Containment assessment, I have bcome increasingly alarmed at the shallowness of understanding that has formed the basis for many of the current designs. It is probable that many more problems will merge with severe consequences, impacting either the safety or the economic viability of the nuclear power program. . . .
>
> In summary, I am no longer convinced of the technical safety of nuclear power and I fear the high risk of political and human factors that will ultimately lead to the mis-

use of its byproducts. This makes it impossible for me to work in an objective manner in my current position and I, therefore, have decided that my only choice is to get out of the nuclear business. . . .

I have come to believe very deeply that we cannot afford nuclear power and I intend to do whatever I can to get the message to the public where the decision on its continuation must ultimately be made.

Minor's letter said:

I am convinced that the reactors, the nuclear fuel cycle, and waste storage systems are not safe. We cannot prevent major accidents or acts of sabotage. I fear that continued nuclear proliferation will quickly consume the limited uranium supply and force us into a plutonium-based fuel economy with even greater dangers of genetic damage and terrorist or weapons activity.

From my earliest days at Hanford, I have been deeply concerned about the dangers of radioactivity. I can still remember my wife's shock at having a container for urine sampling placed on our front doorstep for the use of our family. I wonder now if that police-state atmosphere at Hanford wasn't an omen for all people for the future.

I cannot be a part of an industry that promotes a policy that would lead our generation to consume 30 years of nuclear power for our own selfish purposes and leave behind radioactive wastes that will be a health hazard for thousands of generations to come.

In recent months I have become increasingly dismayed at the industry's opposition to the Nuclear Safeguards Initiative. I have seen the attempts to confuse and whitewash the issues by claiming that there are no unsolvable problems and appealing to individual's fears for their jobs. The public must be told that there are many problems. I am confident that an informed public—given the truth—will decide against continued nuclear proliferation.

I am also sure that there are others in the industry who share my concerns and I hope my decision will cause them to stop and consider the enormous implications and dangers of the nuclear legacy we are creating.

My reason for leaving is a deep conviction that nuclear reactors and nuclear weapons now present a serious danger to the future of all life on this planet.

GE was one of the largest companies in the nuclear-power industry, with orders for 69 U.S. nuclear generating facilities, of which 22 were in operation, and for 48 nuclear-power plants overseas, of which 18 were in operation.

In New York state, the federal safety engineer for three nuclear reactors at Indian Point on the Hudson River, 30 miles north of New York City, announced his resignation, citing the safety factor, at a news conference Feb. 9. The engineer, Robert D. Pollard, project manager for the federal Nuclear Regulatory Commission (NRC), planned to become Washington representative of the Union of Concerned Scientists, a group that opposed nuclear power.

Pollard, 36, chief safety engineer for nuclear reactors in North and South Carolina and Texas as well as at Indian Point, said he could not "in conscience remain silent about the perils associated with the United States nuclear-power program." The Indian Point plants, he said, "have been badly designed and constructed and are susceptible to accidents." He recommended closing down Indian Point Plant No. 2 "at once—it's almost an accident waiting to happen."

Consolidated Edison Co., which owned and operated Indian Point Plant No. 2 and one other plant at Indian Point, rejected Pollard's criticism. Spokesmen for the federal Nuclear Regulatory Commission and the State Power Authority, which recently bought the third Indian Point nuclear reactor from Con Ed, also denied that the plants were unsafe.

NRC chairman William A. Anders and Con Ed chairman Charles F. Luce defended the safety of the plants Feb. 10. Anders described Pollard's safety concerns as nothing more than "generic problems." Pollard showed up at the end of Luce's news conference to deliver a letter requesting shutdown of the No. 2 plant.

Report on atomic plant fire. The official investigation of the 1975 fire at the world's largest nuclear power plant at Browns Ferry, Ala. found Feb. 28, 1976 that federal attention to fire prevention and control there was lacking.

The report, prepared by a seven-man committee of the Nuclear Regulatory Commission, concluded that there were "lapses in quality assurance in design, construction and operation" of the two reactors at Browns Ferry, which were operated by the Tennessee Valley Authority.

TVA was faulted on its design of the plant and the way it managed the fight against the fire; the NRC itself was blamed for failing to properly exercise its responsibility "to assure that the licensee operates the plant safely."

The fire ignited from a small candle flame held by an electrician checking for air leaks in electrical cable. The cable's insulation caught fire and the fire spread to the electrical control room and through it to the room housing the nuclear reactor. There, the fire was put out with water, but only after an effort to contain it with dry chemicals, an incorrect procedure said to have prolonged the fire.

The fire did not damage the uranium cores of the reactor and no radioactivity was released outside.

Smoke detectors in the control room failed to work, the report said, and the design of the detection and control system was sorely deficient. No detectors were installed in the area adjacent to the control room, through which the fire spread, and wires for both the main and back-up detection systems were placed alongside one another. They burned together. Other wires improperly close together short-circuited from the fire's heat and prevented operation of valves and pumps to protect the nuclear cores.

The fire was expected to cost TVA customers $150 million by the time the reactors were back in operation, which was not expected before spring.

Plutonium tested in humans in 1940s. The Energy Research and Development Administration (ERDA) had confirmed that the government injected plutonium into human subjects from 1945 to 1947 to determine what the poisonous, radioactive substance would do to workers manufacturing the atom bomb, The Washington Post reported Feb. 22, 1976.

The injections were administered in varying quantities to eighteen persons (13 males and 5 females) ranging in age from four to 68. All were believed suffering from terminal illnesses.

Officials of the ERDA said records on the experiment were so unclear that it was not known how the subjects were selected or who ordered the tests. The ERDA could only be certain that one person had consented to the injection.

According to a fact sheet prepared by the ERDA, the purpose of the experiment was to gather information that would permit scientists to determine the amount of plutonium to which humans could safely be exposed.

The fact sheet said that seven persons who received injections lived less than one year afterward; three lived between one and three years; two between 14 and 20 years; and one 28 years. The fate of two was unknown, and three were still living.

ERDA confirmed Feb. 28 that its predecessor, the now defunct Atomic Energy Commission (AEC), used 131 prisoners in radiation experiments in the 1960s.

According to the ERDA, the AEC beamed x-rays into the testicles of 131 prisoners in Oregon and Washington to see if heavy radiation could cause sterility. A spokesman for the ERDA said that the prisoners had given written consent to the experiment and later undergone vasectomies to avoid fathering deformed children due to the damage caused by the test on their reproductive systems.

It is not known whether the radiation produced cancer in any of the subjects because no comprehensive follow-up studies had been made.

Staff report OKs offshore A-plant. A staff report of the Nuclear Regulatory Commission April 9, 1976 found no significant environmental risk in plans for a "floating" atomic power plant—the world's first such installation—off the New Jersey coast.

A preliminary environmental-impact statement drafted by the staff said that the risk of radioactivity leak through the air or water was "very low." The staff concluded that it was "unlikely" that the tourist economy of the New Jersey shore would be damaged by such a plant.

The Public Service Electric and Gas Co. projected a $2 billion plant 2.8 miles out in the Atlantic north of Atlantic City. Two reactors, in separate buildings, would be moored in 60 feet of water and would be protected by a breakwater. Electricity transmission lines would be carried inside pipes along both the ocean floor and underground through the wetlands to a switching station about seven miles inland.

Court curbs plutonium. The U.S. Court of Appeals in New York May 26, 1976 barred commercial use of plutonium as a nuclear fuel until the Nuclear Regulatory Commission completed a safety study. The ruling reversed a decision by the commission in November, 1975 to grant interim licenses for the manufacture and use of plutonium pending a decision on permanent licensing.

The NRC ruling had been challenged by environmental organizations, including the Natural Resources Defense Council, and the state of New York. The NRC was supported by 28 companies involved in nuclear power, including the Babcock & Wilcox Co. and Westinghouse Electric Corp.

The court, basing its opinion on the National Environmental Policy Act, found that a draft study on use of plutonium was inadequate in considering both alternative power sources and the problems of theft or sabotage of plutonium.

Atom-waste mixture explodes. A chemical explosion in a 13-liter container of radioactive wastes at the Hanford Nuclear Reservation in Washington State Aug. 30, 1976 injured a workman and contaminated him and nine others with radioactivity.

Eight of them, including two nurses, were decontaminated shortly after the explosion. Two were hospitalized for radiation observation. One of them, Harold McCluskey, 64, was working with a "glove-box chamber," manipulating the material inside it, when the explosion occurred. The blast shattered the box's plexiglas windows, showering McCluskey with shards. Also held for observation was Marvin E. Klundt, 43, who was exposed to radiation when he went to McCluskey's aid.

(A "glove-box chamber" was a device with holes cut into the sides and air-tight gloves fitted into the holes. A workman used the gloves to reach inside the chamber and manipulate the radioactive material inside.)

A medical report Sept. 1 said that McCluskey had received an excessive radiation dose and had problems with his vision, nitric-acid burns on his skin and possible internal radiation contamination.

The explosion occurred in a small building used by the Atlantic Richfield Hanford Co. for radioactive-waste recovery. The waste was produced by nuclear reactors. The material being recovered was americium, a radioactive element used in petroleum exploration to measure ground heat and in medical processes as a source of radiation.

The mixture that blew up was said to be at least 10 years old. George Stocking, president of Atlantic Richfield Hanford Co., reported Aug. 30 that the explosion was caused by a chemical reaction and that, "as far as we know, no radioactivity was released into the atmosphere, but some material was tracked out of the room but remained confined to the building."

Moratorium on A-plant licensing. The Nuclear Regulatory Commission announced Aug. 13, 1976 a moratorium on licensing new A-power plants until the completion of a study of the environmental hazards of fuel reprocessing and disposal of radioactive wastes. A moratorium also was called against any full-power operation of new A-plants.

The action was in compliance with recent court orders requiring the agency to strengthen its licensing procedures concerning reprocessing and waste disposal.

Ending the moratorium, the NRC announced Nov. 5 that it would resume the licensing of new A-plants on a conditional basis.

The commission had issued a staff environmental study Oct. 13 and had proposed at the same time an interim rule to incorporate the environmental impact into its licensing procedures. Licenses issued before adoption of the final rule would be on a conditional basis.

A-waste centers called major hazard—Nuclear-power critic Ralph Nader released a draft report by a nuclear energy expert Sept. 7 concluding that "a major radioactive waste problem already exists" in the U.S. The report was prepared by Mason Willrich for the Energy Research and Development Administration.

The report said that the federal government's past handling of radioactive material had been "marred in a sufficient number of instances to be a cause for concern" and that the system of storing wastes soon "will be unworkable." It said further that the escape of radioactive material into the air and water would "constitute a radiological hazard for hundreds of thousands, perhaps millions of years."

It was estimated that 75 million gallons of high-level radioactive waste and 51 million cubic feet of low-level waste were stored at nine different sites in the U.S. The federal military reactors were said to be producing 7.5 million gallons of liquid high-level waste a year and the commercial reactors were expected to produce a total of 60 million gallons of such waste by the year 2,000.

Federal production of low-level radioactive waste was put at 1.3 million cubic feet a year, the commercial production was estimated at a total of 50 million cubic feet by 2,000.

At the Hanford site, the government's major storage area, 18 leaks had resulted in losses of 430,000 gallons of high-level wastes into the surrounding soil, the Willrich report said.

Health-safety evaluation. An evaluation of the health and safety factors involved in the operation of nuclear power plants was included in the Ford Foundation-sponsored report administered by the Mitre Corp. and released March 21, 1977 by the Nuclear Energy Policy Study Group. The report, entitled "Nuclear Power Issues and Choices," said in regard to these factors:

In principle, one can compare the impact on public health of coal and nuclear power directly in terms of the deaths and illness they cause. In normal operations, a 1,000 MWe nuclear power plant has been estimated to produce roughly one fatality per year from occupational accidents and radiation risks to workers and to the public. A comparable new coal plant, meeting current new source standards, has been estimated to produce from two to twenty-five fatalities per year. Accidents in coal mining and transportation account for roughly two fatalities per year, and the rest of the range is attributed to the health effects of sulfur-related pollutants. This wide range results from the very large uncertainties in the actual effects on human health of the pollution chains resulting from sulfur oxides and from significant differences resulting from plant location with respect to population. The analysis of the risk at specific locations is complicated by uncertainties in meteorology, chemistry, synergistic effects involving other pollutants, and existing backgrounds. In addition to fatalities, pollution from coal plants contributes to large-scale nonfatal illnesses and discomfort for which there is no nuclear counterpart. There may also be significant effects from nitrogen oxides, carcinogenic hydrocarbons, and heavy metals for which a quantitative basis has not yet been established.

Thus, in a comparison of normal operations, nuclear power has smaller adverse health costs than coal. However, in an overall comparison of health effects, the possibility of accidents must also be taken into account. The possibiilty that nuclear accidents could have very serious consequences for public health has long been recognized as a unique problem associated with nuclear power. It is difficult to compare such rare but extremely severe events with the continuous health burden due to fossil fuels, but some perspective is gained from averaging the consequences of estimated accidents over an extended period. This requires knowledge of the probabilities and consequences of a spectrum of possible nuclear accidents.

To date, the safety record of nuclear power reactors in the United States has been excellent, at least as far as public health is concerned. However, the experience of some 200 reactor years of commercial nuclear power does not provide an adequate statistical basis for risk predictions covering the 5,000 reactor years expected during the rest of this century. Probabilistic judgments must be made on related technical experience and theoretical computations. Such an analysis of the current light-water reactors was undertaken in The Reactor Safety Study (WASH-1400, frequently referred to as the Rasmussen Report), published by the Nuclear Regulatory Commission (NRC) in October 1975. This report examined in a systematic fashion a large number of possible paths that could lead to an accident, estimated the overall probability of a nuclear core meltdown and breach of containment, and developed a probabilistic assessment of the consequences of such an accident, averaged over location and weather. Although WASH-1400 is a valuable resource for the study of the safety problem, we believe that it seriously underestimates and has methodological flaws that are discussed in our report.

Without attempting to duplicate the massive analysis of WASH-1400 but taking its uncertainties into account, we have attempted to gain some perspective on the possible social costs of reactor accidents by considering the following questions:

How does the predicted rate of reactor accidents affect the average rate-of-loss comparison between nuclear power and coal?

How serious might the consequences of a reactor accident be?

How likely might an extremely serious nuclear accident be if the associated uncertainties are all viewed pessimistically?

The average rate-of-loss due to reactor accidents calculated in WASH-1400 is only about 0.02 fatalities per year for a 1,000 MWe nuclear power plant. This rate is very low compared with the one fatality per year predicted for normal nuclear operations or the two to twenty-five fatalities per year attributed to a comparable new coal plant. Although we have not made an independent estimate of this average value, for losses due to nuclear accidents, our analysis indicates that with extremely pessimistic assumptions the WASH–1400 estimate might be low by a factor of as much as 500. On the other hand, it could be on the high side as well. . . .

However, even in this extremely unlikely situation, the average fatalities would not exceed the pessimistic end of the range of estimated fatalities caused by coal. Thus, on an average rate-of-loss basis, nuclear power compares favorably with coal even when the possibility of accidents is included.

An extremely serious accident under very adverse conditions is estimated by WASH–1400 to kill as many as three or four thousand people over a few weeks, cause tens of thousands of cancer deaths over thirty years, and cause a comparable number of genetic defects in the next generation, as well as more than $10 billion in property losses. Despite large uncertainties in biological effects, this appears to be a reasonable assessment of the potential consequences. While such an accident would clearly be a major disaster, the consequences would not be out of line with other peacetime disasters that our society has been able to meet without long-term social impact. For example, the United States has experienced a number of hurricanes that have taken over a thousand lives, produced physical damage in the billions of dollars, and required massive evacuation. In such a nuclear accident, the delayed deaths from cancer would not be an immediate effect but might result in a 10 percent increase in the incidence of cancer in the exposed population over a period of thirty years. It must be emphasized that a nuclear accident would probably have much less severe consequences than those estimated for this extremely serious accident and that most nuclear accidents would result in few, if any, fatalities.

The most serious accident considered in WASH–1400 is assigned an exceedingly low probability of occurrence (only one chance in 200 million years of reactor operation). This calculation is based on the combination of a number of low probability estimates for a series of events, most of which are extremely uncertain. When the uncertain factors are viewed in the most pessimistic light, there is a significant chance that such an event might occur during this century if the nuclear program grows at the projected rate. While it is very unlikely that this pessimistic assessment correctly describes the probability of such accidents, it does place an upper bound on the problem.

Having examined nuclear accidents from each of the above perspectives with very pessimistic assumptions, we have concluded that, even when the possibility of reactor accidents is included, the adverse health effects of nuclear power are less than or within the range of health effects from coal. At the same time, this analysis underscores the importance of continuing efforts to reduce the probability and consequences of accidents by improved safety designs and siting policies.

A foreign reactor accident would not necessarily be evidence of risk in this country, since some foreign reactors may be less safely constructed or operated than those in the United States. Nevertheless, a foreign nuclear accident could have a major psychological impact in this and other countries. A high premium should be put on reducing the probability and consequences of reactor accidents wherever they might occur.

A-mishap liability limit voided. U.S. District Court Judge James McMillan March 31, 1977 voided as unconstitutional a provision of the Price-Anderson Act that limited an electric utility company's liability for a nuclear power plant accident to $560 million in damages per accident.

The Price-Anderson Act, passed in 1957 and extended in 1975 until Aug. 1, 1987, had set up a joint private-federal insurance program for the nuclear power industry. It was challenged in 1973 by Ralph Nader's Public Citizens Litigation Group on behalf of the Carolina Environmental Study Group and 40 individuals living near the Duke Power Co.'s McGuire Nuclear Station on Lake Norman in North Carolina. The defendants in the case were Duke Power and the U.S. Nuclear Regulatory Commission (then the Atomic Energy Commission).

Judge McMillan ruled that the act violated constitutional guarantees of equal protection under the laws "because it provides for what Congress deemed to be a benefit to the whole society (the encouragement of the generation of nuclear power) but places the cost of that benefit on an arbitrarily chosen segment of society, those injured by nuclear catastrophe."

William Schultz, attorney for the Nader group, said the nuclear industry had wanted the Price-Anderson Act as protection for its investments. He said that if the judge's ruling stood, the effect could be to slow nuclear power development because it was unlikely that utilities would want to assume unlimited liability.

Industry radiation limit set. The Environmental Protection Agency set a limit

Jan. 6, 1977 for the amount of radiation emissions permitted from the normal operation of nuclear facilities.

The limit, in terms of exposure for any member of the general public, would be a maximum annual radiation dose of 25 millirems to the whole body (75 to the thyroid gland).

The new standard replaced an advisory "radiation guide" with a limit of 500 millirems to the whole body (1,500 to the thyroid gland). Unlike the advisory guide, the new standard was legally enforceable, according to the EPA.

The new standard also called for reduction of allowable emissions of krypton-85 to one-tenth of current levels by 1983.

EPA Administrator Russell E. Train described the new standards as "an important precedent in radiation protection because they consider the long-term potential buildup of radiation in the environment."

The EPA said most nuclear power plants already conformed to the new standards but that milling and other fuel-supply and reprocessing operations needed to be upgraded to meet the new restrictions.

Seabrook, N.H. atom foes protest. About 2,000 demonstrators moved onto the construction site of a nuclear power generating plant in Seabrook, N.H. April 30, 1977. Bringing tents, food and medical supplies, they vowed to "occupy" the site until plans to build the plant were abandoned.

When the protesters refused to leave May 1, about 300 state police officers from New Hampshire and neighboring states began making arrests. The demonstrators did not resist. State officials said 1,414 people had been arrested through early May 2 on charges of criminal trespass.

The well-organized and nonviolent demonstration was planned by a group that called itself the Clamshell Alliance. The alliance was made up largely of members of existing groups who had grown frustrated with what they felt was the predisposition of the federal Nuclear Regulatory Commission (NRC) to favor construction of nuclear power plants.

The proposed $2-billion, twin-tower plant in Seabrook had been embroiled in controversy for the previous four years, as environmentalists and other nuclear energy opponents fought and delayed it in the federal courts. The plant had become a symbol of the widening national debate over whether nuclear energy should be used extensively as a power source in the U.S.

Construction of the plant by Public Service Co. of New Hampshire had been halted by the NRC March 31 because of "uncertainty" over the outcome of a pending U.S. Environmental Protection Agency (EPA) review of plans for the plant's cooling system. The EPA regional office had objected to use of a cooling system that would suck in about 1.2 billion gallons of sea water a day and return it to the ocean 38 degrees warmer, which would be dangerous to seashore and marine life. The NRC action confirmed a Jan. 21 decision by its Atomic Safety and Licensing Board Panel suspending construction permits for the Seabrook facility.

If completed, the 2,300-megawatt plant would provide 80% of New Hampshire's electricity by the mid-1980s, while selling half its power to other New England states.

Power companies, a number of business interests and construction unions in New Hampshire and Massachusetts contended that continued difficulties at the long-delayed plant were another example of government red tape. They said the EPA objections showed an unwarranted concern with fish rather than with producing energy.

The antinuclear forces charged that atomic power in general, and the Seabrook plant in particular, was a dangerous and ill-conceived venture in which faulty construction or maintenance could lead to a disastrous accident. They also objected to the possible danger of cancer from low-level radiation escaping from the plant.

The Clamshell Alliance had organized a demonstration at the Seabrook plant in August 1976 at which 180 of 1,000 protestors had been arrested.

More than 500 imprisoned demonstrators arrested May 1–2 were released from National Guard armories May 13.

Under an agreement with the Rockingham County prosecutor, the demonstrators pleaded not guilty to criminal trespass charges in county district court.

They were then found guilty in mass trials, sentenced to 15-day jail terms and $100 fines and released on their own recognizance pending their automatic appeals to county superior court.

Seventeen protestors had been tried individually in district court May 5. Others had been tried and convicted some days later. In the first cases, a judge had sentenced each of the demonstrators to serve 15 days at hard labor and pay a $100 fine. In a break with local legal practice, he had ordered that the sentences be served immediately, before appeal to superior court.

The judge had suspended the sentence of the first demonstrator convicted but had reversed himself after an appearance in court by state Attorney General David Souter. Souter said a suspended sentence was like no sentence at all. Asking for harsher penalties, he called the Seabrook occupation "one of the most well-planned acts of criminal activity" in the nation's history and said the state police had overheard citizens' band radio messages indicating that the demonstrators planned to reoccupy the construction site.

The total cost of confining the demonstrators was approximately $50,000 a day, state officials said. The National Guard bill alone was $290,566. That figure did not include costs of state or local police, county sheriffs or court facilities.

The occupation was the first massive show of civil disobedience in the nation in opposition to construction of a nuclear power plant.

Proliferation Issue & Nuclear Exports

Exporters agree on safeguards. Senior officials of the seven nations that supplied most of the world's exports of nuclear materials reached agreement in London Nov. 21, 1975 on the basic principles and methods of preventing importing nations from using the acquired technology to make nuclear weapons.

The group, comprising the U.S., the U.S.S.R., Great Britain, France, West Germany, Canada and Japan, had opened secret meetings in London Nov. 14.

Details of the largely unpublicized meetings were not available, but informed British government sources said the countries had agreed in principle to new safeguards governing the sale and use of nuclear materials and facilities. The accord would require each of the seven to elicit a pledge from purchasers to the effect that they would not use the major components of the imported technology—reactors, uranium, enrichment processes and fuel reprocessing plants—to produce weapons, according to a New York Times report Nov. 21.

The seven-nation talks had sprung from a U.S. initiative following the detonation by India in 1974 of an atomic device developed using imported Canadian technology.

After additional secret discussions in London, the seven nations Jan. 27, 1976 exchanged letters agreeing on principles to govern the export of nuclear materials and technology.

The reported agreement aimed at establishing identical security regulations for all countries using nuclear power and would require, in principle, that the seven participants consult among themselves before any one of them concluded a nuclear contract with another party.

Fred C. Ickle, director of the U.S. Arms Control & Disarmament Agency, told the Senate Foreign Relations Subcommittee on Arms Control Feb. 23 that the accord covered not only guidelines for preventing nuclear exports from being turned into atomic weapons, but also follow-up efforts to improve safeguards. A State Department official said later that day that Ikle was prevented from giving more complete testimony because several countries, notably France, had placed restrictions in the understanding on the disclosure of details, including the identity of the participants.

Ickle charged July 23 that the U.S. sold reactors and fuel with little regard for safety standards. While the situation had shown improvement in recent years, Ikle said, the U.S. had supplied reactors to countries which did not need them, such as South Vietnam and Zaire. He also charged that U.S. nuclear aid to Europe had made possible uncontrolled European sales of reactors and reprocessing equipment to less-developed countries.

Ford warns on arms spread—President Gerald Ford told the U.S. Congress July 29 that a "world of many nuclear-weapons states could become extremely unstable and dangerous."

Presenting the annual report of the U.S. Arms Control and Disarmament Agency, Ford said that 20 additional countries had the technical competence and the material to build nuclear weapons. Six nations—the U.S., the Soviet Union, China, Britain, France and India—already had such knowledge and materials. Ford said that all countries supplying nuclear fuel, equipment and assistance should impose International Atomic Energy Agency standards on their sales. These standards included prohibition of all nuclear explosions, insistence upon measures to prevent sabotage and theft, and imposition of the same standards on the equipment if it were transferred to a third country.

Brazilian-West German A-pact. West Germany June 27, 1975 signed an agreement to supply Brazil with a complete nuclear industry by 1990. The accord came in the face of strong opposition by the U.S., which argued that the contract would supply Brazil with the fuel technology to produce nuclear explosives.

Under the pact, worth at least $4 billion, West Germany would sell Brazil eight nuclear reactors which would produce a total of 10,000 kw of electricity, and would deliver a uranium enrichment plant, a fuel fabrication plant and facilities for reprocessing fuel waste products into plutonium. West Germany would also help prospect for and exploit Brazil's uranium deposits in return for access to the reserves.

The agreement was signed in Bonn by West German Foreign Minister Hans-Dietrich Genscher and his Brazilian counterpart, Antonio Azeredo da Silveira.

U.S. State Department spokesman Robert Anderson said June 27 that the U.S. "had concerns about aspects of the Brazilian-West German pact that could have a potential for contributing to the spread of nuclear weapons."

Brazil had not signed the Nuclear Non-Proliferation Treaty; West Germany had signed and ratified it.

As part of the contract, Brazil had agreed to submit to controls by the International Atomic Energy Agency to insure that nuclear fuel material obtained from the German facilities was not diverted for nonpeaceful purposes. U.S. officials had disclosed June 3 that Washington had been instrumental in winning these controls, saying a special treaty incorporating the IAEA safeguards would be signed. The officials, who also admitted the U.S. had sought to dissuade West Germany from making the sale, said Washington had refused to sell nuclear facilities to Brazil because the Latins insisted on inclusion of plutonium and uranium enrichment plants. These would give the Brazilians the capacity to make nuclear weapons. The controls accepted by Brazil did not extend to future Brazilian-built processing facilities.

Brazilian officials vowed that the nuclear facilities would be used solely for peaceful purposes. The nation, which lacked oil and coal resources, had depended mainly on hydroelectric power for industrial development.

Brazilian Energy Minister Shigeaki Ueki said West Germany would lend Brazil $1.5 billion to develop a nuclear industry, it was reported July 2.

Brazil planned to build 63 nuclear-powered electricity generating plants by the year 2000, according to a statement Sept. 9 by Luiz Claudio de Almeida Magalhaes, president of a government-operated electricity company.

The plan, more ambitious than the nuclear energy programs of most European countries, would cost $50 billion at current prices, according to the Washington Post Sept. 11.

Brazil's first atomic power station, built with U.S. technology, was nearing completion near Rio de Janeiro. Eight other plants would be purchased by 1990 from West Germany under the June 27 accord.

Most South American diplomats in Brasilia believed Brazil would build an atomic bomb, for prestige purposes if not actual use, according to the Miami Herald Oct. 19. The Brazilian government denied this.

France July 5 had signed a $2.5 million contract under which Brazil would purchase a French experimental nuclear reactor to be used in research for the construction of a new type of fast neutron, or breeder, reactor.

U.S.-Brazilian dispute—Talks between the U.S. and Brazil on the West German deal were broken off early March 2, 1977 after only one session.

The U.S. delegation, headed by Deputy Secretary of State Warren M. Christopher, left Brasilia before dawn, abruptly closing discussions that had begun the day before. The two countries issued only this communique:

"The two sides exchanged opinions on nuclear matters and energy problems. Each side will now consider the expressed position of the other side. There will be further conversations on these matters."

According to press reports, Christopher had insisted that Brazil not establish the fuel-reprocessing and uranium-enrichment plants provided for in the 1975 agreement with West Germany. In return, Christopher said, the U.S. would provide Brazil with nuclear fuel or arrange to put the uranium-enrichment process under some kind of international control. The Brazilian negotiator, Foreign Minister Antonio Azeredo da Silveira, reportedly rejected Christopher's suggestions on grounds that they would make Brazil dependent on foreign sources for its energy needs. Azeredo also pointed out that the West German agreement already had been signed by both governments and approved by the Brazilian Congress.

The U.S. and several other countries believed that Brazil wanted a uranium-enrichment plant to make plutonium for medium-sized atomic bombs. While the Brazilian government insisted that it wanted only to satisfy the country's energy needs, military officers admitted privately that the agreement with West Germany was designed to raise Brazil to Argentina's level of nuclear development and to enable Brazil to build an atomic bomb, the London newsletter Latin America reported March 11.

U.S. Secretary of State Cyrus Vance, discussing the Brazilian case, had said March 1 that the Carter Administration would consider nuclear proliferation in deciding how much economic aid to give a foreign country. Brazil received little direct U.S. aid, but it received large loans from international financing agencies dominated by the U.S. Brazil already had rejected U.S. military aid because the assistance was linked to greater observance of human rights.

The U.S. also had applied pressure to West Germany, urging it to defer transferring the enrichment plant and reprocessing facility to Brazil at least until arrangements could be negotiated to create regional centers for atomic-fuel manufacture under international control. In response, West Germany postponed the granting of blueprint export licenses for the pilot reprocessing plant until Christopher could visit Bonn for further talks, it was reported March 11.

Other pressure was applied by the Dutch government, which announced in February that it wanted Brazil to return the waste from enriched uranium that would be supplied by the British-Dutch-German consortium URENCO to the first two Brazilian reactors to be constructed under the West German agreement, it was reported March 11.

The Canadian foreign minister, Donald Jamieson, added to the pressure by demanding to know when Brazil would sign the Nuclear Nonproliferation Treaty, it was reported Jan. 28. Since Canada had little influence over Brazil in nuclear matters, it was assumed that Jamieson's statement was directed partly at Argentina, which was building a Canadian-designed natural-uranium nuclear power plant.

Argentina was supporting Brazil in its dispute with the U.S. and other countries over the West German deal. The Argentine ambassador in Brasilia, Oscar Camilion, said Feb. 7, "If Brazil is suffering external pressure against its efforts to develop an independent nuclear policy, it is logical that Argentina, which is in the same situation and, in addition, is Brazil's neighbor, feels the same pressures."

West Germany honors agreement—Despite strong protests from the U.S., the West German government April 5, 1977 approved export licenses for blueprints of a pilot uranium-reprocessing plant and a demonstration uranium-enrichment plant to be built in Brazil.

The plants, to be constructed under the 1975 agreement between Bonn and Brasilia, would produce fissionable uranium and plutonium that could be used in nuclear weapons.

West German Chancellor Helmut Schmidt had held up the export licenses since January, when Vice President Mondale visited Bonn and told him that President Carter had serious objections to the nuclear agreement with Brazil. U.S. Secretary of State Cyrus Vance visited

Bonn March 31 to apply further U.S. pressure, but Schmidt reportedly was adamant, arguing that West Germany must honor its international agreements and that Brazil would use its nuclear technology for peaceful purposes.

(Despite Bonn's public determination, a U.S. State Department spokesman noted April 8 that West Germany's initial exports to Brazil would be limited to blueprints and other data, still leaving the U.S time to persuade Bonn and Brazilia to change their plans. West German officials noted that the actual construction of the enrichment and reprocessing plants would take years, and they said Bonn was willing to confer with the U.S. on possible additional controls in future contracts with other countries.)

The U.S. pressure evoked protests in both West Germany and Brazil. In West Germany, the weekly newspaper Die Zeit charged that the U.S. wanted to destroy the agreement between Bonn and Brasilia to secure the Brazilian contracts for the U.S. nuclear industry, it was reported April 9. Count Otto Lambsdorff, economic spokesman for the Free Democratic Party, said Bonn must honor the Brazilian agreement "to uphold the international credibility of the Federal Republic of Germany as well as to insure the future of German export technology." German labor unions estimated that as many as 200,000 jobs depended on the nuclear industry.

The U.S. pressure had brought U.S.-Brazilian relations to their lowest point in more than a decade, the New York Times reported March 28. Some ranking Brazilian officials believed U.S. pressure against the nuclear deal and in favor of greater observance of human rights in Brazil were part of a larger U.S. effort to prevent Brazil from becoming a major world power, according to the Times.

Brazilian Mines and Energy Minister Shigeaki Ueki said March 26 that Brazil would carry out its nuclear energy program "at all costs." Brazil planned to rely on nuclear and hydroelectric plants because it lacked domestic oil deposits and wanted to cut its purchases of high-priced foreign petroleum.

France bans sale of reprocessing plants. The French government announced Dec. 16, 1976 that it would ban future sales of nuclear fuel-reprocessing plants that could be used to produce plutonium for atomic bombs. This was a change from the previous French policy, which had opposed any interference with its nuclear exports.

The decision followed the French Council for Foreign Nuclear Policy's statement Oct. 11 that France would work against the proliferation of atomic weapons and was ready to study international treaties to effect the policy. Observers believed the council's statement indicated that France might ratify the non-proliferation treaty.

The ban on exporting reprocessing plants was not retroactive. An earlier French agreement to supply Pakistan with a plant remained in effect. It was reported Dec. 16 that the French government wanted to abandon the Pakistan deal but that domestic political pressure from the Gaullist Party made this action difficult. The Gaullists reportedly were extremely sensitive to the appearance of French surrender to pressure from abroad, particularly from the U.S., which had issued vigorous protest over the Pakistan agreement.

President Valery Giscard d'Estaing reportedly was hopeful that the Pakistanis would terminate the agreement, but according to a Dec. 16 report they had decided to continue with plans to purchase the nuclear plant despite U.S. objections.

The French had cancelled an earlier deal for a reprocessing plant sale to South Korea because the guarantees given by the South Koreans against misuse had been regarded by France as inadequate. Agreements concluded earlier with South Africa for the delivery of two nuclear power stations and with Iran for the same number would remain in effect because the French felt there were enough safeguards to prevent diversion of the fuel into weapons.

The French were expected to continue to supply third-world countries with nuclear reactors, fuel and technology. Following the Dec. 16 decision, however, all irradiated fuel produced in French-supplied nuclear plants would be returned to France to be retreated so that the plutonium produced would remain in French hands.

Pakistani Defense & Foreign Affairs Minister Aziz Ahmed Jan. 3, 1977 said

Pakistan would fulfill plans to buy a nuclear reprocessing plant from France despite opposition from the U.S. and Canada. Ahmed said, "No third country [had] any right to demand" otherwise.

The plant, expected to cost $150 million, would give Pakistan the ability to extract plutonium from spent reactor fuel rods used at its nuclear power plant in Karachi. Canada and the U.S. opposed purchases of such plants by Pakistan and other developing nations because of concern that the plutonium extracted could be used for making nuclear weapons.

Canada, which had built the Karachi power plant, had ended its nuclear cooperation with Pakistan in December 1976 because of disagreements over proposed safeguards and regulations.

Foreign Minister Jean Sauvagnargues informed the National Assembly (lower house of Parliament) April 8 that the nation's export of nuclear material and technology would be "subject to controls to ensure that they are used for peaceful purposes."

Sauvagnargues said that the government would demand that "the bulk of our exports of nuclear material and equipment" be placed under the supervision of the United Nations International Atomic Energy Agency in the importing countries. These countries would be required to commit themselves to using the supplies exclusively for peaceful purposes, he said. The French government, Sauvagnargues said, would demand that importing countries rigorously safeguard the transport and use of nuclear material.

French nuclear pacts—French Premier Jacques Chirac and Iraqi Vice President Saddam Hussein had initialed a nuclear cooperation pact Sept. 8, 1975 during a visit to Paris by Hussein.

Under the plan, France would build a nuclear station in Iraq and would train Iraqi scientists and technicians.

France agreed March 22, 1976 to build a 600-megawatt nuclear reactor in Libya. The draft accord was signed at the conclusion of a two-day visit by Premier Chirac to Libya. A spokesman in Paris stressed that the plant could not be turned to military purposes. "This will be a research reactor. There is no question of producing plutonium," the official said. No details on safeguards were cited, however.

Representatives of the French and West German governments signed an agreement in Bonn May 18 on the exchange of information and technology on the development of fast-breeder nuclear reactors. France had an acknowledged lead in the research and development of such reactors which, when operational, were expected to yield both electrical power and additional nuclear fuel.

Under the terms of the accord, the two governments agreed to spend between $84.7 million and $105.9 million annually on reactor development. A French-West German consortium was established to supervise the pooling of information and technology and the sale of production licenses to other nations. The consortium partners were the French Atomic Energy Commission; Navatome, a French builder of atomic reactors that was 40%-owned by Creusot-Loire, and Interatom, a subsidiary of the West German company, Kraftwerk Union AG. The U.S.-based Westinghouse Electric Corp. also participated. It held exclusive licenses for certain nuclear processes involved.

Iran had signed contracts to purchase two nuclear power plants from France, Premier Amir Abbas Hoveida announced May 25. Hoveida made the disclosure in Paris, where he had arrived May 20 to discuss economic cooperation with French officials. The premier met May 25 with President Valery Giscard d'Estaing.

The contract for the two 900-megawatt power stations was said to total $1.2 billion.

Before concluding his state visit May 28, Hoveida said Iran had no intention of using the French nuclear plants to develop nuclear weapons. He insisted that the plants would operate under the control of the International Atomic Energy Agency.

France would build two reactors at Darkuvin at a cost of $1.2 billion under a nuclear-cooperation agreement signed in Teheran Oct. 6 by Shah Mohammed Riza Pahlevi and French President Valery Giscard d'Estaing. The accord also provided for a nuclear research center at Isfahan and the training of Iranian scientists and technicians.

South Africa announced May 29 that it had signed a $1-billion contract May 28 with a French consortium for the purchase of a nuclear-power station to be built at Koeberg, 20 miles north of Cape Town. The plant, comprising two atomic reactors, would be the first on the African continent.

The selection of the French group, announced by South Africa's Electricity Supply Commission (Escom), reversed a prior decision to award the contract to an American-Dutch-Swiss consortium headed by the U.S.' General Electric Co. (GE).

Jan H. Smith, president of Escom, said the state-owned utility had given GE and its partners—Brown, Boveri & Cie. of Switzerland and Rijn-Schelde-Vorolme Machinefabrieken en Scheepswerven N.V. of the Netherlands—a "letter of intent" for the contract two months earlier, stipulating that the three firms supply the required licensing and financial guarantees by May 21. However, talks with the tri-national group had been suspended with the passage of this deadline.

Smith said the breakdown of the talks resulted from the failure of the Dutch government to approve the sale by the cutoff date.

Although he cited only the Dutch obstacle to completion of the deal, opposition was also vigorous in the U.S. There, as in the Netherlands, critics of South Africa's racial policy of apartheid and its failure to have signed the nuclear non-proliferation treaty demanded that the government not approve the contract.

The Paris-based consortium to which the contract was awarded comprised Framatome, a subsidiary of the Creusot-Loire group, which would supply two pressurized-water nuclear reactors; Alsthom, which would provide twin-turbine generators capable of producing 925 megawatts of power each, and Spie-Batignolles S.A., which would provide general engineering services. The first reactor was scheduled to go into operation in November 1982 and the second a year later.

The Paris consortium would be paid 70% in French francs and 30% in South African rands. The Credit Lyonnais, France's second largest bank, headed the financing group which guaranteed the contract for the firms.

With a combined output of nearly two million kilowatts, the two nuclear reactors would supply most of the electricity needs of the western Cape region, which had traditionally relied for its electrical power on supplies from coal-fired stations in the Traansvaal, 1,000 miles north, or on power generated by local stations fed by coal brought from the Traansvaal mines by rail.

U.S. to provide uranium enrichment—Under a contract signed in 1975, the U.S. would enrich the uranium for the nuclear reactors to be built in South Africa by the French consortium which won the contract, the Wall Street Journal reported June 1.

Along with its application to the Nuclear Regulatory Commission (NRC) May 10 to build the South African reactors, the General Electric Co., leader of the three-nation consortium which had been considered certain to secure the contract, had also applied for licenses to export 1.4 million pounds of enriched uranium to South Africa as fuel for the plants.

The State Department May 27 indicated its approval of the licenses during hearings before the Senate Foreign Relations Committee. Myron B. Kratzer, deputy assistant secretary of state for scientific affairs, told the committee that approval of the contract was the best way to assure that South Africa would not divert plutonium from the spent fuel to make atomic weapons.

Congressional opponents of the sale said it would undermine the U.S.' recently-enunciated policy of support for black majority rule in southern Africa. The senate subcommittee on African affairs held hearings on the proposed sale May 26–29. The subcommittee chairman, Sen. Dick Clark (D, Iowa) opposed the sale, citing South Africa's policy of apartheid.

The Washington Post reported May 26 that the U.S. had three years earlier sold South Africa two large computers which were being used at a uranium enrichment plant at Valindaba. Details of the sale by the Foxboro Co. of Foxboro, Mass. appeared in previously unpublished correspondence between the White House and the Senate Government Operations Committee.

A-plant deal split Dutch Cabinet—The Dutch Cabinet May 21 let pass a deadline for providing credit guarantees for the construction by a U.S.-led consortium of a nuclear-power plant in South Africa.

The coalition government had come under strong pressure by groups within the Netherlands who opposed South Africa's policy of apartheid and who feared South Africa would divert nuclear wastes to the construction of atomic bombs.

The government was to have provided a $260-million credit guarantee to Rijn-Schelde-Verolme Machinefabrieken Scheepswerven N.V., one of the consortium partners.

Although the deadline had passed, the Cabinet had continued talks on the nuclear-reactor deal through May 27, suggesting that South Africa might have been willing to extend the cutoff date in order to conclude the contract with the American-led group. The sudden award of the contract to the French consortium came as a surprise in Paris and elsewhere, according to reports in the French daily Le Monde June 1 and other newspapers.

U.K. nuclear pacts. An agreement was signed in Teheran Oct. 19, 1975 in which Britain was to assist Iran's nuclear energy development program. Iranians were to be trained in Britain and the British were to help develop a nuclear research program at Teheran University. The accord was signed by Stanley Marshall of Britain and Akbar Etemad, head of Iran's Atomic Energy Authority.

James Stewart, deputy chairman of the British Nuclear Power Company, said Nov. 9 that an enabling agreement with the Soviet Union had been signed, providing for the exchange of experience, techniques and, possibly, materials which would benefit British activities in the development of nuclear power stations due to come into operation in the 1980s. Stewart, who had returned that day from a tour of Soviet atomic plants, said the U.S.S.R. was substantially "further ahead" of Britain in several areas, including production of new alloys, engineering techniques and the scaling up of small prototypes and reactors into larger reactors.

Australia sells uranium. The Australian government said Nov. 11, 1976 that it would lift its 1972 embargo on uranium exports and would permit firms to fill their existing orders with Japan, West Germany and the U.S. The existing Australian contracts with those countries provided for the supply of 9,000 tons of uranium before 1986.

Environment Minister Kevin Newman indicated that despite the new export policy no new mining operations would be considered until the second part of the Fox commission report on uranium mining had been received.

Newman said an export permit would be issued to the Rio Tinto-Zinc Corp. Ltd.'s Mary Kathleen mine in North Queensland, the country's only producing uranium mine. Two other Australian firms that had export contracts but no production permission, would be able to arrange exports from the government's stockpile.

The Mary Kathleen mine had been stockpiling its own uranium since it had resumed operations in March. The mine had borrowed uranium oxide from the British Atomic Energy Commission since August to meet its overseas contracts because it had lacked the export permit and because major trade unions had banned the transport of uranium. According to Newman, no further uranium supplies would be available through Great Britain.

Australia's nuclear power development had sparked widespread controversy in the country. Conservationists opposed the mining of uranium, while other groups argued that uranium exports would provide an important boost to Australia's foreign earnings. The report estimated that export earnings could rise from $125 million annually to more than $1.25 billion yearly in the 1990s. Development of the uranium industry had been virtually halted for the previous four years as a result of stringent government limitations on exports.

Uranium deposits in Australia were estimated at 20%-25% of the world's economically recoverable reserves.

The major trade unions in Australia Nov. 12 called for a continued union ban on the mining and transport of uranium in defiance of the government's decision to permit its export.

In addition to union opposition, the government was faced with environmental objections to the development of a large uranium industry and the prospect of the material's misuse by overseas buyers.

But scarcely a full month later, the Australian Council of Trade Unions (ACTU) Dec. 8 voted 13-5 to endorse the government's decision to honor existing uranium contracts and urged the trade unions involved to accept work in mining and transport operations.

The council's decision reversed the existing policy of the trade unions and backed the position taken by the Labor Party in Parliament.

Australia's first uranium shipment since 1963 was held up several times by demonstrators, according to press reports Jan. 2, 1977. The 130-ton cargo moved 1,200 miles by train to Brisbane, from where it was to be shipped to the Hamburg Electrical works in West Germany. The demonstrators reportedly stopped the train three times during the journey.

Fox report—An inquiry group sponsored by the government had issued a report Oct. 28 endorsing the mining and selling of uranium under strict conditions. The inquiry, headed by Justice Russel Walter Fox, had been commissioned 18 months previously by the former Labor government.

The report recommended that the government postpone any decision on mining uranium reserves in the Northern Territory, where the major deposits were located, until a final report.

The inquiry expressed concern that the "nuclear power industry is unintentionally contributing to an increased risk of nuclear war," but concluded that this danger could be minimized by government-imposed safeguards. The inquiry suggested that the government:

- Retain the right to immediately terminate uranium mining and sales.
- Bar uranium sales to nations that were not signatories of the Nuclear Non-Proliferation Treaty.
- Conclude back-up agreements "applying to the entire civil nuclear industry in the country supplied."
- Establish a permanent council to advise on the export and use of Australian uranium.

U.S.-Australian correspondence—Prime Minister Malcolm Fraser told Australia's Parliament March 23, 1977 that he had exchanged letters with U.S. President Jimmy Carter on international safeguards for nuclear facilities. Fraser also said Australia would probably become a major supplier of uranium in the world market if the government authorized uranium development, but that such development would be subject to strict controls.

Fraser's letter to Carter, written Feb. 11, pledged Australia to stringent control over uranium exports. It was intended to align Australia's policies with those of the U.S., Canada and other uranium producers.

Carter's reply to Fraser included a proposal to guarantee fuel supplies to nations that accepted non-proliferation safeguards and did not acquire sensitive nuclear facilities that could be used to produce weapons. Carter said that Australia's potential as a major supplier of uranium gave the country a particular interest in collaborating with like-minded countries on supply policies.

The Australian government was under continuous pressure from other nations that wanted its uranium deposits developed and exported. A delegation of U.S. utility executives held talks Feb. 24 with the government and the mining industry on the subject. The executives said the U.S. nuclear power industry would go ahead "with or without Australian uranium." However, the group was clearly anxious to encourage the pro-uranium faction in Australia by holding up the prospect of Australia being left out of the lucrative uranium market.

The Philippines had sent a representative to inquire about the purchase of uranium, it was reported March 8, and a European Community (EC) delegation held talks March 14 and 16 on Australian mineral resources. In particular, the EC representatives were interested in uranium, coal and natural gas.

In a related development, the Australian uranium company, Queensland Mines Ltd., told two Japanese power utilities March 24 that it would be unable to honor export contracts in 1977 for 400 tons of uranium. The company said that the government had prevented development of the rich Narbarlek deposit in the Northern Territory and had refused to supply the uranium from the government stockpile. The government had promised foreign utilities that contracts signed in 1972 would be honored whatever the final decision on full-scale development of the reserves.

Whitlam urges uranium safeguards—Federal opposition leader Gough Whitlam March 29 outlined a Labor Party plan for a system of rigorous safeguards for the export of uranium. The major element of his program would be the creation of an international nuclear police force to prevent the use of Australian uranium for nuclear weapons or plutonium-based fast-breeder reactors.

Under other proposals in the plan:

- Uranium would be sold only to signatories of the Nuclear Nonproliferation Treaty;
- Sales would be subject to a United Nations International Atomic Energy Agency safeguards agreement;
- Buyers would have to guarantee that Australian uranium would not be reprocessed to yield plutonium;
- Australia would participate in research to find safe storage procedures for nuclear waste.

In giving his qualified approval to the mining and export of uranium, Whitlam declared in Parliament, "It is clear that the government will proceed to establish new uranium mines in Australia and to export uranium before adequate international safeguards have been developed and irrespective of the public debate which takes place on the final findings of the Fox commission."

Following Whitlam's speech, the deputy leader of the Labor Party, Thomas Uren, rose to oppose the plan. Uren and the party's left wing were trying to persuade the party to adopt a proposal for a moratorium on uranium mining and export for at least five years.

Canada tightens rules. The Canadian government Dec. 22, 1976 instituted greater restrictions on the sale of nuclear-reactor fuels and technology to other countries.

Announcing the new policy, External Affairs Minister Donald Jamieson said: "Shipment to non-nuclear weapon states under future contracts will be restricted to those which ratify the nuclear non-proliferation treaty or otherwise accept international safeguards on their entire nuclear program." Previous Canadian restrictions had applied only to items supplied by Canada, leaving receiving countries free to develop weapons with materials imported from other sources.

The new regulations placed Canadian exports under "the most stringent conditions in the world," Jamieson said, adding: "For all practical purposes, nuclear cooperation with Pakistan is effectively at an end." Pakistan had rejected placing its nuclear reactor, purchased from Canada, under the new restrictions.

Canada bars A-aid to India. Canadian Foreign Minister Allan MacEachen May 18, 1976 announced a permanent ban on sales of nuclear materials to India.

The sales had been suspended in May 1974 when India exploded a nuclear device made from plutonium produced by a Canadian-supplied reactor. Indian-Canadian negotiations since the suspension had failed to reach agreement on Ottawa's demands for safeguards against the use of Canadian material for atomic explosions.

MacEachen said Canada would resume nuclear shipments to India only on condition that they "not be used for the manufacture of a nuclear device."

India May 20 assailed Canada for having "turned its back" on a long-negotiated settlement of nuclear matters between the two countries. Addressing parliament, Foreign Minister Yeshwantaro B. Chavan said his government was "disappointed that after two years of strenuous negotiations, when a detailed understanding had been reached, the Canadian government should have unilaterally taken the step to terminate nuclear cooperation."

U.S. & Soviet sales to India—The U.S. Nuclear Regulatory Commission July 21, 1976 approved an initial sale of 13.5 tons of uranium fuel to India.

The Soviets agreed to sell India 200 metric tons of heavy water for India's nuclear reactors, U.S. officials reported Dec. 8. (A metric ton is 2,024.6 pounds.)

The International Atomic Energy Agency (IAEA) was notified of the deal, and Soviet officials indicated that international inspection and safeguard procedures would be observed. The sale was considered unusual because it was a case of Soviet willingness to supply nuclear material to a country with bomb-making capability.

Heavy water, used to control nuclear reactions in some reactors, is water in which the hydrogen atoms are replaced by deuterium, a hydrogen isotope twice the weight of ordinary hydrogen.

Uranium 'lost' at sea. The European Community (EC) Commission May 2, 1977 officially confirmed press reports that 200 metric tons of uranium ore had mysteriously disappeared from a freighter bound from Antwerp, Belgium to Genoa, Italy in November 1968. The incident had been reported April 28 by the Los Angeles Times and the New York Times and on April 29 by Paul Leventhal, a former Senate Government Operations Committee staff expert on the spread of nuclear weapons.

Leventhal, in a speech to the Conference for a Non-Nuclear Future in Salzburg, Austria, said that several weeks after the ship had failed to make its scheduled arrival in Genoa, it had "reappeared with a new name, new registry, new crew, but no uranium." (The conference was timed to coincide with a two-week conference on nuclear power, also in Salzburg, sponsored by the International Atomic Energy Agency.)

Though investigations by at least four nations had never officially resolved the mystery of the disappearance, some U.S. and European intelligence officials reportedly were convinced that the uranium had found its way to Israel. EC officials privately indicated May 2 that this was their understanding as well.

The Israeli Atomic Energy Commission April 29 denied that Israel had any connection with the disappearance of the uranium.

The reports of the missing uranium again raised speculation as to whether Israel had the capability to produce nuclear weapons and, if so, whether it had already done so. The Israeli Atomic Energy Commission spokesman, in making the denial April 29, refused to say where Israel had obtained the uranium for its top-secret nuclear reactor at Dimona in the Negev Desert. Built in the 1950s with French assistance, the Dimona reactor had never been opened to inspection by foreigners.

EC officials also privately alleged that the leak of the nine-year-old story in the U.S. press had been a deliberate tactic by U.S. officials to give substance to their current campaign to get Europeans to accept stricter international controls on the sale and transport of nuclear materials. One EC official said that if the story had indeed been leaked, "it was a dirty blow."

Time magazine reported May 30 that the apparent disappearance of the uranium ore was the culmination of a carefully orchestrated plot by Israeli intelligence agents.

Time said that after several weeks of investigation by a team of correspondents, it had learned that the plot had been devised to disguise a secret purchase of uranium by Israel for Dimona. Secrecy was necessary, Time said, because an open purchase might have caused the Soviet Union to supply nuclear arms to the Arab nations.

According to Time, Israel received assurances from the West German government of Chancellor Kurt Georg Kiesinger that it would be permitted to disguise the purchase of uranium as a private commercial transaction. In return, Time said, Israel promised Germany access to its advanced uranium separation process that could be used to produce nuclear weapons. (West German officials now refuse either to confirm or deny government involvement in such a deal, Time reported.)

Some $3.7 million worth of uranium ore was purchased, in what later became Zaire, from the Belgian firm Societe General des Minerais, according to the magazine. It was bought by a now-defunct German petrochemical company, Asmara Chemie, which Time said had never before bought uranium. The two companies told the European Community's atomic energy agency, Euratom, in August 1968 that the ore would be shipped to SAICA, a paint company in Milan, which Time said also had never been known to use uranium. (The destination had been shifted from the original choice, a Moroccan firm, because Morocco was not in the European Community and nuclear material could not be shipped there without a special permit.)

SAICA was sent $12,000 to buy equipment to mix the uranium and paint, Time said. But a few days after the uranium was shipped, Asmara reportedly told SAICA the ore ship had been lost and to keep the money.

Time sources said the uranium was shipped from Antwerp, Belgium aboard the *Scheersberg A*, a tramp steamer secretly owned by the Israeli intelligence agency, Mossad. The uranium was transferred, under cover of darkness, to an Israeli ship in the Mediterranean Sea between Cyprus and Iskenderun, Turkey. The Israeli boat then took the uranium to Haifa, according to Time. Port records showed that the *Scheersberg A* arrived empty in Iskenderun Dec. 2.

The *Scheersberg A*, according to Time, "was almost certainly involved" in the refueling of five French-built gunboats seized by Israeli agents from Cherbourg harbor in December 1969. *The Scheersberg A* had since been sold and renamed twice and currently was making a regular run between Greece and Libya under the name *Kerkyra*.

A Euratom investigation of the missing uranium had been hampered by the agency's lack of police powers, Time said, and after a few months Euratom had asked for help from the security forces of Western nations. A West German investigation, Time said, had been "abruptly—and mysteriously—halted" soon after it began in 1969.

Israel reported to have A-bomb—Israel had assembled 13 atomic bombs at the start of the 1973 war, and then stored them when the tide of war turned in its favor, Time magazine reported in its April 12, 1976 issue. Israel denied the report.

According to the Time account, Premier Golda Meir had ordered Defense Minister Moshe Dayan to activate the A-weapons after receiving a report Oct. 8 of marked deterioration of the Israeli military position on the northern front with Syria. "As each bomb [of 20 kiloton yield] was assembled, it was rushed off to waiting air force units," the magazine said, in preparation for dropping "on enemy forces from specially equipped Kfir and Phantom fighters or Jericho missiles."

According to Time, the Israelis believed the Soviet Union as well as the U.S. had learned of the Israeli nuclear arsenal, possibly through their spy satellites, prompting the Soviet Union Oct. 13 to ship nuclear warheads to be fitted on their Scud missiles in Egypt. The U.S. detected the Soviet nuclear shipment Oct. 15, which lead to an American global military alert 10 days later, the article said.

The magazine reported that in early 1968 the Israeli government had vetoed a project for work on a separation plant to produce the fissionable material necessary for an A-bomb, but that Dayan had secretely ordered its construction and the program went ahead. Dayan was quoted as having told Time correspondent Marlin Levin recently: "Israel has no choice" but to build atomic weapons. "With our manpower we cannot physically, financially or economically go on acquiring more and more tanks and more and more planes."

In denying the Time report, the Israeli government April 5 reiterated its previously stated position that it was not a nuclear power, and would not be the first to introduce nuclear weapons into the Middle East conflict.

U.S. approves reactor sale to Spain. The U.S. Nuclear Regulatory Commission June 21, 1976 approved the sale of a reactor to Spain despite objections of one member of the four-man panel.

Victor Gilinsky registered the first protest of a commission member against granting a reactor-export license. He said that the deal did not provide sufficient safeguards to prevent Spain from manufacturing nuclear weapons with the reactor fuel. He had argued for a license modification to restrict Spain to using only U.S. uranium as fuel. Such a restriction would prevent Spain from reprocessing the fuel to produce plutonium, an important fuel for nuclear weapons, without U.S. permission. Spain had previously signed deals with U.S. companies to build reactors for generating electricity.

Report says Taiwan secretly reprocessing nuclear fuel. U.S. intelligence over the first six months of 1976 found that the Republic of China (Taiwan) was secretly reprocessing used uranium fuel to obtain plutonium, the Washington Post reported Aug. 29. With plutonium Taiwan could make a nuclear bomb. A Taiwan Embassy spokesman in Washington Aug. 28 denied the report.

The Post said that a hot-cell reprocessing center had been used to produce about one pound of plutonium. American, Canadian and European experts were quoted as saying that this amount was far short of the minimum of 18 pounds needed for a sophisticated nuclear device, but enough to provide knowledge of plutonium handling and explosive fabrication.

The hot-cell plant had been built near Taipei with American aid and with parts obtained from countries around the world. The installation had been constructed in 1970 with the understanding that it would be subject to inspections by the Vienna-based U.N. International Atomic Energy Agency (IAEA). Taiwan was ousted by the IAEA in 1971.

Taiwan had a research reactor and four other American-built nuclear power plants in addition to the recycling plant, and hoped to acquire two additional power plants from the U.S. by 1980. But U.S. Arms Control and Disarmament Agency officials said they were stalling on the application for the two additional plants in an effort to stop Taiwan's secret uranium reprocessing, the Post report said.

Nuclear conference held in Salzburg. Most of the more than 60 nations represented at the International Conference on Nuclear Power and Its Fuel Cycle in Salzburg, Austria May 2–13, 1977 showed little willingness to retreat from their plans to use nuclear power or use plutonium fuel to supply that power. U.S. President Jimmy Carter previously had appealed for deferral of the use of plutonium worldwide to help curb the spread of nuclear weapons.

Attending the conference, which was sponsored by the International Atomic Energy Agency (IAEA), were some 2,000 government delegates, industrial representatives and academic specialists. Among the issues discussed were the difficulties of reactor safety, the spread of nuclear weapons, the development of new technologies and the risks of wide use of nuclear power.

The issue of nuclear nonproliferation, another primary concern of President Carter, was a major one at the conference. Much attention was paid to making nuclear bomb-grade fuels inaccessible, either by limiting fuel-processing technology or by imposing international controls. A problem debated by uranium enrichment specialists, for example, was that new methods of enrichment might make production of bomb-grade uranium too simple. The making of bomb-grade materials currently was dependent on relatively slow, highly sophisticated processes.

(Raw uranium consisted mostly of the isotope Uranium-238, which could be used neither as a fuel nor in a bomb, and about .7% of the isotope Uranium-235. Uranium enrichment was the process by which the concentration of Uranium-235 was increased. Enrichment to a 3%–5% concentration of Uranium-235 was sufficient for reactor fuels. Bomb-grade uranium had approximately a 90% concentration of Uranium-235.

(Two methods of enrichment currently in use were gaseous diffusion and centrifugation. Both methods relied on the slight difference in weight between Uranium-235 and Uranium-238 in order to separate the two isotopes.

(Uranium-238, when bombarded with neutrons, was converted into radioactive Plutonium-239, which could be used as a reactor fuel or a nuclear explosive. The chief motive for using plutonium as a fuel was that it could be derived in a fast-breeder reactor from the otherwise useless Uranium-238. Since most natural uranium was in the form of Uranium-238, use of plutonium-fueled breeders would greatly extend the life and usefulness of existing uranium reserves. A breeder reactor could create more plutonium from Uranium-238 than it used as a fuel.)

French delegates to the conference announced May 6 that France had developed a new method of uranium enrichment that could not practically be used to make bomb-grade material. The process involved the separation of Uranium-235 and Uranium-238 by means of chemical exchange. Two safeguards against proliferation cited by the French were the danger of an explosion in the plant if an attempt were made to reach 90% enrichment and the fact that, in any case, it would take about 30 years to achieve that level of enrichment. Other scientists, familiar with the theory but not the details of the French technique, said such a process would involve a very large plant and a very large uranium inventory. Both of these factors, they said, would make it difficult for the process to be commercially competitive.

Another major question for the future of nuclear energy was that of public acceptance and assessment of the risk. Throughout the 12-day meeting, leaders of national atomic energy programs conceded that public hostility was their chief handicap.

A central issue in public acceptance was the difficulty of accurately assessing the risk of nuclear energy use. At the conference, delegates cited a 1975 report by the U.S. Nuclear Regulatory Commission that estimated the chances of a major accident causing several thousand deaths at one in a billion for each year of operation of a power plant. Dr. Leonard D. Hamilton of Brookhaven National Laboratory and Dr. A.S. Manne of Stanford University presented a paper in which they said coal was more hazardous than uranium as an energy source. Using estimates of mortality attributable to air pollution and including hazards of mining and transport of fuel, they blamed coal for between 1,900 and 15,000 deaths in 1975, compared with between 18 and 42 deaths due to nuclear energy programs.

In other conference developments:

■ In his nation's first public response to President Carter's curtailment of the U.S. breeder-reactor research program, a delegate from the Soviet Union told the conference May 9 that "plutonium must be used." The Soviet Union would continue its development of breeder reactors, he said.

■ An article by IAEA staff member Morris Rosen, distributed to the delegates May 3, said that developing nations might be buying and installing nuclear power plants that were less safe than those operating in supplier nations. Rosen's article, which was widely discussed, pointed to the danger of earthquakes in developing countries with nuclear programs; to "sub-minimal" local watchdog organizations, and to the failure of nuclear suppliers to update design and safety features of exported reactors.

■ Dr. Hannes Alfven, winner of the Nobel Prize and a member of Sweden's Atomic Energy Commission, told the conference May 9 that nuclear energy should be abandoned. He gave four main objections: that it was associated with the production of radioactive substances; that the link between nuclear energy and weapons was so strong that "we have to accept both or neither"; that it was becoming difficult to support the claim that nuclear energy was cheap, and that there were "several other and more attractive" ways of solving the energy problem. Earlier in the day, a leading U.S. proponent of nuclear energy use, Dr. Hans Bethe, also a Nobel Prize-winner, had said nuclear energy was "a necessity, not an option."

'Counterconference' held—A group of about 100 scientists, environmental specialists and antinuclear activists from 20 countries May 1 condemned the use of plutonium as a source of energy and called for a stronger fight against conventional nuclear power plants. The group, which called itself the Salzburg Conference for a Non-Nuclear Future, met in Salzburg from April 29 to May 1. It advocated international efforts to develop such alternative technologies as the large-scale harnessing of solar energy.

Carter approves uranium shipments. President Carter had approved the shipment of approximately 827 pounds of highly enriched uranium to five countries for use in nuclear research reactors, it was reported May 6, 1977.

Applications for shipments of about 2,866 pounds of enriched uranium had been pending for as long as 18 months. Officials in Western Europe had been complaining for nearly a year about the delay, asserting recently that without swift U.S. action, some of their nuclear facilities would have to cut back operations by summer. Some U.S. allies also had complained that Carter's slowness in approving the applications seemed to contradict his assertion that the U.S. would remain a reliable uranium supplier.

The shipments approved by Carter were destined for Belgium, Canada, Japan, the Netherlands and West Germany. In addition, the State Department had cleared the shipment of as much as 496 more pounds of high-grade uranium in amounts too small to require presidential review. All the applications still had to be reviewed by the Nuclear Regulatory Commission.

Cuba to get Comecon nuclear plant. Comecon, the Communist bloc's economic association, would build the first atomic power station in Cuba, the Soviet news agency Tass reported Jan. 6, 1977.

Under an agreement signed in Moscow in April 1976, the first stage of the plant would have "an output capacity of 440 megawatts." Cuban President Fidel Castro had said the plant eventually would have twice that capacity.

Carter signs Latin pact. Fulfilling a promise he made to the Organization of American States in April, President Carter May 26, 1977 signed the Treaty of Tlateloco, which banned atomic weapons in Latin America and the Caribbean.

If the U.S. Senate ratified the treaty, the U.S. would have to keep nuclear weapons out of its territories in the area, including Puerto Rico, the Virgin Islands, the Panama Canal Zone and the Guantanamo naval base in Cuba.

Other Developments

Conference asks waste-disposal data exchange. Representatives of seven leading nuclear energy users—Britain, Canada, France, Japan, Sweden, the U.S. and West Germany—agreed at a five-day conference in Denver July 12, 1976 to increase their exchange of information on the disposal of nuclear-waste material.

The conferees agreed that without a safe system for storing radioactive wastes, public apprehension would seriously curtail nuclear-energy programs. (The waste material was the residue of nuclear fuels which were reprocessed after a reaction to extract any additional useful uranium 235 and plutonium.)

The conferees agreed to set up an international committee of waste-handling experts. Under discussion were plans such as storing the wastes in steel canisters and burying them in deep-lying salt deposits in the U.S. and West Germany and in crystalline rocks in Canada.

U.K. accepts A-wastes—Britain resumed the receipt of nuclear wastes at Barrow-in-Furness Dec. 1, 1975, with the delivery of a consignment of used nuclear fuel from Italy which was scheduled for transport to the reprocessing plant of British Nuclear Fuels at Cumbria where radioactive waste was extracted and stored.

It was the first consignment of used fuel from an overseas atomic power station since unions at Barrow were instructed by leaders of the Confederation of Shipbuilding and Engineering Unions not to handle the materials in a controversy over whether the waste products could be safely stored.

The government had decided Oct. 22 to review the arrangement under which Britain reprocessed the nuclear materials of other countries.

U.S. OKs Japan's spent-fuel export—The U.S. had agreed to a resumption of Japan's transfer of used nuclear fuel to Great Britain for reprocessing, the Japan Atomic Power Co. reported March 28, 1977. This decision would authorize a renewal of shipments halted in October 1976, when President Ford asked the Japanese government to suspend such shipments. Ford had called for a three-year ban on exports of spent nuclear fuel and uranium enrichment technology.

Carter proposals criticized—Proposals of President Carter regarding the reprocessing of spent nuclear fuel were criticized severely April 28–29, 1977 at the 15-nation Conference of Atomic Energy Supplier Nations in London. The main feature of the proposals under attack was that licenses for the re-exporting of U.S.-supplied enriched uranium fuel should be severely restricted. If accepted, such a measure could effectively kill British-French plans to reprocess other nations' spent nuclear fuel on a large scale. It could also thus prevent some nations from obtaining plutonium to fuel fast-breeder reactors and to lessen their dependence on U.S. uranium. Japan, for instance, was almost totally dependent on U.S.-supplied uranium fuel.

The conferees reportedly worked out a code to regulate the export of nuclear equipment for peaceful purposes. But meetings on additional safeguards were inconclusive. Sources said this occurred in part because it was felt that U.S. policy was not sufficiently formulated to make definitive reaction possible.

Maritime transport pact in force. The international convention relating to civil liability in the field of maritime transport of nuclear materials entered into force July 15, 1975 after five nations had submitted instruments of ratification of the pact. The five nations were France, Spain, Denmark, Sweden and Norway.

The convention had been adopted in December 1971 by an international conference in Brussels, held under the joint auspices of the Intergovernmental Maritime Consultative Organization, the Nuclear Energy Agency of the Organization for Economic Cooperation and Development and the International Atomic Energy Agency.

The new convention would free shipowners from liability under international maritime law in the case of nuclear damage involving the transport of nuclear materials to or from a nuclear installation. The nuclear installation would be liable for all damages.

France adopts new A-plan. The French cabinet Aug. 6, 1975 approved a major new program to restructure France's civilian nuclear energy industry, granting a monopoly to the U.S.-designed pressurized water reactor manufactured by Framatome, a French subsidiary of the French-Belgian Creusot-Loire group and the U.S.-based Westinghouse Electric group. Westinghouse held the license for the pressurized water reactor.

The decision meant that France's Compagnie Generale d'Electricite (CGE), which produced boiling water reactors under license from General Electric of the U.S., was forced out of the reactor field. But CGE would be compensated with a monopoly for supplying turbine equipment for Framatome's installations. Until now, Framatome had obtained 18 firm orders and four options for reactors from the French state-owned electricity board, while CGE had received only two firm orders and six options.

Framatome was owned 45% by Westinghouse and 51% by Creusot-Loire. In order to increase French control over the industry, the state Atomic Energy Authority (CEA) would purchase a major share of Westinghouse's stake in Framatome.

The government would also require Westinghouse to convert its licensing system into a partnership with Framatome and work out joint research and development programs. The current license agreement would expire in 1982.

As part of the new policy, the CEA would be reorganized so that research and production would be split into two autonomous divisions, with the commissioner to provide only general supervision. In addition, the government authorized high-level talks with West Germany to discuss advanced reactor research, joint security issues and uranium enrichment.

The reorganization of the civilian nuclear power industry coincided with a government decision to retreat from plans for a long-term atomic energy program designed to shift the burden of France's power needs away from imported oil. Under the new policy, France would continue with its plans to build 12 nuclear stations of 1,000 megawatts each in 1976–77, but would decide on future expansion gradually. Underlying the gradualist approach was the belief that planners could not predict French energy needs or the price of oil years in advance.

Strasbourg A-plant canceled—Plans of the French electricity authorities to build a nuclear power plant near Strasbourg were dropped Nov. 20, 1976 at a meeting of the Franco-German regional committee that oversaw the area's development. The location of the planned station at Marckholsheim was one of 46 sites in France chosen for nuclear facilities.

Environmentalists had opposed French officials and industrialists over the issue and had demonstrated Oct. 10 in Strasbourg against the project.

Uranium mines bombed—Five bombs exploded in the surface buildings of the French atomic authority's uranium mines at Margnac (reported Nov. 15, 1976). The explosions caused millions of dollars in damages to the structures and the extracting plant. The identity of the bombers was not discovered.

Margnac was located in the south-west corner of France. Its mine produced about 40% of the uranium obtained in the area.

British to make thermonuclear fuel. The British Defense Ministry announced April 27, 1976 that it had contracted with British Nuclear Fuels Ltd. to produce tritium. Tritium, a radioactive isotope of hydrogen gas, is an essential component in the manufacture of hydrogen bombs.

A defense official said that security prohibited release of detailed figures about the deal. The Defense Ministry statement emphasized, however, that domestic production of tritium was "more convenient and saves dollars."

In the past, Britain had imported its tritium exclusively from the U.S. under a 1958 agreement on the exchange of nuclear materials and information. The pact, renegotiated in 1973, was set to expire in 1977. Under the pact's terms, either party could terminate the exchange after giving a year's notice. The British government did not say how its new domestic contract would affect its dealings with the U.S.

London sources quoted April 28 in the New York Times said that the high cost of tritium had been a factor behind the domestic deal with British Nuclear Fuels. The recent decline in the value of the pound against the dollar further increased the cost of importing tritium, they noted.

Other reports April 28 maintained that the move was part of an effort by the British government toward developing self-sufficiency in the production of thermonuclear weapons. British officials rejected contentions that the decision was one step in a British attempt to end nuclear dependence on the U.S.

Bonn, Moscow drop Soviet A-plant plan. West Germany and the Soviet Union March 30, 1976 dropped plans to build in the U.S.S.R. the world's largest nuclear power station. Planned for a site at Kaliningrad in western Russia, the $1.6 billion station was to have supplied electricity to West Germany and West Berlin. It was to have been constructed by the West German firm Kraftwerk Union AG, using West German financing.

Both parties to the project sought to emphasize that its collapse did not represent a setback to West German-Soviet relations. Bonn's economics minister, Hans Friderichs, said in Moscow upon the announcement of the deal's cancellation that the two sides had been unable to agree on economic terms.

However, a West German spokesman, Klaus Bolling, said in Bonn March 31, that the Soviet Union "was having trouble agreeing on the route for the power lines." Western news reports indicated that the East German government, through whose territory the wires would have led to West Berlin, objected to the project on political grounds. According to a London Times report April 1, the Democratic Republic had opposed the direct provision of electricity to West Berlin and had wanted only a "branch line" to the city.

Kosygin opens Finnish plant. Soviet Premier Alexei Kosygin May 23, 1977 formally opened Finland's first nuclear power station. The ceremony came on the second day of his five-day official visit to Finland. The plant, built at Loviisa, a small town on the coast near Helsinki, was constructed with Soviet aid. A second plant in Loviisa would be completed in 1979. In addition to the Loviisa plants, Finland had two other nuclear stations under construction.

The power plant was a 440-megawatt pressurized water reactor. Half the work on the $260 million plant was done by Finnish companies and foreign subcontractors and half by the Soviets. Canada supplied the uranium, which was enriched by the Soviet Union.

230 injured in W. German protest. A demonstration by 30,000 persons Nov. 13, 1976 at the Brokdorf nuclear power plant building site, 25 miles northwest of Hamburg on the Elbe River, erupted into violence when about 3,000 of the more militant protesters attempted to storm the enclosed area. Police used fire hoses, tear gas and clubs to repel the assault. At least 151 demonstrators and 79 policemen were hurt.

The protest underlined a growing popular resistance to the West German nuclear-power program, which had begun in 1974 as an answer to the oil crisis. The government had projected a nuclear electrical capacity of 45,000-50,000 megawatts (mw) by 1985. Local protests and increased safety requirements had cut the government's projections by 10,000 mw, it was reported Nov. 11.

Current megawatt capacity was 6,400, with about 12,000 under construction and an additional 9,000 planned.

Opinion polls, made public Nov. 11, indicated that 20% of the West German population was concerned over expansion of nuclear-energy facilities. The opposition in Bonn had accused the government of planning the growth of the nuclear energy industry, then backing away from specific projects after those projects were criticized, it was reported Nov. 11.

Other nuclear protests had been centered in Lower Saxony, where feelings were reportedly more intense. The government planned to build West Germany's first full-scale atomic fuel reprocessing and disposal plant in Lower Saxony. The area had large underground salt strata where the government hoped to bury radioactive waste from all West German nuclear power stations. Lower Saxony residents opposed the plan and had protested repeatedly.

Earlier in the year Interior Minister Werner Maihofer had announced that no new nuclear-plant construction permits would be issued until plans existed for the disposal of spent fuel. Three Bonn government ministers visited the premier of Lower Saxony Nov. 11 to pressure him into reaffirming his support for the disposal project.

North Rhine bans atomic plants—West Germany's most industrialized and populated state, North Rhine-Westphalia, announced Jan. 13, 1977 that it was banning the construction of more nuclear power plants until the problem of spent atomic fuel was settled. Heinz Kuhn, premier of the state, said there was no chance that plants already in operation would be shut down. Four of West Germany's 13 operating atomic-power plants were located in North Rhine-Westphalia.

Public protests in West Germany over the disposal of radioactive waste and worry over the safety of the plants themselves had led the federal government in 1976 to cut back the number of power stations planned for the country to 25 from 35.

The violence that had accompanied the November 1976 anti-atomic power demonstration in Brokdorf on the Elbe River had a special impact on Bonn politicians and had led to a debate on the value of nuclear development itself. Hans Matthofer, technology minister, said the Brokdorf protest represented "not just resistance to a nuclear power plant. The Elbe is a polluted river," Matthofer said, ". . . and people are protesting against all of this. Nuclear energy is just the symbol."

Public concern over the safety of the atomic plants was increased Jan. 14 when the ministry for protection of the environment in the southern state of Bavaria shut down an atomic-power station near Gundremminger after radioactive steam escaped. No one was injured or endangered by the leak, but the plant was closed indefinitely.

It was the third incident at the Gundremminger plant. Two men had been killed a year earlier by hot steam while working on a burst valve, and the plant had been closed three days in December 1976 because of a leak in the circulation system.

The events leading to the Gundremminger closing were similar to circumstances that led to the shutdown of the power station in Biblis, which had been a showpiece of the West German nuclear industry. The plant was closed for months in 1976 because of a problem in its main coolant circuit.

In addition to plant safety, the problem of fuel disposal had become more serious since the Brokdorf demonstration. Government experts had said that a central underground dump would be needed within 10 or 20 years to handle the mounting waste from the reactors currently in operation. Lower Saxony, the only West German state with the salt caverns necessary for disposal, had resisted attempts to locate the dump within its border.

The limits on domestic development put pressure on the nuclear power industry to increase exports in order to make up for the closed domestic market. To fully employ the nation's nuclear work force, eight power stations had to be produced each year.

A $4-billion nuclear deal with Brazil, in which West Germany agreed to supply eight nuclear-power stations and a reprocessing plant, had been regarded as the major export hope and a justification for the billions spent by the government on nuclear research and development.

Government plans new breeder research—The West German cabinet April 27 proposed spending the equivalent of $2.7 billion for a four-year comprehensive energy research program. Some 70% of that, or about $1.88 billion, would be spent on nuclear energy research, including work on the fast-breeder reactors opposed by U.S. President Jimmy Carter and on reprocessing of spent uranium fuel.

The program would seek to safeguard economic growth by formulating a comprehensive answer to the nation's energy needs. It would, for the first time, tie together all types of energy research in one funding plan.

A government official said West Germany could not delay work on the reprocessing of spent nuclear fuel because of its lack of raw materials and its dense population. Asked whether the research plans contradicted the U.S. policies on nuclear research, he said, "The U.S. recognizes that there must be different strategies for different situations."

Chancellor Helmut Schmidt had said April 26 that West Germany considered reactor exports its major future industry and would sell reactors to "any country that wants one or more than one." In a television address April 21, Schmidt had said the Bonn government would reject U.S. calls to stop selling sensitive atomic technology abroad unless all rival nuclear exporters accepted similar restrictions. In the address, he restated his opposition to Carter's request that West Germany cancel the sale of potential nuclear-bomb technology to Brazil.

Japan starts up first breeder. Japan's first experimental fast-breeder reactor April 24 reached criticality, or the point at which a nuclear reaction was virtually self-sustaining. It was the first time Japan had ever used plutonium to spark a critical nuclear reaction. The exercise made Japan the fifth nation to breed plutonium for peaceful purposes. The other countries were the U.S., the Soviet Union, Great Britain and France.

The reactor, named "Joyo," was built by the government's Power Reactor and Nuclear Fuel Development Corp. at a cost of $100 million. It was the first of two planned prototypes. There were 12 similar reactors in the other four countries.

Index

D

N

V

W